BIOGRAPHICAL INDEX TO AMERICAN SCIENCE

BIOGRAPHICAL INDEX TO AMERICAN SCIENCE

The Seventeenth Century to 1920

Compiled by
Clark A. Elliott

Bibliographies and Indexes in American History, Number 16

GREENWOOD PRESS
NEW YORK • WESTPORT, CONNECTICUT • LONDON

Library of Congress Cataloging-in-Publication Data

Biographical index to American science : the seventeenth century to
   1920 / compiled by Clark A. Elliott.
      p.   cm.—(Bibliographies and indexes in American history,
ISSN 0742-6828 ; no. 16)
   ISBN 0-313-26566-6 (lib. bdg. : alk. paper)
   1. Scientists—United States—Biography—Dictionaries.
I. Elliott, Clark A.   II. Series.
Q141.B533   1990
509.2'2—dc20   90-31735
[B]

British Library Cataloguing in Publication Data is available.

Library of Congress Catalog Card Number: 90-31735
ISBN: 0-313-26566-6
ISSN: 0742-6828

First published in 1990

Greenwood Press, 88 Post Road West, Westport, CT 06881
An imprint of Greenwood Publishing Group, Inc.

Printed in the United States of America

The paper used in this book complies with the
Permanent Paper Standard issued by the National
Information Standards Organization (Z39.48-1984).

10 9 8 7 6 5 4 3 2

To my father
Leroy Albert Elliott

and to the memory of my mother
Bertha Lyons Elliott

Contents

Preface

This work is conceived consciously as a supplement to my _Biographical Dictionary of American Science: The Seventeenth Through the Nineteenth Centuries_ (Greenwood Press, 1979). Although the information on any one individual is considerably briefer, some three times as many names are included here as are in the _Biographical Dictionary_.

I will be happy if this work serves the needs of harried students and scholars seeking information on the individual scientists listed here. My more ambitious hope is that it will help to open up research on a much wider array of persons active in the formative years of science in America. The history of science, as in so many spheres, tends to concentrate on a select group of persons, problems, or phenomenon at the center of the historiographic vision. A truer picture of diversity on a wider and deeper stage is in order; and the promotion of attention to that diversity or plurality is a central motivation for this work. It is particularly appropriate for American science, in which action--the formulating of a career, the building of resources for education and research, the construction of the bonds of community, the promotion of political support--is as much a part of the history of early science as is the work in the field and laboratory or the preparation of research reports. Actions arise from actors, and the names and the lives represented here are the elements for a fuller and more realistic reconstruction of the history of science in America. As a personal motivation for the work as well, a collecting instinct is not to be denied. Accumulating and studying the lives of deceased scientists grows out of an endless fascination with the patterns and personalities, the strategies, successes and failures that these lives represent.

One accumulates many obligations in the course of such a project, not the least of which are owed to the legions of librarians formerly and now in the Harvard libraries who make search and research a matter of repeatedly taken-for-granted success. I am happy as well to acknowledge the

support of a Douglas W. Bryant Fellowship from the Harvard
University Library. It came at a crucial time when I
needed the encouragement as well as the funds.   My family
has endured the clutter of files and my hours at the library and at the computer. For their patient acceptance of
my obsession with these "other lives," I thank Priscilla
and our sons Andy and Glenn with gratitude that is beyond
words.   Priscilla worked as well at the task of typing the
initial list of names, dates, and fields of science. Glenn
earned spending money by assisting early with routine but
crucial work; he has grown to young manhood and inde-
pendence since the project began in earnest five years ago.
Cynthia Harris at Greenwood Press has been responsive and
encouraging, but also willing to place a degree of faith in
the judgement of the author. To all of them I am immeasur-
ably thankful.

Introduction

PURPOSE

The purpose of this <u>Biographical Index to American Science</u> (<u>B.I.A.S.</u>) is to assist in locating basic biographical information on some 2850 American scientists who died prior to 1921. Bibliographic references (under "Works About" in the Biographical Index entries) are to standard biographical directories, dictionaries, indexes, etc., or to obituaries and similar notices for lesser known persons not covered by the standard works. A secondary but also important goal is the presentation of a substantial (but by no means complete) census of early American scientists. Thus, <u>B.I.A.S.</u> can serve those who know a name and who seek sources of biographical data on the person. It also can serve those who are interested in a wider sweep of personnel, either by scanning the pages of the Biographical Index, or by consulting the Index of Names by Scientific Field at the end of the volume. The entries in the Biographical Index ordinarily include full name, year of birth and death, scientific field(s), occupational category, manuscript sources, and works about the person.

BIOGRAPHICAL SOURCES

The biographical sources that are indexed represent an inevitable hierarchy in regard to their authority, selectivity, or completeness. A hierarchy similarly exists among the scientists themselves so that some are in the most authoritative, selective, or commonly available reference works while others are represented only in works of less accessibility or authority. "Works About" a scientist, which is the heart of <u>B.I.A.S.</u>, reflects this differentiation.

All of the works cited in <u>B.I.A.S.</u> are given in the section labelled Key To Biographical Sources, which follows this introduction. With an important exception, the works listed immediately below always are included under "Works About" when a name in B.I.A.S. appears there.[1]

Am Men Sci   [American Men of Science]

Barr   [E. S. Barr, Index to Biographical Fragments in Unspecialized Scientific Journals]

Biog Geneal Master Ind   [Biography and Genealogy Master Index]

Dict Am Biog   [Dictionary of American Biography]

Dict Am Med Biog   [Dictionary of American Medical Biography]

Dict Sci Biog   [Dictionary of Scientific Biography]

Elliott   [C. A. Elliott, Biographical Dictionary of American Science]

Ireland Ind Sci   [N. O. Ireland, Index to Scientists]

Natl Ac Scis Biog Mem   [National Academy of Sciences, Biographical Memoirs]

Natl Cycl Am Biog   [National Cyclopaedia of American Biography]

Notable Am Wom   [Notable American Women]

Ogilvie   [M. B. Ogilvie, Women in Science]

Pelletier   [P. A. Pelletier, Prominent Scientists: An Index to Collective Biographies]

Roy Soc Cat   [Royal Society of London, Catalogue of Scientific Papers 1800-1900]

Siegel and Finley   [P. J. Siegel and K. T. Finley, Women in the Scientific Search]

Who Was Who Am   [Who Was Who in America]

In preparing this work, substantial benefit has come from four of the works with a mission somewhat similar to that of B.I.A.S. These are Barr, Biog Geneal Master Ind, Ireland Ind Sci, and Pelletier. These are other indexes rather than direct sources of biographical information. All four of these indexes were crucial for locating bio- graphical sources of information for the American scien- tists. Additionally, four of the biographical works were considered to be particularly valuable because of their authoritativeness and wide availability and also because

---

[1] The initial abbreviated form of the citation can be used to find more complete bibliographical information in the Key To Biographical Sources.

they not only include substantive biographical information
but   in turn routinely cite further sources of information.
Those   four biographical works are Dict Am Biog, Dict Sci
Biog, Elliott, and Notable Am Wom.

If one of the four biographical works just mentioned
appears under a scientist's name, the indexing sources--
Barr, Biog Geneal Master Ind, Ireland Ind Sci, or
Pelletier--simply are listed for benefit of those readers
who want to consult them. If none of the above mentioned
four biographical works are found under a scientist's name,
then any biographical references cited in Barr or the other
indexes are extracted and listed directly in B.I.A.S. For
example, references to the American Journal of Science or
to Science are largely due to Barr, while sources such as
Adams O, Allibone, Appleton's, and Wallace (see Key To Biographical Sources) are derived from the Biog Geneal Master
Ind. Pelletier indexes such discipline-related works as Am
Chem and Chem Eng and Mallis (see Key), and both Ireland
Ind Sci and Pelletier have indexed a number of special publications that would not be referenced here without their
diligence and foresight. I am pleased to acknowledge my
indebtedness to these sources, but also recognize that reference works by their nature have an interlocking relation-
ship and that few are complete without the others.

Although every reasonable effort at accuracy has been
exerted, a certain caution is warranted. As explained
above, B.I.A.S. draws upon previous efforts of other
authors or projects, and not every reference has been examined and verified. A small possibility exists that in some
cases a source listed here may in fact not relate to the
person under whose name it appears. All those familiar
with historical research (or modern telephone directories)
will appreciate the problem of persons with similar names.

COMPILATION OF NAMES

It will be helpful to note that the names in B.I.A.S.
were compiled from a number of sources as a step toward a
census of scientists. Thus, unlike many similar works, it
does not index completely and systematically a stated and
limited number of publications. The B.I.A.S. list of scientists' names from multiple sources was a starting point
and that list in turn was checked against other published
indexes or biographical reference works. Emphasis is on
the list of names and not on the works referenced. In compiling the list of names, certain of the works listed above
were of particular value. Barr, Dict Sci Biog (using the
Concise edition, New York 1981), Elliott, Ogilvie,
Pelletier, Siegel and Finley, and Who Was Who Am (using Who
Was Who Am Hist [see Key To Biographical Sources]) were
systematically reviewed. The notices of scientists
deceased that were listed in early Am Men Sci volumes also
were systematically examined. The Dict Am Biog was consulted by reference to its Concise edition (New York 1964)
and its occupational index. Notable Am Wom was examined

through its classified list of selected biographies. The Roy Soc _Cat_ was systematically examined and references to biographical works as well as names of authors of a significant number of articles whose names had not emerged from other sources was compiled. The latter group subsequently were identified, insofar as possible, in standard reference works. Those not readily identifiable were eliminated from the list.

Also significant in compiling the initial list were the following (which are not in the Key To Biographical Sources): Margaret Ann Young, "Early American Scientists, 1607-1818" (masters thesis, University of Oklahoma, 1957); and Clark A. Elliott, "The American Scientist 1800-1863: His Origins, Career, and Interests" (doctoral dissertation, Case Western Reserve University, 1970).

The scientific population listed here is a very substantial portion of the total for its time period, hopefully including all of those of the first rank, a very large percentage of the second level, and at least a representation of some of the personnel nearer to the periphery of the scientific community. In the course of this project, however, I have come to appreciate the degree to which the scientific population in the period covered tended to blend, at its outer edges, with the American populace at large and the foolhardiness of making anything but qualified statements about the percentage of coverage in any list of names. As illustration, among my initial attempts at a more-or-less complete census of the scientific population was a systematic review of all the entries in the Roy Soc _Cat_ to the end of the nineteenth century. This humbling exercise was informative about how widespread scientific writing was in the late years of the nineteenth century, the difficulty of the task of identifying the authors of an appreciable portion of that literature, and the subsequent inability to determine what proportion was written by "professionals" (however defined).

I have chosen to do a work that draws upon a range of biographical sources, and one that accepts a less than systematic or rigid definition of who was and was not a scientist. For the most part, however, I have tended to avoid the direct use of collective biographical sources limited to particular disciplines because I did not want to skew the overall picture of the scientist. (My general impression is that natural history is fairly well attended to in such works while the mathematical, astronomical, and physical sciences have been relatively neglected by the authors of biographical or bibliographical reference works.) In spite of my general wish to present here a reasonably representative view of the scientific population, the somewhat diverse means of compiling the names prompts caution in any use of the list for statistical purposes. For readers interested in the larger body of reference works on American scientists, some of which may supplement _B.I.A.S._, see my "Collective Lives of American Scientists: An Introductory Essay and a Bibliography" in _Beyond History of Science:_

Essays   in   Honor   of   Robert   E. Schofield,   ed.   Elizabeth
Garber   (Bethlehem,   Penn.:   Lehigh   University   Press,   in
press).

DISCIPLINARY COVERAGE

B.I.A.S. relates to persons in mathematics, astronomy,
physics,   chemistry,   geology,   botany,   zoology,   and related
sciences,   and to some persons in psychology and anthropology.   Occasionally patrons or   others not contributors   to
science are included.   Medicine, engineering, and invention
are included to some extent, in recognition of   how   diffi-
cult   it is to separate these fields from the general definition of science in America during the period   covered   by
B.I.A.S.   However, I have tended to include medical persons
and   engineers when they were known for their contributions
to   knowledge rather than strictly as practitioners.   Sometimes   my criterion was   as simple as   whether or not   they
taught,   while in other   cases inclusion was   based on   the
initial   source of the name   (e.g., that a medical   practitioner was listed in Am Men Sci).   Inventors are recognized
particularly when they seemingly dealt with general princi-
ples   rather than simply tinkering.   One of the reasons for
doing   this work has been to define by example the American
scientist involved in the development, manipulation, compi-
lation,   dissemination, or application of knowledge, and to
distinguish   that person more clearly from the inspired but
uninformed inventor so often pointed to as the scientist in
early   America.   Some of the latter undoubtedly have gained
entry, but they are not the focus of the work.

CHRONOLOGICAL COVERAGE

A word should be said about the ending date   of   1920.
This   is   largely   arbitrary,   but   defensible   on   several
fronts.   That year marked (in round numbers) the end of the
World   War I   period and   what would   begin   a   substantial
growth   in American science.   For example, I initially compiled   a list of   all American scientists   deceased   before
1941 and found that by a significant factor more scientists
lived (and died) between 1921 and 1940 than had   lived   and
died   in all of American history prior to 1921.   Twentieth-
century American science has been well-served by the directory   American   Men   of   Science   (now, more   accurately and
justly, American Men and Women of Science), which first was
published in 1906 with a second edition in 1910.   The third
edition did not appear until 1921 but   subsequent   editions
have   appeared with regularity.   That work and its recently
published index[2]   serve as a basic source for biographical
data on recent scientists.   Thus, B.I.A.S. can stand   as   a
complement to that work, for early American science.

---

[2]American Men and Women of Science Editions 1-14
Cumulative Index (New York and London: R. R. Bowker,
1983).

NOTES ON ENTRIES IN THE BIOGRAPHICAL INDEX

Following are some final notes on the several categories of information given under each entry.

Name. Standard biographical sources sometimes differ in the
form of a person's name, dates, or other basic information.
Ordinarily, B.I.A.S. adopts the form used in what appears
to be the most authoritative source. When the differences
may have significance (e.g., in identifying or locating a
biography) a statement is included indicating that the form
or spelling of the the name differs, or that birth or death
dates vary.

Dates. See "Name," above. If a person was born before 1851
but the death date is not known, the person is included
here based on the fact that he or she would have been 7C
years or older in 1920.

Fields. The overall coverage of the various fields or disciplines is referred to above. I have tended to adopt the
characterization of a person's interests as given in the
biographical sources cited, rather than using a precon-
ceived classification scheme. Some small adjustments have
to be made in compiling the Index of Names by Scientific
Field, in order to give that list a reasonable coherency.

Occupation. The brief, generalized occupational titles are
included both to characterize the scientist and to help the
user in determining whether the entry does in fact refer to
the person in whom they are interested.   Although an a
priori classification of occupational categories was not
used, a loose scheme was developed as the project pro-
ceeded.   There are inherent difficulties in the creation
and consistent application of such a classification during
a project of this size which is compounded by the historic
changes themselves over three centuries. In addition to
certain inevitable variations in application, the use of a
standardized terminology may in certain cases veil somewhat
the specific work the person was doing.   Nonetheless, for
purposes of comparability and as a means of understanding
the overall occupational character of the scientists, a
degree of consistent usage seemed preferable to a free and
uncontrolled vocabulary. A fuller understanding of an individual's occupation is possible by considering both the
occupation and his or her field(s) of science. An explana-
tion of the use of a couple of occupational terms is in
order. "Industrial" is used to encompass a wide range of
commercial or (generally) for-profit employers, from manufacturing to insurance concerns (e.g., actuary). "Consultant" sometimes is used in its accepted sense of offering
advice or assistance to a client, but in certain cases it
is used here to indicate an independent practitioner of a
science (e.g., a person whose occupation is described as
chemist).   Up to three occupational categories are given
for each scientist.

Manuscripts. Only the National Union Catalog of Manuscript Collections (to 1984) is referenced here. (See Ind NUCMC in Key To Biographical Sources). Although the limitations of this source are well-known, it is a standard and reasonably widely available reference work and a useful beginning point for research in unpublished sources.

Works About. The rationale and conventions in compiling and presenting this section are described above. When reference is made to biographical indexes rather than directly to biographies themselves, the indexes are listed as "See also."

Key to Biographical Sources

Ac Nat Scis Philadelphia <u>Proc</u> =
<u>Proceedings of Academy of Natural Sciences of</u>
<u>Philadelphia</u>.

Ac Sci Saint Louis <u>Bull</u> =
<u>Bulletin of Academy of Science of Saint Louis</u>.

Adams O =
Adams, Oscar Fay. <u>A Dictionary of American Authors</u>. 5th
ed., revised and enlarged. New York: Houghton Mifflin Co.,
1904. Reprinted: Detroit: Gale Research Co., 1969.

Adams R =
Adams, Russell. <u>Great Negroes Past and Present</u>. Chicago:
Afro-Am Publishing, 1969.

<u>Albany Med Ann</u> =
<u>Albany Medical Annals</u>.

Albert =
Albert, George D. <u>History of the County of Westmoreland,</u>
<u>Pennsylvania</u>. Philadelphia: L. H. Everts & Co., 1882.

Allibone =
Allibone, S. Austin. <u>A Critical Dictionary of English</u>
<u>Literature and British and American Authors Living and</u>
<u>Deceased from the Earliest Accounts to the Latter Half of</u>
<u>the Nineteenth Century</u>. 3 vols. Philadelphia: J. B.
Lippincott & Co., 1858-1871. Reprinted: Detroit: Gale
Research Co., 1965.

Allibone <u>Supp</u> =
Kirk, John Foster. <u>A Supplement to Allibone's Critical</u>
<u>Dictionary of English Literature and British and American</u>
<u>Authors</u>. 2 vols. Philadelphia: J.B. Lippincott & Co., 1891.
Reprinted: Detroit: Gale Research Co., 1965.

Am Ac Arts Scis <u>Proc</u> =
<u>Proceedings of American Academy of Arts and Sciences</u>.

Am Assoc Advancement Sci Proc =
Proceedings of American Association for the Advancement of
Science.

Am Chem and Chem Eng =
Miles, Wyndham, ed. American Chemists and Chemical
Engineers. Washington, D.C.: American Chemical Society,
1976.

Am Chem Journ =
American Chemical Journal.

Am Chem Soc Journ =
Journal of American Chemical Society.

Am Geol =
American Geologist.

Am Inst Mining Eng Trans =
Transactions of American Institute of Mining Engineers.

Am Journ Pharm =
American Journal of Pharmacy.

Am Journ Sci =
American Journal of Science.

Am Lit Yearbook =
Traub, Hamilton, ed. The American Literary Yearbook. Vol.
1, 1919. Henning, Minn.: Paul Traub, Publisher, 1919.
Reprinted: Detroit: Gale Research Co., 1968.

Am Machinist =
American Machinist.

Am Med Assoc Journ =
Journal of American Medical Association.

Am Men Sci =
American Men of Science: A Biographical Directory. New
York, N.Y.; Garrison, N.Y.; Lancaster, Penn.: The Science
Press, 1906- Note: 2nd edition 1910; 3rd ed. 1921; 4th ed.
1927; 5th ed. 1933. "d" indicates edition containing death
notice.

Am Meteorol Journ =
American Meteorological Journal.

Am Micr Soc Proc =
Proceedings of American Microscopical Society.

Am Micr Soc Trans =
Transactions of American Microscopical Society.

Am Mo Micr Journ =
American Monthly Microscopical Journal.

Am Nat =
American Naturalist.

Am Pharm Assoc Proc =
Proceedings of American Pharmaceutical Association.

Am Phil Soc members =
Biographical data files on members of the American
Philosophical Society (MS). American Philosophical Society
Library, Philadelphia, Penn.

Am Phil Soc Proc =
Proceedings of American Philosophical Society.

Am Phil Soc Proc (Memorial Vol) =
American Philosophical Society. Proceedings...Memorial
Volume 1. Philadelphia: American Philosophical Society,
1900.

Am Soc Civil Eng Trans =
Transactions of American Society of Civil Engineers.

Am Soc Micr Proc =
Proceedings of American Society of Microscopists.

Am Wom =
Willard, Frances E., and Mary A. Livermore, eds. American
Women. A revised edition of Women of the Century, 1500
biographies with over 1400 portraits; a comprehensive
encyclopedia of the lives and achievements of American
women during the nineteenth century. 2 vols. New York:
Mast, Crowell & Kirkpatrick, 1897. Reprinted: Detroit:
Gale Research Co., 1973.

Amherst Biog Rec 1871 =
Amherst College. Biographical Record of the Alumni of
Amherst College during Its First Half Century, 1821-1871.
Edited by W. L. Montague. Amherst, Mass., 1883.

Amherst Biog Rec 1973 =
Amherst College. Amherst College Biographical Record 1973.
Amherst, Mass.: Trustees of Amherst College, 1973.

Anderson =
Anderson, George B. Our County and Its People, a
Descriptive and Biographical Record of Saratoga County, New
York. [Boston]: Boston History Co., 1899.

Andreas =
Andreas, A. T. History of Chicago from the Earliest Period
to the Present Time. 3 vols. Chicago: The A. T. Andreas
Co., Publishers, 1885-1886.

Appleton's =
Wilson, James Grant, and John Fiske, eds. Appleton's
Cyclopaedia of American Biography. 6 vols. New York: D.
Appleton & Co., 1888-1889. Reprinted: Detroit: Gale
Research Co., 1968.

Appleton's Supp 1901 =
Wilson, James Grant, ed. Appleton's Cyclopaedia of American
Biography. Vol. 7, Supplement. New York: D. Appleton & Co.,
1901. Reprinted: Detroit: Gale Research Co., 1968.

Appleton's Supp 1918-31 =
Dearborn, L. E., ed. A Supplement to Appleton's Cyclopaedia
of American Biography. 6 vols. New York: Press Association
Compilers, Inc., 1918-1931. Originally published as The
Cyclopaedia of American Biography, Supplementary edition.

Arnold =
Arnold, Howard P. Memoir of Jonathan Mason Warren, M.D.
Boston: Printed for private distribution, 1886.

Assoc Am Eng Soc Journ =
Journal of Association of American Engineering Societies.

Atkinson =
Atkinson, William B., ed. Biographical Dictionary of
Contemporary American Physicians and Surgeons. 2nd enlarged
edition. Philadelphia: D.G. Brinton, 1880.

Auk =
Auk. Note: Some biographies from this journal are
reprinted in: T. S. Palmer and others. Biographies of
Members of the American Ornithological Union. Washington,
D.C., 1954.

Baldwin C =
Baldwin, Charles C. Baldwin Genealogy from 1500 to 1881.
Cleveland, Ohio: Printed by Leader Printing Co., 1881.

Baldwin T =
Baldwin, Thomas W. Patten Genealogy, William Patten of
Cambridge, 1635, and His Descendants. Boston: Thomas W.
Baldwin, 1908.

Barnhart =
Barnhart, John, comp. Biographical Notes upon Botanists,
Maintained in the New York Botanical Garden Library. 3
vols. Boston: G. K. Hall, 1965.

Barr =
Barr, Earnest Scott. An Index to Biographical Fragments in
Unspecialized Scientific Journals. University: University
of Alabama Press, 1973.

Bartonia =
Bartonia.

Bell =
Bell, Whitfield J., Jr. Early American Science: Needs and
Opportunities for Study. Williamsburg, Va.: Institute of
Early American History and Culture, 1955.

Benson =
Benson, Maxine. _Martha Maxwell: Rocky Mountain Naturalist_.
Lincoln and London: University of Nebraska Press, 1986.

Biog Dict Am Educ =
Ohles, John F., ed. _Biographical Dictionary of American
Educators_. 3 vols. Westport, Conn.: Greenwood Press, 1978.

Biog Dict Parapsychology =
_Biographical Dictionary of Parapsychology with Directory
and Glossary, 1964-1966_. New York: Garrett Publications,
Helix Press, 1964-1966-

Biog Dict Synopsis Books =
Warner, Charles Dudley, ed. _Biographcial Dictionary and
Synopsis of Books Ancient and Modern_. Akron, Ohio: Werner
Co., 1902. Reprinted: Detroit: Gale Research Co., 1965.

Biog Dir Am Congress =
Biographical Directory of the American Congress 1774-1971.
Washington, D.C.: U.S. Government Printing Office, 1971.
Note: Biographies begin on p.487.

Biog Geneal Master Ind =
Herbert, Miranda C., and Barbara McNeil, eds. _Biography and
Genealogy Master Index: A Consolidated Index to More than
3,200,000 Biographical Sketches..._ 2nd ed. 8 vols. Gale
Biographical Index Series no. 1. Detroit: Gale Research
Co., 1980. Note: There also are supplementary volumes.
Issued also in a cumulative microfiche form under the title
Bio-Base and on-line by way of DIALOG Information Services,
Inc., as _Biography Master Index_ (_BMI_). These latter forms
were not consulted for _B.I.A.S._ but users are referred to
them for possible additional or updated references.

Biol Soc Washington _Proc_ =
Proceedings of Biological Society of Washington.

Blake =
Blake, Samuel. _Blake Family: A Genealogical History of
William Blake of Dorchester, and His Descendants_. Boston:
Ebenezer Clapp, Jr., 1857.

Boatner =
Boatner, Mark M. _The Civil War Dictionary_. New York: David
McKay Co., Inc., 1959.

Boston Evening Transcript =
Boston Evening Transcript.

Boston Med Surg Journ =
Boston Medical and Surgical Journal.

Boston Soc Nat Hist _Proc_ =
Proceedings of Boston Society of Natural History.

Bot Gaz =
Botanical Gazette.

Bot Zentralblatt =
Botanisches Zentralblatt.

Bradley =
Bradley, Francis. "Extracts from an Annual Register Kept at
New Haven, Connecticut from May 1842 - September 1854."
Transactions of Connecticut Academy of Arts and Sciences
1(1866-1871):139-54.

Brigham =
Brigham, Willard I. The Tyler Genealogy: The Descendants of
Job Tyler, of Andover, Massachusetts, 1619-1700. 2 vols.
Plainfield, N.J.: Cornelius B. Tyler; Tylerville, Conn.:
Rollin U. Tyler, 1912.

Brooklyn Ent Soc Bull =
Bulletin of Brooklyn Entomological Society.

Brown Hist Cat 1904 =
Brown University. Historical Catalogue of Brown University,
1764-1904. Providence, R.I.: Published by the University,
1905.

Brown Hist Cat 1934 =
Brown University. Historical Catalogue of Brown University,
1764-1934. Providence, R.I.: Published by the University,
1936.

Browne and Weeks =
Browne, Charles Albert, and Mary Elvira Weeks. A History of
the American Chemical Society--Seventy-five Eventful Years.
Washington, D.C.: American Chemical Society, 1952.

Bulloch =
Bulloch, William. The History of Bacteriology. London and
New York: Oxford University Press, 1938.

Burke and Howe =
Burke, William J., and Will D. Howe. American Authors and
Books, 1640 to the Present Day. 3d revised edition. Revised
by Irving Weiss and Anne Weiss. New York: Crown Publishers,
Inc., 1972.

California Ac Scis Proc =
Proceedings of California Academy of Sciences.

Canadian Ent =
Canadian Entomologist.

Carolyn Sherwin Bailey Collection =
Davis, Dorothy R., ed. and comp. The Carolyn Sherwin Bailey
Historical Collection of Children's Books: A Catalogue. New
Haven, Conn.: Southern Connecticut State College, 1966.

Carwell =
Carwell, Hattie. Blacks in Science. Hicksville, N.Y.:
Exposition, 1977.

Cassinia =
Cassinia: Proceedings of the Delaware Valley Ornithological
Club.

Chamberlain =
Chamberlain, Joshua L., ed. Universities and Their Sons. 5
vols. Boston: R. Herndon Co., 1898-1900.

Chem Indus Contribution =
Haynes, William, and Edward L. Gory, eds. Chemical
Industry's Contribution to the Nation: 1635-1935; A Record
of Chemical Accomplishments... [New York, 1935].   "A
supplement to Chemical Industries, May 1935."

Chem News =
Chemical News and Journal of Physical Science.

Chicago Ac Scis Bull =
Bulletin of Chicago Academy of Sciences.

Chymia =
Chymia: Annual Studies in the History of Chemistry.

Cincinnati Soc Nat Hist Journ =
Journal of Cincinnati Society of Natural History.

Claghorn =
Claghorn, Charles Eugene. Biographical Dictionary of
American Music. West Nyack, N.Y.: Parker Publishing Co.,
Inc., 1973.

Cleveland Ac Nat Scis Proc =
Proceedings of the Cleveland Academy of Natural Sciences,
1845 to 1859. Cleveland, Ohio: Published by a gentleman of
Cleveland, 1874.

Coffin =
Coffin, Seldon J. The Men of Lafayette, 1826-1893:
Lafayette College, Its History, Its Men, Their Record.
Easton, Penn.: George W. West, 1891.

Cohen =
Cohen, Rose N. The Men Who Gave Us Wings. Aviation Readers.
New York: Macmillan, 1944.

Colby Obit Rec =
Colby College. An Obituary Record of Colby University,
1822-1870. Waterville, Maine: Printed for the Alumni
[Lewiston, Journal Press], 1870.

Conch Exchange =
Conchologists' Exchange.

Cordell =
Cordell, Eugene F. Historical Sketch of the University of
Maryland School of Medicine (1807-1890). Baltimore: Press
of Isaac Friedenwald, 1891.

Cordell Med Ann =
Cordell, Eugene F. The Medical Annals of Maryland, 1799-1899. Baltimore: [Press of Williams and Wilkins Co.], 1903.

Country Gentleman =
Country Gentleman.

Croom =
Croom, H. B. Catalogue of Plants, Native or Naturalized, in the Vicinity of Newbern, North Carolina. New York: George P. Scott and Co., Printers, 1837.

Culin =
Culin, Stewart. "The Dickeson Collection of American Antiquities." Bulletin of University Museum of the University of Pennsylvania 2(1900):113-68.

Cullum =
Cullum, George W. Biographical Register of the Officers and Graduates of the U.S. Military Academy at West Point, New York. 3d edition revised and extended. Boston and New York: Houghton Mifflin Co.; Riverside Press, 1891-

Cutright =
Cutright, Paul Russell. Theodore Roosevelt: The Naturalist. New York: Harper, 1956.

Davenport Ac Nat Scis Proc =
Proceedings of Davenport Academy of Natural Sciences.

Davison =
Davison, Charles. Founders of Seismology. Cambridge: Cambridge University Press, 1927.

Dawdy =
Dawdy, Doris Ostrander. Artists of the American West: A Biographical Dictionary. Chicago: Swallow Press, Inc., 1974.

Deutsche Akademie Nat Nova Acta =
Nova Acta Leopoldina [of] Deutsche Akademie der Naturforscher.

Dexter Biog Notices =
Dexter, Franklin B. Biographical Notices of Graduates of Yale College Including Those Graduated in Classes Later than 1815, Who Are Not Commemorated in the Annual Obituary Records. Issued as a Supplement to the Obituary Record. New Haven: [Yale University?], 1913.

Dexter Biog Sketches =
Dexter, Franklin B. Biographical Sketches of the Graduates of Yale College with Annals of the College History. 6 vols. New York: Henry Holt and Co.; New Haven: Yale University Press, 1885-1912.

Dict Am Biog =
Dictionary of American Biography. Published under the
auspices of the American Council of Learned Societies. 20
vols. New York: C. Scribner, 1928-1937. Supplement 1 (vol.
21), 1944.

Dict Am Lib Biog =
Wynar, Bohdan S., ed. Dictionary of American Library
Biography. Littleton, Col.: Libraries Unlimited, Inc.,
1978.

Dict Am Med Biog =
Kaufman, Martin, Stuart Galishoff, and Todd L. Savitt, eds.
Dictionary of American Medical Biography. 2 vols.
Westport, Conn.: Greenwood Press, 1984.

Dict Natl Biog =
Stephen, Sir Leslie, and Sir Sidney Lee, eds. Dictionary of
National Biography, from the Earliest Times to 1900. 21
vols. London: Oxford University Press, H. Milton Co., 1937-
1966.

Dict Sci Biog =
Gillispie, Charles Coulston, ed. Dictionary of Scientific
Biography. 16 vols. New York: Scribner, 1970-1980.

Drake =
Drake, Francis S. Dictionary of American Biography
including Men of the Time. Boston: James R. Osgood & Co.,
1872. Reprinted: Detroit: Gale Research Co., 1974.

Druggists Circ =
Druggists' Circular.

Dunlap =
Dunlap, Orrin E., Jr. Radio's 100 Men of Science. New York
and London: Harper and Brothers, 1944.

Durfee =
Durfee, Calvin. Williams [College] Biographical Annals.
Boston: Lee and Shepard; New York: Lee, Shepard, and
Dillingham, 1871.

Duyckinck =
Duyckinck, Evert A., and George L. Duyckinck. Cyclopaedia
of American Literature. 2 vols. Philadelphia: William
Rutter & Co., 1875. Reprinted: Detroit: Gale Research Co.,
1965.

Electrician =
Electrician.

Elisha Mitchell Sci Soc Journ =
Journal of Elisha Mitchell Scientific Society.

Elliott =
Elliott, Clark A. _Biographical Dictionary of American Science: The Seventeenth through the Nineteenth Centuries_. Westport, Conn.: Greenwood Press, 1979. Note: "(s)" indicates short entry, generally for persons living after 1906.

Ellis =
Ellis, Franklin. _History of Columbia County, New York_. Philadelphia: Everts and Ensign, 1878.

Ency Occultism =
Shepard, Leslie, ed. _Encyclopedia of Occultism and Parapsychology_. 2 vols. and supplement. Detroit: Gale Research Co., 1978.

Ent Am =
Entomologica Americana.

Ent Mo Mag =
Entomologist's Monthly Magazine.

Ent News =
Entomological News.

Ent Rec =
Entomologist's Record and Journal of Variation.

Ent Soc Ontario _Rep_ =
Report of Entomological Society of Ontario.

Ent Soc Philadelphia _Proc_
Proceedings of Entomological Society of Philadelphia.

Erythea =
Erythea: A Journal of Botany, West American and General.

Essex Inst. _Bull_ =
Bulletin of Essex Institute.

Essig =
Essig, E. O. _A History of Entomology_. New York: Macmillan, 1931.

Field and Lab =
Field and Laboratory.

Fisher =
Fisher, George P. _Life of Benjamin Silliman_. 2 vols. New York: Charles Scribner, 1866.

Franklin Inst _Journ_ =
Journal of Franklin Institute.

Futhey and Cope =
Futhey, J. Smith, and Gilbert Cope. _History of Chester County, Pennsylvania with Genealogical and Biographical Sketches_. Philadelphia: Louis H. Everts, 1881.

Ganter =
Ganter, Herbert L. "William Small, Jefferson's Beloved Teacher." William and Mary Quarterly 3d series 4(1947):501-11.

Gard and Forest =
Garden and Forest.

Gard Chronicle =
Gardeners' Chronicle.

Geog Jahrbuch =
Geographisches Jahrbuch.

Geol Mag =
Geological Magazine.

Geol Soc Am Bull =
Bulletin of Geological Society of America.

Geol Soc London Q Journ =
Quarterly Journal of Geological Society of London.

Gies =
Gies, Joseph, and Frances Gies. The Ingenious Yankees. New York: Crowell, 1976.

Goss =
Goss, Charles F. Cincinnati--The Queen City 1788-1912. 4 vols. Chicago and Cincinnati: The S. J. Clarke Publishing Co., 1912.

Groce and Wallace =
Groce, George C., and David H. Wallace. The New York Historical Society's Dictionary of Artists in America, 1564-1860. New Haven, Conn.: Yale University Press, 1957.

Gruson =
Gruson, Edward S. Words for Birds: A Lexicon of North American Birds with Biographical Notes. New York: Quadrangle, 1972.

Haber =
Haber, Louis. Black Pioneers of Science and Invention. New York: Harcourt, Brace and World, 1970.

Hall =
Hall, Henry, ed. America's Successful Men of Affairs: An Encyclopedia of Contemporaneous Biography. 2 vols. New York: New York Tribune, 1895-1896.

Hampden Sidney Gen Cat =
Hampden-Sidney College. General Catalogue of the Officers and Students of Hampden-Sidney College, Virginia, 1776-1906. Richmond, Va.: Whittet and Shepperson, Printers, [1908].

Hanley =
Hanley, Wayne. <u>Natural History in America</u>. New York:
Quadrangle, 1977.

Harrington =
Harrington, Thomas F. <u>The Harvard Medical School: A
History</u>. 3 vols. New York and Chicago: Lewis Publishing
Co., 1905.

Harshberger =
Harshberger, John W. <u>The Botanists of Philadelphia and
Their Work</u>. Philadelphia: [Press of T. C. Davis and Sons],
1899.

Harvard Class of 1893 <u>Rep</u> =
Harvard College, Class of 1893. <u>Secretary's Seventh Report</u>.
Cambridge: Crimson Printing Co., printed for the Class,
1923.

Harvard <u>Hist Register</u> =
Harvard University. <u>Historical Register of Harvard
University, 1636-1936</u>. Cambridge: Harvard University,
1937.

Harvard <u>Quinquennial Cat</u> =
Harvard University. <u>Quinquennial Catalogue of Officers and
Graduates 1636-1930</u>. Cambridge: Published for the
University, 1930.

Hawks =
Hawks, Ellison. <u>Pioneers of Wireless</u>. London: Methuen and
Co., 1927.

Haynes =
Haynes, William. <u>Chemical Pioneers</u>. vol. 1. New York: D.
Van Nostrand Co., 1939.

Heinmuller =
Heinmuller, John. <u>Man's Fight to Fly</u>. New and London: Funk
& Wagnalls, 1944.

Hinkel =
Hinkel, Edgar J., ed. <u>Biographies of California Authors and
Indexes to California Literature</u>. Published by the Alameda
County Library, Oakland, California ... under the auspices
of the Works Projects Administration. 2 vols. Oakland,
Calif., 1942.

Holloway =
Holloway, Lisabeth M. <u>Medical Obituaries: American
Physicians' Biographical Notices in Selected Medical
Journals Before 1907</u>. New York and London: Garland, 1981.

Howard =
Howard, Arthur V. <u>Chambers's Dictionary of Scientists</u>. New
York: Dutton, 1958.

Howard F =
Howard, Fred. Wilbur and Orville: A Biography of the Wright
Brothers. New York: Knopf, 1987.

Humming Bird =
Humming Bird.

Ibis =
Ibis: A Quarterly Journal of Ornithology.

Ind NUCMC =
Index to Personal Names in the National Union Catalog of
Manuscript Collections 1959 - 1984. 2 vols. Alexandria:
Chadwyck-Healey, 1988. Note: The numbers that sometimes
follow this reference are National Union Catalog of
Manuscript Collections (i.e., NUCMC) numbers for
collections of the person's own papers or those in which he
or she is a major component. The absence of NUCMC numbers
means that all references in the Ind NUCMC are for
manuscripts found among the papers of another person.
"(?)" indicates some degree of uncertainty as to whether
the person listed in Ind NUCMC and in B.I.A.S. are the
same.

Indiana Authors 1816 =
Banta, R. E., comp. Indiana Authors and Their Books, 1816-
1916. Crawfordsville, Ind.: Wabash College, 1949.

Indiana Authors 1917 =
Thompson, Donald E., comp. Indiana Authors and Their Books,
1917-1966. A continuation of Indiana Authors and Their
Books, 1816-1916, and containing additional names from the
earlier period. Crawfordsville, Ind.: Wabash College,
1974.

Insect Life =
Insect Life [U.S. Department of Agriculture. Division of
Entomology. Periodical bulletin].

Iowa Ac Scis Proc =
Proceedings of Iowa Academy of Sciences.

Ireland Ind Sci =
Ireland, Norma Olin. Index to Scientists of the World from
Ancient to Modern Times: Biographies and Portraits. Boston:
F. W. Faxon Co., Inc., 1962.

Ireland Ind Wom =
Ireland, Norma Olin. Index to Women of the World from
Ancient to Modern Times: Biographies and Portraits.
Westwood, Mass.: F. W. Faxon Co., Inc., 1970.

Jessup =
Jessup, Philip C. Elihu Root. 2 vols. New York: Dodd, Mead
and Co., 1938.

Jewett and McCausland =
Jewett, Frances L., and Clare L. McCausland. Plant Hunters.
Boston: Houghton Mifflin Co., 1958.

Johns Hopkins Biol Lab Mem =
Memoirs from Johns Hopkins University Biological
Laboratory.

Johns Hopkins Circ =
Johns Hopkins University Circulars.

Johns Hopkins Half Cent Dir =
Johns Hopkins University. Johns Hopkins Half-Century
Directory...1876-1926. Baltimore: The Johns Hopkins
University, 1926.

Johnson =
Johnson, Thomas C. Scientific Interests in the Old South.
New York: D. Appleton Century Co., Inc., for the Institute
for Research in the Social Sciences, University of
Virginia, 1936.

Journ Bot =
Journal of Botany, British and Foreign.

Journ Conch =
Journal of Conchology.

Journ Mycology =
Journal of Mycology.

Kagan =
Kagan, Solomon R. Modern Medical World. Boston: Medico-
Historical Press, 1945.

Kansas Ac Sci Trans =
Transactions of Kansas Academy of Science.

Kansas Univ Q =
Kansas University Quarterly.

Kelly =
Kelly, Howard A. A Cyclopedia of American Medical
Biography. 2 vols. Philadelphia and London: W. B. Saunders,
Co., 1912.

Kelly and Burrage =
Kelly, Howard A., and Walter L. Burrage. Dictionary of
American Medical Biography. New York and London: Appleton
and Co., 1928.

Klein =
Klein, Aaron E. The Hidden Contributors. New York:
Doubleday, 1971.

Knight =
Knight, Lucian Lamar, comp. Biographical Dictionary of
Southern Authors. Detroit: Gale Research Co., 1978.
Originally published as Library of Southern Literature,
vol. 15, Biographical Dictionary of Authors (Atlanta:
Martin & Hoyt Co., 1929).

LaBorde =
LaBorde, M. History of the South Carolina College.
Columbia, S.C.: Peter B. Glass, 1859.

Lanman =
Lanman, Charles. Biographical Annals of the Civil
Government of the United States, During Its First Century.
Washington, D.C.: James Anglim, Publisher, 1876.
Reprinted: Detroit: Gale Research Co., 1976.

Leonardo =
Leonardo, Richard. Lives of Master Surgeons. New York:
Froben, 1948.

Leopoldina =
Leopoldina [Deutsche Akademie der Naturforscher].

Lincoln Lib Language Arts =
The Lincoln Library of Language Arts. 3d ed. 2 vols.
Columbus, Ohio: Frontier Press Co., 1978. Note:
Biographies begin vol. 1, p.345.

Lincoln Lib Social Studies =
The Lincoln Library of Social Studies. 8th ed. 3 vols.
Columbus, Ohio: Frontier Press Co., 1978. Note:
Biographies begin vol. 3, p.865.

Macmillan Dict Canadian Biog =
Wallace, W. Stewart, ed. The Macmillan Dictionary of
Canadian Biography. 4th ed., revised, enlarged and updated
by W. A. McKay. Toronto: Macmillan of Canada, 1978.

Maitland =
Maitland, Lester J. Knights of the Air. Garden City, N.Y.:
Doubleday, Doran & Co., 1929.

Major =
Major, Ralph H. A History of Medicine. 2 vols. Springfield,
Ill.: Charles Thomas, 1954.

Mallis =
Mallis, Arnold. American Entomologists. New Brunswick,
N.J.: Rutgers University Press, 1971.

Martin =
Martin, John H. Martin's Bench and Bar of Philadelphia,
Together with Other Lists of Persons Appointed to
Administer the Laws in the City and County of Philadelphia,
and the Province and Commonwealth of Pennsylvania.
Philadelphia: R. Welsh and Co., 1883.

Med Soc St California <u>Trans</u> =
<u>Transactions of Medical Society of the State of
California</u>.

Merrill =
Merrill, George P. <u>The First One Hundred Years of American
Geology</u>. New Haven: Yale University Press, 1924.

<u>Meteorol Zeitschrift</u> =
<u>Meteorologische Zeitschrift</u>.

Minnesota Ac Nat Scis <u>Bull</u> =
<u>Bulletin of Minnesota Academy of Natural Sciences</u>.

Morris and Morris =
Morris, Don, and Inez Morris. <u>Who Was Who in American
Politics. A Biographical Dictionary of Over 4,000 Men and
Women Who Contributed to the United States Political Scene
from Colonial Days Up to and Including the Immediate Past</u>.
New York: Hawthorn Books, Inc., Publishers, 1974.

Natl Ac Scis <u>Biog Mem</u> =
National Academy of Sciences. <u>Biographical Memoirs</u>.
Washington, D.C.: National Academy of Sciences, 1-, 1877-
Note: Vol. 10 and 11 of the <u>Biographical Memoirs</u> were
published as the Academy's <u>Memoirs</u> vol. 17 and 21
respectively.

Natl Am Soc <u>Am Families</u> =
National Americana Society, New York. <u>American Families of
Historic Lineage, Long Island Edition</u>. Edited by William S.
Pelletreau and John H. Brown. Edition Etoile D'Argent. 2
vols. New York: National Americana Society, [1913].

Natl Cycl Am Biog =
<u>National Cyclopaedia of American Biography</u>. New York: James
T. White & Co., 1892-1984.

<u>Natl Union Cat</u> =
<u>National Union Catalog, Pre-1956 Imprints</u>. 685 vols.
London: Mansell, 1968-1980.

<u>Nature</u> =
<u>Nature</u>.

<u>Nautilus</u> =
<u>Nautilus</u>.

New Jersey Hist Soc <u>Proc</u> =
<u>Proceedings of New Jersey Historical Society</u>.

New Orleans Med Surg Journ =
<u>New Orleans Medical and Surgical Journal</u>.

New York Ac Scis <u>Ann</u> =
<u>Annals of New York Academy of Sciences</u>.

New York Ac Scis <u>Trans</u> =
<u>Transactions of New York Academy of Sciences</u>.

New York Ent Soc <u>Journ</u> =
<u>Journal of New York Entomological Society</u>.

<u>New York Med Journ</u> =
<u>New York Medical Journal</u>.

New York St Ag Soc <u>Trans</u>
<u>Transactions of New York State Agricultural Society</u>.

<u>New York Times</u> =
<u>New York Times</u>.

New York Univ <u>Gen Cat</u> =
New York University. <u>General Alumni Catalogue of New York</u>
<u>University, 1833-1907, Medical Alumni</u>. New York: General
Alumni Society, 1908.

Nicholls =
Nicholls, Peter. <u>The Encyclopedia of Science Fiction: An</u>
<u>Illustrated A to Z</u>. London: Grenada Publishing Ltd., 1979.

North =
North, Dexter. <u>John North of Farmington, Connecticut and</u>
<u>His Descendants, with a Short Account of Other Early North</u>
<u>Families</u>. Washington, D.C., 1921.

Notable Am Wom =
James, Edward T., ed. <u>Notable American Women 1607-1950</u>. 3
vols. Cambridge: The Belknap Press of Harvard University
Press, 1971.

<u>Notarisia</u> =
<u>Notarisia</u>.

Ogilvie =
Ogilvie, Marilyn Bailey. <u>Women in Science: Antiquity</u>
<u>through the Nineteenth Century: A Biographical Dictionary</u>
<u>with Annotated Bibliogrpahy</u>. Cambridge, Mass.: MIT Press,
1986. Note: "(app)" indicates that the entry is in the
volume's appendix.

<u>Ohio Authors</u> =
Coyle, William H. ed. <u>Ohio Authors and Their Books:</u>
<u>Biographical Data and Selective Bibliographies for Ohio</u>
<u>Authors, Native and Resident, 1796-1950</u>. Cleveland and New
York: World Publishing Co., 1962

<u>Ornith Monatsberichte</u> =
<u>Ornithologische Monatsberichte</u>.

<u>Ornith Ool</u> =
<u>Ornithologist and Oologist</u>.

Osborn Hen =
Osborn, Henry Fairfield, ed. A Naturalist in the Bahamas:
John I. Northrop, October 12, 1861 - June 25, 1891: A
Memorial Volume. New York: Columbia University Press,
1910.

Osborn Herb =
Osborn, Herbert. Fragments of Entomological History.
Columbus, Ohio: The author, 1937.

Ottawa Nat =
Ottawa Naturalist.

Oxford Canadian Hist =
The Oxford Companion to Canadian History and Literarure.
Toronto: Oxford University Press, 1967.   Supplement, 1973.

Papilio =
Papilio.

Pelletier =
Pelletier, Paul A. Prominent Scientists: An Index to
Collective Biographies. New York: Neal-Schuman Publishers,
Inc., 1980.

Pennsylvania Biog Cat =
University of Pennsylvania, Society of the Alumni.
Biographical Catalogue of the Matriculates of the College,
Together with Lists of the Members of the College Faculty
and Trustees, Officers and Recipients of Honorary Degrees,
1749-1893. Philadelphia: Printed for the Society, 1894.

Pennsylvania Cat Med =
University of Pennsylvania, Society of the Alumni of the
Medical Department. Catalogue of the Alumni of the Medical
Department of the University of Pennsylvania, 1765-1877.
Philadelphia: Published by the Society of the Alumni of the
Medical Department; Collins Printers, 1877.

Pennsylvania Gen Cat =
University of Pennsylvania. General Alumni Catalogue of the
University of Pennsylvania, 1922. [Philadelphia: Alumni
Association of the University of Pennsylvania], 1922.

Phil Soc Washington Bull =
Bulletin of Philosophical Society of Washington.

Philadelphia College Pharm First Cent =
Philadelphia College of Pharmacy and Science. The First
Century of the Philadelphia College of Pharmacy, 1821-1921.
[Philadelphia]: Philadelphia College of Pharmacy and
Science, 1922.

Poggendorff =
Poggendorff, J. C. Biographisch-Literarisches
Handworterbuch zur Geschichte de Exacten Wissenschaften.
Leipzig: Verlag von Johann Ambrosius Barth, 1863-

Pop Ast =
Popular Astronomy.

Pop Sci Mo =
Popular Science Monthly.

Powell =
Powell, William H., ed. Officers of the Army and Navy
(Volunteers) Who Served in the Civil War. Philadelphia: L.
R. Hamersly & Co., 1893.

Preston =
Preston, Wheeler. American Biographies. New York: Harper &
Brothers Publishers, 1940. Reprinted: Detroit: Gale
Research Co., 1974.

Princeton Theol Sem Biog Cat =
Princeton Theological Seminary. Biographical Catalogue of
the Princeton Theological Seminary, 1815-1932. Princeton,
N.J.: Trustees of the Theological Seminary of the
Presbyterian Church, 1933.

Princeton Theol Sem Necrological Rep =
Princeton Theological Seminary, Alumni Association.
Necrological Reports and Annual Proceedings of the Alumni
Association. Princeton, N.J., 1891-19[32?].

Princeton Univ Gen Cat =
Princeton University. General Catalogue of Princeton
University, 1746-1906. Princeton, N.J.: Published for the
University, [1908].

Psyche =
Psyche, a Journal of Entomology.

Quinlan =
Quinlan, James E. History of Sullivan County Embracing an
Account of Its Geology, Climate, Aborigines, Early
Settlement, Organization, the Formation of Its Towns, with
Biographical Sketches of Prominent Residents. Liberty,
N.Y.: G. M. Beebe and W. T. Morgans, 1873.

Rail Eng Journ =
Railroad and Engineering Journal.

Rail Gaz =
Railroad Gazette.

Rensselaer Biog Rec =
Rensselaer Polytechnic Institute. Biographical Record of
the Officers and Graduates of the Rensselaer Polytechnic
Institute, 1824-1886. Edited by Henry B. Nason. Troy, N.Y.:
William H. Young, 1887.

Roy Micr Soc Journ =
Journal of Royal Microscopical Society.

Roy Soc <u>Cat</u> =
Royal Society of London. <u>Catalogue of Scientific Papers,</u>
<u>1800-1900</u>. 19 vols. London: Clay, 1867-1902; Cambridge:
University Press, 1914-1925.

Roy Soc Edinburgh <u>Proc</u> =
<u>Proceedings of Royal Society of Edinburgh</u>.

Roy Soc <u>Proc</u> =
<u>Proceedings of Royal Society of London</u>.

Samuels =
Samuels, Peggy, and Harold Samuels. <u>The Illustrated</u>
<u>Biographical Encyclopedia of Artists of the American West</u>.
Garden City, N.Y.: Doubleday & Co., Inc., 1976.

Sargent =
Sargent, C. S. <u>Silva of North America</u>. 14 vols. Boston and
New York: Houghton Mifflin, 1891-1902.

Savage =
Savage, George. <u>Dictionary of Nineteenth Century Antiques</u>
<u>and Later Objets d'Art</u>. London: Barrie & Jenkins Ltd.,
1978.

Sch Mines Q =
<u>School of Mines Quarterly: A Journal of Applied Science</u>.

Schapsmeier =
Schapsmeier, Edward L., and Frederick H. Schapsmeier.
<u>Encyclopedia of American Agricultural History</u>. Westport,
Conn.: Greenwood Press, 1975.

<u>Sci</u> =
<u>Science</u>.

Shaw =
Shaw, John Mackay. <u>Childhood in Poetry</u>. A catalogue, with
biographical and critical annotations, of the books of
English and American poets comprising the Shaw Childhood in
Poetry Collection in the Library of Florida State
University, with lists of the poems that relate to
childhood, notes, and index. Detroit: Gale Research Co.,
1967, 1972, 1976, 1980.

<u>Sidereal Messenger</u> =
<u>Sidereal Messenger</u>.

Siegel and Finley =
Siegel, Patricia Joan, and Kay Thomas Finley. <u>Women in the</u>
<u>Scientific Search: An American Bio-Bibliography, 1724-1979</u>.
Metuchen, N.J. and London: The Scarecrow Press, Inc.,
1985.

Simpson =
Simpson, Henry. <u>The Lives of Eminent Philadelphians, Now</u>
<u>Deceased</u>. Philadelphia: William Brotherhead, 1859.

Sinclair =
Sinclair, Bruce. Philadelphia's Philosopher Mechanics: A
History of the Franklin Institute 1824-1865. Baltimore:
Johns Hopkins University Press, 1974.

Smith A =
Smith, Anna W. Genealogy of the Fisher Family, 1682-1896.
Philadelphia, 1896.

Smith E =
Smith, Edgar F. Chemistry in America. New York: D. Appleton
& Co., 1929.

Smithsonian Bur Eth Rep =
Annual Report of Bureau of American Ethnology to the
Secretary of the Smithsonian Institution.

Smithsonian Rep =
Annual Report of Board of Regents of the Smithsonian
Institution.

Soc Belge Micr Bull =
Bulletin [of] Societe Belge de Microscopie.

Soc Chem Indus Journ =
Journal of Society of Chemical Industry.

Sonnedecker =
Sonnedecker, Glenn. "The History of American Pharmaceutical
Education to 1900." Ph.D. dissertation, University of
Wisconsin, 1953.

Staten Island Assoc Arts Scis Proc =
Proceedings of Staten Island Association of Arts and
Sciences.

Stearns =
Stearns, Raymond P. Science in the British Colonies of
America. Urbana: University of Illinois Press, 1970.

Stephens =
Stephens, Lester D. Joseph LeConte: Gentle Prophet of
Evolution. Baton Rouge and London: Louisiana State
University Press, 1982.

Stewart =
Stewart, Frank H. History of the First United States Mint:
Its People and Its Operations. [Camden, N.J.]: Privately
printed, 1924.

Stiles =
Stiles, Henry R., ed. The Civil, Political, Professional,
and Ecclesiastical History and Commercial and Industrial
Record of the County of Kings and the City of Brooklyn, New
York, from 1683 to 1884. New York: W. W. Munsell and Co.,
[1884].

Struik =
Struik, Dirk. Yankee Science in the Making. Boston: Little,
1948.

Sullivan =
Sullivan, Thomas P. Sketches of the Lives of John L.
Sullivan and Thomas Russell. Boston, 1861.

Sylvester =
Sylvester, Nathaniel B. History of Saratoga County, New
York. Philadelphia: Everts & Ensign, 1878.

Thomas =
Thomas, John P. The History of the South Carolina Military
Academy. Charleston, S.C.: Walker, Evans & Cogswell Co.,
1893.

Three Am =
Three American Microscope Builders. Buffalo, N.Y.: American
Optical Co., Scientific Instrument Division, 1945.

Tobey and Thompson =
Tobey, Walter L., and William O. Thomson. The Diamond
Anniversary Volume ... 1824-1899 Miami University, Oxford,
Ohio. Hamilton, Ohio: The Republican Publishing Co.,
[1900?].

Torrey Bot Club Bull =
Bulletin of Torrey Botanical Club.

Tulane Graduates Mag =
Tulane Graduates' Magazine.

Twentieth Cent Biog Dict =
Johnson, Rossiter, ed. The Twentieth Century Biographical
Dictionary of Notable Americans. 10 vols. Boston: The
Biographical Society, 1904. Reprinted: Detroit: Gale
Research Co., 1968.

U S Natl Lib Med Ind Cat =
United States National Library of Medicine. Index Catalogue
of the Surgeon General's Office, U.S. Army (Army Medical
Library), Authors and Subjects. Washington, D.C.:
Government Printing Office, 1880-, series 1-

U S Natl Mus Bull =
Bulletin of United States National Museum.

U S Natl Mus Rep =
United States National Museum. Annual Report of the Board
of Regents of the Smithsonian Institution...Report of the
U.S. National Museum. Washington, D.C.: Government Printing
Office, 1883-

U S Weather Bur Mo Weather Rev =
United States Weather Bureau. Monthly Weather Review.

Upham and Dunlap =
Upham, Warren, and Rose B. Dunlap, comp. <u>Minnesota</u>
<u>Biographies, 1655-1912</u>. Collections of the Minnesota
Historical Society, vol. 14. St. Paul, Minn.: The Society,
1912.

<u>Victorian Nat</u> =
<u>Victorian Naturalist</u>.

Waite =
Waite, Frederick C. <u>Western Reserve University: The Hudson</u>
<u>Era</u>. Cleveland, Ohio: Western Reserve University, 1943.

Wakelyn =
Wakelyn, Jon L. <u>Biographical Dictionary of the Confederacy</u>.
Westport, Conn.: Greenwood Press, 1977.

Wallace =
Wallace, W. Stewart, comp. <u>A Dictionary of North American</u>
<u>Authors Deceased before 1950</u>. Toronto: Ryerson Press, 1951.
Reprinted: Detroit: Gale Research Co., 1968.

Wallbridge =
Wallbridge, William G. <u>Descendants of Henry Wallbridge Who</u>
<u>Married Anna Amos, December 25, 1688</u>. Philadelphia: Press
of Franklin Printing Co., 1898.

Wesleyan <u>Alumni Rec</u> =
Wesleyan University. <u>Alumni Record of Wesleyan University,</u>
<u>Middletown, Conn.</u>   3d ed. Hartford, Conn.: Press of the
Case, Lockwood and Brainard Co., 1883.   4th ed. New Haven,
Conn.: The Tuttle, Morehouse, and Taylor Co., 1911.

<u>West Am Sci</u> =
<u>West American Scientist</u>.

<u>Who Was Who Am</u> =
<u>Who Was Who in America</u>. Chicago: Marquis, 1943- Note:
"(add)" indicates that an entry is in the volume's
addendum.
vol. h: Historical volume, 1607-1896 (1963/1967).
vol. 1: Volume one, 1897-1942 (1943).
vol. 2: Volume two, 1943-1950 (1963).
vol. 3: Volume three, 1951-1960 (1966).
vol. 4: Volume four, 1961-1968 (1968).
vol. 5: Volume five, 1969-1973 (1973).

<u>Who Was Who Am Hist</u> =
<u>Who Was Who in American History: Science and Technology</u>. A
component of Who's Who in American History. Chicago:
Marquis's Who's Who, 1977.

<u>Who Was Who Lit</u> =
<u>Who Was Who in Literature, 1906-1934</u>. 2 vols. Gale
Composite Biographical Dictionary Series, no. 5. Detroit:
Gale Research Co., 1979.

Who Was Who North Am Authors =
Who Was Who among North American Authors, 1921-1939. 2
vols. Gale Composite Biographical Dictionary Series, no.
1. Detroit: Gale Research Co., 1976.

Whos Who Am =
Who's Who in America. Chicago: A. N. Marquis and Co., 1899-

Williams =
Williams, Trevor. A Biographical Dictionary of Scientists.
2nd ed. New York: Wiley, 1974.

Williams Gen Cat =
Williams College. General Catalogue of the Officers and
Graduates of Williams College 1910. Williamstown, Mass.:
The College, 1910.

Wimmer =
Wimmer, Curt P. College of Pharmacy of the City of New
York, Included in Columbia University in 1904: A History.
[Baltimore: Printed by Read-Taylor], 1929.

Wolf =
Wolf, Ralph F. India Rubber Man: The Story of Charles
Goodyear. Caldwell, Idaho: The Caxton Printers, 1939.

Wom Whos Who Am =
Leonard, John William, ed. Woman's Who's Who of America: A
Biographical Dictionary of Contemporary Women of the United
States and Canada, 1914-1915. New York: American
Commonwealth Co., 1914. Reprinted: Detroit: Gale Research
Co., 1976.

Woods Hole Biol Lectures =
Woods Hole, Mass. Marine Biological Laboratory. Biological
Lectures.

World Whos Who Sci =
World Who's Who in Science: A Biographical Dictionary of
Notable Scientists from Antiquity to the Present. Chicago:
Marquis-Who's Who, 1968.

Wright =
Wright, P. G., and E. Q. Wright. Elizur Wright [Jr.], the
Founder of Life Insurance. Chicago: University of Chicago
Press, 1937.

Yale Cat =
Yale University. Yale Catalogue of Officers and Graduates,
1701-1924. New Haven: Published by the University, 1924.

Yale Obit Rec =
Yale University. Obituary Record of Graduates of Yale
University... New Haven, 1860-

Young =
Young, Herman, and Barbara Young. Scientists in the Black
Perspective. Louisville, Ky.: Lincoln Foundation, 1974.

<u>Zoologist</u> =
<u>Zoologist: A Monthly Journal of Natural History</u>.

BIOGRAPHICAL INDEX

A

ABBE, CLEVELAND (1838-1916). Astronomy; Meteorology. OCCUP: U.S. government scientist. MSS: <u>Ind NUCMC</u> (64-1556, 75-1935). WORKS ABOUT: <u>Am Men Sci</u> 1-2 d3; <u>Dict Am Biog</u>; <u>Dict Sci Biog</u>; Elliott (s); Natl Ac Scis <u>Biog Mem</u> 8; <u>Natl Cycl Am Biog</u> 8; <u>Who Was Who Am</u> 1. See also: Barr; <u>Biog Geneal Master Ind</u>; Ireland <u>Ind Sci</u>; Pelletier.

ABBOT, JOEL (1766-1826). Physics (magnetism). OCCUP: Physician; Legislator. MSS: <u>Ind NUCMC</u> (?). WORKS ABOUT [some as Abbott; birth dates may differ]: Appleton's; <u>Biog Dir Am Congress</u>; Lanman; <u>Natl Union Cat</u>; <u>Twentieth Cent Biog Dict</u>; Wallace; <u>Who Was Who Am</u> h.

ABBOT, JOHN (1751-1840 or 1841). Entomology; Ornithology. OCCUP: Natural history collector. MSS: <u>Ind NUCMC</u> (79-765). WORKS ABOUT [Some as Abbott; birth or death dates may differ]: Elliott. See also: Barr; <u>Biog Geneal Master Ind</u>; Pelletier.

ABBOTT, ARTHUR VAUGHAN (1854-1906). Electrical engineering. OCCUP: Industrial engineer. WORKS ABOUT: Adams O; <u>Sci</u> 24(1906):830; Wallace; <u>Who Was Who Am</u> 1.

ABBOTT, CHARLES CONRAD (1843-1919). Natural history; Zoology; Archeology. OCCUP: Engaged in independent study. MSS: <u>Ind NUCMC</u> (66-2). WORKS ABOUT: <u>Am Men Sci</u> 1-2 d3; <u>Dict Am Biog</u>; Elliott (s); <u>Natl Cycl Am Biog</u> 10; <u>Who Was Who Am</u> 1. See also: Barr; <u>Biog Geneal Master Ind</u>.

ABBOTT, SAMUEL WARREN (1837-1904). Statistics; Demography; Medicine. OCCUP: Physician; Public official. MSS: <u>Ind NUCMC</u> (62-4515). WORKS ABOUT: <u>Dict Am Biog</u>; <u>Natl Cycl Am Biog</u> 20; <u>Who Was Who Am</u> 1. See also: <u>Biog Geneal Master Ind</u>.

ABERT, JOHN JAMES (1788-1863). Topographical engineering. OCCUP: Army officer; Engineer. MSS: <u>Ind NUCMC</u>. WORKS ABOUT: <u>Dict Am Biog</u> supp 1; Elliott; <u>Natl Cycl Am Biog</u> 4; <u>Who Was Who Am</u> h. See also: <u>Biog Geneal Master Ind</u>.

ABERT, SILVANUS THAYER (1828-1903). Civil engineering. OCCUP: U.S. government engineer. WORKS ABOUT: Adams O; Allibone Supp; Appleton's; Wallace; Who Was Who Am 1.

ACKER, CHARLES ERNEST (1868-1920). Electrochemistry. OCCUP: Inventor; Manufacturer. WORKS ABOUT: Am Men Sci 2 d3; Dict Am Biog; Natl Cycl Am Biog 13; Who Was Who Am 1. See also: Biog Geneal Master Ind.

ACKHURST, JOHN (1816-1902). Taxidermy. OCCUP: Taxidermist. WORKS ABOUT: Sci 15(1901):358.

ADAMS, ARTHUR KINNEY (1883-1920). Geology. OCCUP: Professor. WORKS ABOUT: Am Men Sci 2 d3.

ADAMS, CHARLES BAKER (1814-1853). Natural history; Conchology. OCCUP: Professor; State government geologist. MSS: Ind NUCMC. WORKS ABOUT: Dict Am Biog; Elliott; Natl Cycl Am Biog 5; Who Was Who Am h. See also: Barr; Biog Geneal Master Ind; Ireland Ind Sci.

ADAMS, CHARLES FRANCIS (1854-1914). Physics. OCCUP: School science teacher. WORKS ABOUT: Sci 40(1914):780, 42(1915):826; Wallace.

ADAMS, DANIEL (1773-1864). Mathematics. OCCUP: Physician; Writer and editor. WORKS ABOUT: Dict Am Biog; Natl Cycl Am Biog 20; Who Was Who Am h. See also: Biog Geneal Master Ind.

ADAMS, ISAAC, JR. (1836-1911). Chemical engineering. OCCUP: Physician; Industrialist. WORKS ABOUT: Am Chem and Chem Eng; Sci 21(1905):875-76.

ADLER, ISAAC (1849-1918). Medicine. OCCUP: Physician; Professor. WORKS ABOUT: Am Men Sci 1-2 d3; Natl Cycl Am Biog 11; Who Was Who Am 4.

ADLUM, JOHN (1759-1836). Agriculture. OCCUP: Farmer; Horticulturist. MSS: Ind NUCMC (82-1786). WORKS ABOUT: Dict Am Biog; Who Was Who Am h. See also: Biog Geneal Master Ind.

ADRAIN, ROBERT (1775-1843). Mathematics. OCCUP: School teacher; Professor. WORKS ABOUT: Dict Am Biog; Dict Sci Biog; Elliott; Natl Cycl Am Biog 1; Who Was Who Am h. See also: Biog Geneal Master Ind; Ireland Ind Sci.

AGASSIZ, ALEXANDER EMMANUEL RODOLPHE (1835-1910). Natural history; Zoology; Oceanography; Engineering. OCCUP: Museum administrator; Businessman. MSS: Ind NUCMC. WORKS ABOUT: Am Men Sci 1 d3; Dict Am Biog; Dict Sci Biog; Elliott (s); Natl Ac Scis Biog Mem 7; Natl Cycl Am Biog 3; Who Was Who Am 1. See also: Barr; Biog Geneal Master Ind; Ireland Ind Sci; Pelletier.

AGASSIZ, ELIZABETH CABOT CARY (1822-1907). Natural history; Science writing. OCCUP: College president. MSS: Ind NUCMC (61-1779). WORKS ABOUT: Natl Cycl Am Biog 12; Notable Am Wom; Ogilvie; Siegel and Finley; Who Was Who Am 1. See also: Barr; Biog Geneal Master Ind.

AGASSIZ, JEAN LOUIS RODOLPHE (1807-1873). Natural history; Zoology; Ichthyology; Geology; Paleontology. OCCUP: Professor. MSS: Ind NUCMC (81-399). WORKS ABOUT: Dict Am Biog; Dict Sci Biog; Elliott; Natl Ac Scis Biog Mem 2; Natl Cycl Am Biog 2; Who Was Who Am h. See also: Barr; Biog Geneal Master Ind; Ireland Ind Sci; Pelletier.

AGASSIZ, LOUIS. See AGASSIZ, JEAN LOUIS RODOLPHE.

AGNEW, CORNELIUS REA (1830-1888). Ophthalmology. OCCUP: Physician; Professor. MSS: Ind NUCMC (74-258, 76-1456, 79-756). WORKS ABOUT: Dict Am Biog; Natl Cycl Am Biog 8; Who Was Who Am h. See also: Barr; Biog Geneal Master Ind.

AIKIN, WILLIAM EDWARD AUGUSTIN (1807-1888). General science; Chemistry. OCCUP: Professor; Academic administrator. Scientific work: see Roy Soc Cat. MSS: Ind NUCMC (?). WORKS ABOUT: Cordell; Rensselaer Biog Rec.

ALBERTSON, MARY A. (-1914). Astronomy; Botany. OCCUP: Librarian; Curator. WORKS ABOUT: Ogilvie (app); Sci 40(1914):341.

ALDEN, CHARLES HENRY (1836-1910). Medicine. OCCUP: Army surgeon; Academic administrator. WORKS ABOUT: Am Men Sci 1 d3; Wallace; Who Was Who Am 1.

ALDEN, TIMOTHY, JR. (1771-1839). Electrical phenomena; Diving bell. OCCUP: Clergyman; Professor (language and history); College president. Scientific work: see Roy Soc Cat. MSS: Ind NUCMC (66-1010). WORKS ABOUT: Dict Am Biog; Natl Cycl Am Biog 13; Who Was Who Am h. See also: Biog Geneal Master Ind.

ALDRICH, CHARLES (1828-1909). Natural history; Zoology; Ornithology. OCCUP: Editor and publisher; Legislator; Collector. MSS: Ind NUCMC (62-1511, 62-4036, 77-1125). WORKS ABOUT: Am Men Sci 1 d3; Appleton's Supp 1901; Natl Cycl Am Biog 9; Sci 29(1909):610; Twentieth Cent Biog Dict; Wallace; Who Was Who Am 1.

ALEXANDER, A. B. (-1916). Statistics. OCCUP: U.S. government employee. WORKS ABOUT: Sci 44(1916):707.

ALEXANDER, JAMES (1691-1756). Mathematics. OCCUP: State government surveyor; Lawyer; Public official. MSS: Ind NUCMC (60-650, 60-1945). WORKS ABOUT: Dict Am Biog; Who Was Who Am h. See also: Biog Geneal Master Ind.

ALEXANDER, JOHN HENRY (1812-1867). Applied science; Mathematics; Geology. OCCUP: State government engineer-surveyor; Businessman; Professor. MSS: Ind NUCMC. WORKS ABOUT: Dict Am Biog; Elliott; Natl Ac Scis Biog Mem 1; Natl Cycl Am Biog 9; Who Was Who Am h. See also: Barr; Biog Geneal Master Ind.

ALEXANDER, SAMUEL (1837?-1917). Botany. OCCUP: City government forester. MSS: Ind NUCMC (67-25). WORKS ABOUT: Sci 45(1917):498-99.

ALEXANDER, STEPHEN (1806-1883). Astronomy. OCCUP: Professor. MSS: Ind NUCMC. WORKS ABOUT: Dict Am Biog; Elliott; Natl Ac Scis Biog Mem 2; Natl Cycl Am Biog 11; Who Was Who Am h. See also: Barr; Biog Geneal Master Ind.

ALEXANDER, WILLIAM (LORD STIRLING) (1726-1783). Mathematics; Astronomy. OCCUP: Public official; State government surveyor. MSS: Ind NUCMC (60-2736, 70-1692). WORKS ABOUT: Dict Am Biog; Natl Cycl Am Biog 1; Who Was Who Am h. See also: Biog Geneal Master Ind; Ireland Ind Sci.

ALEXANDER, WILLIAM DeWITT (1833-1913). Science; Ethnology; Geography; History. OCCUP: Educator; Surveyor. WORKS ABOUT: Am Men Sci 1-2 d3; Dict Am Biog; Who Was Who Am 1. See also: Biog Geneal Master Ind.

ALGER, FRANCIS (1807-1863). Mineralogy. OCCUP: Businessman; Manufacturer; Mineralogical explorer. WORKS ABOUT: Elliott. See also: Barr.

ALLEN, ALFRED REGINALD (1876-1918). Neurology. OCCUP: Neurologist. WORKS ABOUT: Sci 48(1918):467; Who Was Who Am 1.

ALLEN, DUDLEY PETER (1852-1915). Surgery. OCCUP: Surgeon; Professor. WORKS ABOUT: Leonardo p11-12; Natl Cycl Am Biog 14, 16; Sci 41(1915):163; Who Was Who Am 1.

ALLEN, HARRISON (1841-1897). Zoology; Anatomy; Medicine. OCCUP: Physician; Professor. MSS: Ind NUCMC (68-1459). WORKS ABOUT: Dict Am Biog; Elliott; Natl Cycl Am Biog 9; Who Was Who Am h. See also: Barr; Biog Geneal Master Ind.

ALLEN, JOHN ROBINS (1869-1920). Mechanical engineering. OCCUP: Professor. WORKS ABOUT: Am Men Sci 1-3 d4; Natl Cycl Am Biog 19; Wallace.

ALLEN, JONATHAN ADAMS (1787-1848). Botany; Chemistry; Geology. OCCUP: Physician; Professor. WORKS ABOUT: Kelly and Burrage; Wallace

ALLEN, OSCAR DANA (1836-1913). Chemistry. OCCUP: Professor. WORKS ABOUT: Elliott. See also: Barr; Biog Geneal Master Ind.

ALLEN, RICHARD HINCKLEY (1838-1908). Astronomy. OCCUP: Not determined; Author. WORKS ABOUT: <u>Sci</u> 27(1908):276; Wallace.

ALLEN, TIMOTHY FIELD (1837-1902). Botany. OCCUP: Physician; Professor; Academic administrator (medicine). WORKS ABOUT: <u>Dict Am Biog</u>; Elliott; <u>Natl Cycl Am Biog</u> 7:282; <u>Who Was Who Am</u> 1. See also: <u>Biog Geneal Master Ind</u>.

ALLEN, WALTER MORRISON (1867?-1909). Optics. OCCUP: Instrument maker. WORKS ABOUT: <u>Sci</u> 29(1909):294.

ALLEN, WILLIAM FREDERICK (1846-1915). Metrology; Railway economics. OCCUP: Engineer; Administrator. MSS: <u>Ind NUCMC</u> (66-1895, 70-1693). WORKS ABOUT: <u>Am Men Sci</u> 1-2 d3; <u>Dict Am Biog</u>; <u>Natl Cycl Am Biog</u> 14, 23; <u>Who Was Who Am</u> 1. See also: <u>Biog Geneal Master Ind</u>.

ALLEN, ZACHARIAH (1795-1882). Science; Invention. OCCUP: Businessman; Inventor. WORKS ABOUT: <u>Dict Am Biog</u>; <u>Natl Cycl Am Biog</u> 8; <u>Who Was Who Am</u> h. See also: <u>Biog Geneal Master Ind</u>; Ireland <u>Ind Sci</u>.

ALLIN, ARTHUR (1869-1903). Psychology. OCCUP: Professor. WORKS ABOUT: Adams O; <u>Sci</u> 18(1903):703; <u>Who Was Who Am</u> 1.

ALPERS, WILLIAM CHARLES (1851-1917). Chemistry; Pharmaceutical chemistry. OCCUP: Industrialist; Academic administrator. MSS: <u>Ind NUCMC</u> (?). WORKS ABOUT: <u>Am Men Sci</u> 1-2 d3; <u>Natl Cycl Am Biog</u> 22; Wallace; <u>Who Was Who Am</u> 1.

ALTER, DAVID (1807-1881). Physics. OCCUP: Physician; Manufacturer. WORKS ABOUT: <u>Dict Am Biog</u>; Elliott; <u>Who Was Who Am</u> h. See also: <u>Biog Geneal Master Ind</u>; Ireland <u>Ind Sci</u>; Pelletier.

ALVORD, BENJAMIN (1813-1884). Mathematics; Biology. OCCUP: Army officer. MSS: <u>Ind NUCMC</u>. WORKS ABOUT: <u>Dict Am Biog</u>; <u>Natl Cycl Am Biog</u> 4; <u>Who Was Who Am</u> h. See also: <u>Biog Geneal Master Ind</u>.

ALVORD, HENRY ELIJAH (1844-1904). Agriculture. OCCUP: Farmer; Academic administrator; U.S. government scientist. MSS: <u>Ind NUCMC</u>. WORKS ABOUT: <u>Dict Am Biog</u>; <u>Natl Cycl Am Biog</u> 22; <u>Who Was Who Am</u> 1. See also: Barr; <u>Biog Geneal Master Ind</u>

AMEND, BERNHARD GOTTWALD (1821?-1917). Chemistry. OCCUP: Businessman; Banker. WORKS ABOUT: <u>Sci</u> 33(1911):575; <u>Who Was Who Am</u> 3.

AMES, MARY L. PULSIFER (1845?-1902). Botany. OCCUP: Science writer. MSS: <u>Ind NUCMC</u> (?). WORKS ABOUT: <u>Sci</u> 15(1902):517.

AMES, NATHANIEL (1708-1764). Almanac making; Mathematics. OCCUP: Physician; Publisher (almanacs). MSS: Ind NUCMC. WORKS ABOUT: Dict Am Biog; Natl Cycl Am Biog 8; Who Was Who Am h. See also: Biog Geneal Master Ind.

AMON, FRANK (-1918). Chemistry. OCCUP: Academic researcher. WORKS ABOUT: Sci 49(1919):42.

AMORY, ROBERT (1842-1910). Physiology. OCCUP: Physician. WORKS ABOUT: Allibone Supp; Appleton's; Sci 32(1910):303; Twentieth Cent Biog Dict; Wallace; Who Was Who Am 1.

ANDERSON, FREDERICK W. (1866-1891). Botany. OCCUP: U.S. government scientist; Agricultural publishing associate. MSS: Ind NUCMC (?). WORKS ABOUT: Bot Gaz 17(1892):78-81.

ANDERSON, MALCOLM PLAYFAIR (1879-1919). Zoology. OCCUP: Zoological explorer. MSS: Ind NUCMC (?). WORKS ABOUT: Am Men Sci 2 d3.

ANDREWS, EBENEZER BALDWIN (1821-1880). Astronomy; Geology. OCCUP: Clergyman; Professor; State government scientist. WORKS ABOUT: Allibone Supp; Am Journ Sci 120(1880):255; Appleton's; Ohio Authors; Wallace.

ANDREWS, EDMUND (1824-1904). Archeology; Surgery. OCCUP: Surgeon; Professor (surgery). WORKS ABOUT: Allibone Supp; Am Men Sci 1 d3; Appleton's; Dict Am Med Biog; Leonardo p13-14; Twentieth Cent Biog Dict; Wallace; Who Was Who Am 1.

ANDREWS, WILLIAM HOLMES (-1905). Chemistry. OCCUP: State government scientist. WORKS ABOUT: Sci 22(1905):726.

ANTHONY, JOHN GOULD (1804-1877). Zoology. OCCUP: Businessman; Museum curator. MSS: Ind NUCMC (65-1221). WORKS ABOUT: Dict Am Biog; Elliott; Natl Cycl Am Biog 10; Who Was Who Am h. See also: Barr; Biog Geneal Master Ind.

ANTHONY, WILLIAM ARNOLD (1835-1908). Physics. OCCUP: Professor. MSS: Ind NUCMC. WORKS ABOUT: Am Men Sci 1 d3; Dict Am Biog supp 1; Elliott (s); Natl Cycl Am Biog 11; Who Was Who Am 1. See also: Barr; Biog Geneal Master Ind.

ANTISELL, THOMAS (1817-1893). Chemistry; Geology. OCCUP: Physician; Professor; U.S. government scientist. MSS: Ind NUCMC (?). WORKS ABOUT: Allibone; Allibone Supp; Am Chem and Chem Eng; Merrill p313; Natl Cycl Am Biog 19; Phil Soc Washington Bull 13(1900):367-70; Wallace.

APGAR, AUSTIN CRAIG (1838-1908). Natural history. OCCUP: Teacher; Educator. WORKS ABOUT: Adams O; Sci 27(1908):518; Wallace; Who Was Who Am 1.

APGAR, ELLIS A. (1836-1905). Botany. OCCUP: Educational administrator; Author. WORKS ABOUT: Biog Dict Am Educ; Sci 22(1905):319; Wallace.

APPLE, ANDREW THOMAS GEIGER (1858-1918). Astronomy. OCCUP: Clergyman; Professor; Observatory director. WORKS ABOUT: Sci 47(1918):236-37; Who Was Who Am 1.

ARENTS, ALBERT (1840-1914). Metallurgy. OCCUP: Industrial metallurgist. WORKS ABOUT: Dict Am Biog; Who Was Who Am 4. See also: Biog Geneal Master Ind.

ARMSTRONG, JOSEPH R. (-1919?). Exploration. OCCUP: African explorer. WORKS ABOUT: Sci 50(1919):565.

ARNOLD, DELOS (-1909). Geology. OCCUP: Geological collector and donor. WORKS ABOUT: Sci 30(1909):437.

ARNOLD, JOHN WILLIAM SCHMIDT (1846-1888). Medicine; Physiology. OCCUP: Physician; Professor. WORKS ABOUT: Atkinson; Boston Med Surg Journ 119(1888):442; New York Med Journ 48(1888):463. See also: Holloway.

ASHBURNER, CHARLES ALBERT (1854-1889). Geology. OCCUP: State government scientist; Mining manager. WORKS ABOUT: Dict Am Biog; Elliott; Natl Cycl Am Biog 11; Who Was Who Am h. See also: Barr; Biog Geneal Master Ind; Pelletier.

ASHLEY, HARRISON EVERETT (1876-1911). Chemistry. OCCUP: U.S. government scientist. WORKS ABOUT: Am Men Sci 1-2 d3.

ASHMEAD, WILLIAM HARRIS (1855-1908). Entomology. OCCUP: Museum curator. MSS: Ind NUCMC (?). WORKS ABOUT: Am Men Sci 1 d3; Dict Am Biog; Elliott (s); Natl Cycl Am Biog 20; Who Was Who Am 1. See also: Barr; Biog Geneal Master Ind; Ireland Ind Sci; Pelletier.

ATKINS, H. A. (1821-1885). Ornithology. OCCUP: Physician. WORKS ABOUT: Ornith Ool 10(1885):120.

ATKINSON, GEORGE FRANCIS (1854-1918). Botany. OCCUP: Professor; Academic researcher. MSS: Ind NUCMC (62-2338). WORKS ABOUT: Am Men Sci 1-2 d3; Dict Am Biog; Elliott (s); Natl Ac Scis Biog Mem 29; Natl Cycl Am Biog 13; Who Was Who Am 1. See also: Barr; Biog Geneal Master Ind.

ATWATER, CALEB (1778-1867). Natural history. OCCUP: Lawyer; Legislator. MSS: Ind NUCMC (68-1212, 75-1882). WORKS ABOUT: Dict Am Biog; Elliott; Natl Cycl Am Biog 22; Who Was Who Am h. See also: Biog Geneal Master Ind.

ATWATER, WILBUR OLIN (1844-1907). Chemistry; Agricultural chemistry; Physiology; Scientific administration. OCCUP: Professor. WORKS ABOUT: Am Men Sci 1 d3; Dict Am Biog; Dict Am Med Biog; Dict Sci Biog; Elliott (s); Natl Cycl Am Biog 6; Who Was Who Am 1. See also: Barr; Biog Geneal Master Ind; Ireland Ind Sci; Pelletier.

ATWOOD, LUTHER (1826-1868). Chemistry. OCCUP: Industrial scientist. WORKS ABOUT: Natl Cycl Am Biog 13.

ATWOOD, NATHANIEL ELLIS (1807-1886). Ichthyology. OCCUP:
Seaman; Manufacturer; Legislator. WORKS ABOUT: Am Ac Arts
Scis <u>Proc</u> 22(1887):522-23; Essex Inst <u>Bull</u> 19(1888):68-71.

ATWOOD, WILLIAM (1830-1884). Chemistry. OCCUP: Industrial
scientist and manager. WORKS ABOUT: <u>Natl Cycl Am Biog</u> 13.

AUDUBON, JOHN JAMES (1785-1851). Ornithology. OCCUP:
Merchant; Artist. MSS: <u>Ind NUCMC</u> (60-859, 60-3108, 61-849,
62-1920, 66-8, 81-406). WORKS ABOUT: <u>Dict Am Biog</u>; <u>Dict Sci
Biog</u>; Elliott; <u>Natl Cycl Am Biog</u> 6; <u>Who Was Who Am</u> h. See
also: Barr; <u>Biog Geneal Master Ind</u>; Ireland <u>Ind Sci</u>;
Pelletier.

AUGHEY, SAMUEL (1831-1912). Geology. OCCUP: Clergyman;
Professor; Mining expert. MSS: <u>Ind NUCMC</u>. WORKS ABOUT: <u>Sci</u>
35(1912):414; Wallace.

AUSTEN, PETER TOWNSEND (1852-1907). Chemistry. OCCUP:
Professor; Consultant; Inventor. MSS: <u>Ind NUCMC</u> (?). WORKS
ABOUT: <u>Am Men Sci</u> 1 d3; <u>Dict Am Biog</u>; Elliott (s); <u>Natl
Cycl Am Biog</u> 13; <u>Who Was Who Am</u> 1. See also: Barr; <u>Biog
Geneal Master Ind</u>; Pelletier.

AUSTIN, COE FINCH (1831-1880). Bryology; Muscology. OCCUP:
Herbarium curator. WORKS ABOUT: Adams O; Allibone <u>Supp</u>; <u>Am
Journ Sci</u> 119(1880):423, 123(1882):332, 127(1884):242;
Appleton's; Torrey Bot Club <u>Bull</u> 7(1880):38-39; Wallace.

AVERY, CHARLES ELLERY (1848?-1916). Pharmacology. OCCUP:
Professor. WORKS ABOUT: <u>Sci</u> 44(1916):743.

AVERY, WILLIAM CUSHMAN (ca.1832-1894). Ornithology. OCCUP:
Physician. MSS: <u>Ind NUCMC</u> (60-1165). WORKS ABOUT: <u>Auk</u>
11(1894):263-64.

AYRES, BROWN (1856-1919). Physics. OCCUP: Professor;
College president. WORKS ABOUT: <u>Am Men Sci</u> 1-2 d3; <u>Dict Am
Biog</u>; <u>Natl Cycl Am Biog</u> 18; <u>Who Was Who Am</u> 1. See also:
Barr; <u>Biog Geneal Master Ind</u>.

AYRES, WILLIAM ORVILLE (1817-1887). Zoology; Ichthyology.
OCCUP: Physician; Professor; Academic administrator
(medicine). WORKS ABOUT: Elliott. See also: Barr.

B

BABCOCK, HENRY HOMES (1832-1881). Botany. OCCUP: Teacher; Educator. WORKS ABOUT: Chicago Ac Scis <u>Bull</u> 2(1891-1901):vi-viii.

BABCOCK, JAMES FRANCIS (1844-1897). Chemistry. OCCUP: Professor; Public official. WORKS ABOUT: <u>Dict Am Biog</u>; Elliott; <u>Natl Cycl Am Biog</u> 10; <u>Who Was Who Am</u> h. See also: Barr; <u>Biog Geneal Master Ind</u>; Pelletier.

BACHE, ALEXANDER DALLAS (1806-1867). Physics; Geophysics. OCCUP: Professor; U.S. government scientist. MSS: <u>Ind NUCMC</u> (61-701, 62-4858, 72-1228). WORKS ABOUT: <u>Dict Am Biog</u>; <u>Dict Sci Biog</u>; Elliott; Natl Ac Scis <u>Biog Mem</u> 1; <u>Natl Cycl Am Biog</u> 3; <u>Who Was Who Am</u> h. See also: Barr; <u>Biog Geneal Master Ind</u>; Ireland <u>Ind Sci</u>; Pelletier.

BACHE, FRANKLIN (1792-1864). Chemistry. OCCUP: Physician; Professor. MSS: <u>Ind NUCMC</u> (60-2022). WORKS ABOUT: <u>Dict Am Biog</u>; <u>Dict Am Med Biog</u>; Elliott; <u>Natl Cycl Am Biog</u> 5; <u>Who Was Who Am</u> h. See also: <u>Biog Geneal Master Ind</u>; Pelletier.

BACHE, THOMAS H. (1826?-1912). Medicine. OCCUP: Physician. MSS: <u>Ind NUCMC</u> (?). WORKS ABOUT: <u>Sci</u> 36(1912):79.

BACHMAN, JOHN (1790-1874). Natural history; Zoology. OCCUP: Clergyman. MSS: <u>Ind NUCMC</u>. WORKS ABOUT: <u>Dict Am Biog</u>; Elliott; <u>Who Was Who Am</u> h. See also: <u>Biog Geneal Master Ind</u>; Ireland <u>Ind Sci</u>; Pelletier.

BACON, CHARLES A. (1860?-1901). Astronomy. OCCUP: Professor; Observatory director. WORKS ABOUT: <u>Sci</u> 14(1901):780.

BACON, JOHN, JR. (1817-1881). Chemistry; Physiological chemistry. OCCUP: Hospital scientist; Professor. MSS: <u>Ind NUCMC</u> (?). WORKS ABOUT: Am Ac Arts Scis <u>Proc</u> 18(1883):419-22.

BAILEY, BERT HEALD (1875-1917). Ornithology. OCCUP: Professor; Museum curator. WORKS ABOUT: Sci 46(1917):14, 450-51; Wallace; Who Was Who Am 1.

BAILEY, FRANK KELTON (-1909). Physics. OCCUP: Academic instructor. WORKS ABOUT: Sci 30(1909):111.

BAILEY, JACOB WHITMAN (1811-1857). Botany; Chemistry; Geology. OCCUP: Professor. MSS: Ind NUCMC (70-1316). WORKS ABOUT: Dict Am Biog; Elliott; Natl Cycl Am Biog 10; Who Was Who Am h. See also: Barr; Biog Geneal Master Ind; Ireland Ind Sci; Pelletier.

BAILEY, JAMES SPENCER (1830-1883). Entomology. OCCUP: Physician. WORKS ABOUT: Brooklyn Ent Soc Bull 6(1883-1884):48; Ent Soc Ontario Rep (1883):82; Leopoldina 20(1884):57; Papilio 3(1883):166-67; Sci 1st ser 2(1883):148.

BAILEY, WILLIAM WHITMAN (1843-1914). Botany. OCCUP: Professor. MSS: Ind NUCMC (70-1317). WORKS ABOUT: Adams O; Allibone Supp; Am Journ Sci 187(1914):366; Am Men Sci 1-2 d3; Appleton's; Natl Cycl Am Biog 10, 29; Sci 39 (1914):323; Twentieth Cent Biog Dict; Wallace; Who Was Who Am 1, 4.

BAIN, SAMUEL McCUTCHEN (1869-1914). Botany. OCCUP: Professor. WORKS ABOUT: Am Men Sci 1-2 d3.

BAIRD, JOHN WALLACE (1873-1919). Psychology. OCCUP: Professor. MSS: Ind NUCMC. WORKS ABOUT [birth dates may differ]: Am Journ Sci 197(1919):454; Am Men Sci 1-2 d3; Natl Cycl Am Biog 22; Sci 49(1919):213, 393-94; Who Was Who Am 1

BAIRD, JULIAN WILLIAM (1859-1911). Chemistry. OCCUP: Professor; Academic administrator. WORKS ABOUT: Sci 34(1911):12; Who Was Who Am 1.

BAIRD, SPENCER FULLERTON (1823-1887). Natural history; Zoology. OCCUP: Museum curator; Administrator; U.S. government scientist. MSS: Ind NUCMC (72-1229). WORKS ABOUT: Dict Am Biog; Dict Sci Biog; Elliott; Natl Ac Scis Biog Mem 3; Natl Cycl Am Biog 3; Who Was Who Am h. See also: Barr; Biog Geneal Master Ind; Ireland Ind Sci; Pelletier.

BAKER, FRANK (1841-1918). Zoology; Anatomy. OCCUP: Professor; Zoological park administrator. WORKS ABOUT: Am Men Sci 1-2 d3; Dict Am Biog; Natl Cycl Am Biog 19; Who Was Who Am 1.

BAKER, GEORGE LORIMER (-1909). Medicine. OCCUP: Physician(?); Researcher. WORKS ABOUT: Sci 29(1909):542.

BAKER, MARCUS (1849-1903). Geography; Cartography; Mathematics. OCCUP: U.S. government scientist. WORKS ABOUT: Dict Am Biog; Elliott; Natl Cycl Am Biog 11; Who Was Who Am 1. See also: Biog Geneal Master Ind.

BAKER, PHILIP S. (-1901). Chemistry. OCCUP: Professor. WORKS ABOUT: Sci 14(1901):501.

BALBACH, EDWARD, JR. (1839-1910). Metallurgy. OCCUP: Industrialist. WORKS ABOUT: Dict Am Biog; Natl Cycl Am Biog 17; Who Was Who Am 4.

BALDWIN, DAVID DWIGHT (1831-1912). Botany; Conchology. OCCUP: Museum curator; Educator; Industrial officer. WORKS ABOUT: Am Men Sci 2 d3; Wallace.

BALDWIN, LOAMMI (1740-1807). Engineering. OCCUP: Civil engineer. MSS: Ind NUCMC (81-411). WORKS ABOUT [birth dates may differ]: Dict Am Biog; Natl Cycl Am Biog 10; Who Was Who Am h. See also: Biog Geneal Master Ind; Pelletier.

BALDWIN, LOAMMI, JR. (1780-1838). Engineering. OCCUP: Lawyer; Civil engineer. MSS: Ind NUCMC (84-2068). WORKS ABOUT: Dict Am Biog; Natl Cycl Am Biog 10; Who Was Who Am h. See also: Biog Geneal Master Ind; Pelletier.

BALDWIN, WILLIAM (1779-1819). Botany. OCCUP: Physician; Natural history explorer/ collector. MSS: Ind NUCMC. WORKS ABOUT: Dict Am Biog; Elliott; Natl Cycl Am Biog 10; Who Was Who Am h. See also: Barr.

BANCROFT, EDWARD (1744-1821). Chemistry. OCCUP: Writer; Researcher. MSS: Ind NUCMC. WORKS ABOUT: Dict Am Biog; Who Was Who Am h. See also: Biog Geneal Master Ind; Pelletier.

BANDELIER, ADOLPH FRANCIS ALPHONSE (1840-1914). Anthropology; Archeology. OCCUP: Explorer; Museum associate. MSS: Ind NUCMC. WORKS ABOUT: Am Men Sci 1-2 d3; Dict Am Biog; Natl Cycl Am Biog 26; Who Was Who Am 1. See also: Barr; Biog Geneal Master Ind; Ireland Ind Sci.

BANISTER, JOHN (1650-1692). Anthropology; Botany; Entomology; Malacology. OCCUP: Clergyman; Natural history explorer/ collector. WORKS ABOUT: Dict Am Biog; Dict Sci Biog; Elliott; Who Was Who Am h. See also: Biog Geneal Master Ind.

BANNEKER, BENJAMIN (1731-1806). Astronomy; Mathematics. OCCUP: Farmer; Surveyor. MSS: Ind NUCMC. WORKS ABOUT: Elliott; Natl Cycl Am Biog 5; Who Was Who Am h. See also: Biog Geneal Master Ind; Pelletier.

BANNISTER, HENRY MARTYN (1844-1920). Geology; Medicine. OCCUP: Museum employee; Physician; Editor. MSS: Ind NUCMC (?). WORKS ABOUT: Am Men Sci 2 d3; Who Was Who Am 4.

BARBER, EDWIN AtLEE (1851-1916). Archeology; Botany. OCCUP: U.S. government scientist; Museum curator. WORKS ABOUT: Am Men Sci 1-2 d3; Dict Am Biog; Natl Cycl Am Biog 22; Who Was Who Am 1. See also: Biog Geneal Master Ind.

BARBOUR, VOLNEY G. (-1901). Engineering. OCCUP: Professor. WORKS ABOUT: Sci 13(1901):956.

BARCLAY, L. P. (-1908). Optics. OCCUP: Industrial employee. WORKS ABOUT: Sci 28(1908):370.

BARD, JOHN (1716-1799). Anatomy. OCCUP: Physician. MSS: Ind NUCMC (60-1236). WORKS ABOUT: Dict Am Biog; Dict Am Med Biog; Who Was Who Am h. See also: Biog Geneal Master Ind; Ireland Ind Sci.

BARD, SAMUEL (1742-1821). Medicine. OCCUP: Physician; Professor. MSS: Ind NUCMC. WORKS ABOUT: Dict Am Biog; Dict Am Med Biog; Natl Cycl Am Biog 8; Who Was Who Am h. See also: Biog Geneal Master Ind; Ireland Ind Sci; Pelletier.

BARDWELL, ELIZABETH M. (1832?-1899). Astronomy. OCCUP: Professor. WORKS ABOUT: Sci 9(1899):821-22.

BARKER, FRANKLIN LUTHER (1871-1920). Economic geology. OCCUP: Professor. WORKS ABOUT: Am Men Sci 2 d3.

BARKER, GEORGE FREDERICK (1835-1910). Chemistry; Physics. OCCUP: Professor. MSS: Ind NUCMC. WORKS ABOUT: Am Men Sci 1 d3; Dict Am Biog; Elliott (s); Natl Cycl Am Biog 4; Who Was Who Am 1. See also: Barr; Biog Geneal Master Ind; Ireland Ind Sci; Pelletier.

BARNARD, EDITH ETHEL (1880-1914). Chemistry. OCCUP: Academic instructor. WORKS ABOUT: Am Men Sci 2 d3; Sci 39(1914):421.

BARNARD, FREDERICK AUGUSTUS PORTER (1809-1889). Astronomy; Mathematics; Physics. OCCUP: Professor; College president. MSS: Ind NUCMC (61-3232). WORKS ABOUT: Dict Am Biog; Elliott; Natl Ac Scis Biog Mem 20; Natl Cycl Am Biog 6; Who Was Who Am h. See also: Barr; Biog Geneal Master Ind; Ireland Ind Sci.

BARNARD, JOHN (1681-1770). Mathematics. OCCUP: Clergyman. MSS: Ind NUCMC (?). WORKS ABOUT: Dict Am Biog; Natl Cycl Am Biog 7:305; Who Was Who Am h. See also: Biog Geneal Master Ind.

BARNARD, JOHN GROSS (1815-1882). Engineering; Mathematics; Physics. OCCUP: Army officer; Engineer. MSS: Ind NUCMC. WORKS ABOUT: Dict Am Biog; Elliott; Natl Ac Scis Biog Mem 5; Natl Cycl Am Biog 4; Who Was Who Am h. See also: Barr; Biog Geneal Master Ind.

BARNARD, WILLIAM STEBBINS (1849-1887). Natural history; Entomology. OCCUP: Academic instructor; U.S. government scientist. MSS: Ind NUCMC (?). WORKS ABOUT: Am Nat 21 (1887):1136-37; Appleton's; Natl Cycl Am Biog 12; Osborn Herb p179-80; Twentieth Cent Biog Dict.

BARNES, CHARLES REID (1858-1910). Botany; Plant physiology. OCCUP: Professor. MSS: Ind NUCMC. WORKS ABOUT: Am Men Sci 1 d3; Dict Am Biog; Elliott (s); Natl Cycl Am Biog 13; Who Was Who Am 1. See also: Barr; Biog Geneal Master Ind.

BARNES, DANIEL HENRY (1785-1828). Zoology; Conchology. OCCUP: Clergyman; Educator. WORKS ABOUT [death dates may differ]: Allibone; Am Journ Sci 15(1829):401-02; Appleton's; Drake; Twentieth Cent Biog Dict.

BARRELL, JOSEPH (1869-1919). Geology. OCCUP: Professor. MSS: Ind NUCMC. WORKS ABOUT: Am Men Sci 1-2 d3; Dict Am Biog; Dict Sci Biog; Natl Ac Scis Biog Mem 12; Who Was Who Am 1. See also: Biog Geneal Master Ind.

BARRIS, WILLIS HERVEY (1821?-1901). Geology. OCCUP: Professor (theology?). MSS: Ind NUCMC (?). WORKS ABOUT: Sci 14(1901):117.

BARROWS, CHARLES CLIFFORD (1857-1916). Gynecology. OCCUP: Physician; Professor. WORKS ABOUT: Appleton's Supp 1918-31; Natl Cycl Am Biog 3, 36; Sci 43(1916):64-65, 528; Who Was Who Am 1.

BARTLETT, JOHN RUSSELL (1843-1904). Physical geography. OCCUP: Naval officer. WORKS ABOUT: Appleton's; Twentieth Cent Biog Dict; Who Was Who Am 1.

BARTLETT, WILLIAM HOLMS CHAMBERS (1804-1893). Astronomy; Mathematics. OCCUP: Army engineer; Professor; Actuary. MSS: Ind NUCMC. WORKS ABOUT: Dict Am Biog supp 1; Elliott; Natl Ac Scis Biog Mem 7. See also: Biog Geneal Master Ind.

BARTLETT, WILLIAM PITT GREENWOOD (1837-1865). Mathematics. OCCUP: U.S. government scientist. Scientific work: see Roy Soc Cat. WORKS ABOUT: Appleton's.

BARTON, BENJAMIN SMITH (1766-1815). Natural history; Botany; Zoology; Ethnography; Medicine. OCCUP: Physician; Professor. MSS: Ind NUCMC (60-2224, 61-316, 61-721, 76-892). WORKS ABOUT: Dict Am Biog; Dict Am Med Biog; Dict Sci Biog; Elliott; Natl Cycl Am Biog 8; Who Was Who Am h. See also: Barr; Biog Geneal Master Ind; Ireland Ind Sci.

BARTON, WILLIAM PAUL CRILLON (1786-1856). Botany. OCCUP: Physician; Professor. MSS: Ind NUCMC. WORKS ABOUT: Dict Am Biog; Dict Am Med Biog; Elliott; Natl Cycl Am Biog 13; Who Was Who Am h. See also: Biog Geneal Master Ind.

BARTRAM, JOHN (1699-1777). Botany. OCCUP: Farmer; Natural history explorer/ collector. MSS: <u>Ind NUCMC</u> (60-1467, 61-729, 66-11). WORKS ABOUT: <u>Dict Am Biog</u>; <u>Dict Sci Biog</u>; Elliott; <u>Natl Cycl Am Biog</u> 7:153; <u>Who Was Who Am</u> h. See also: Barr; <u>Biog Geneal Master Ind</u>; Ireland <u>Ind Sci</u>; Pelletier.

BARTRAM, WILLIAM (1739-1823). Natural history; Botany; Ornithology. OCCUP: Farmer; Natural history explorer/ collector. MSS: <u>Ind NUCMC</u> (76-893). WORKS ABOUT: <u>Dict Am Biog</u>; <u>Dict Sci Biog</u>; Elliott; <u>Natl Cycl Am Biog</u> 7:154; <u>Who Was Who Am</u> h. See also: Barr; <u>Biog Geneal Master Ind</u>; Ireland <u>Ind Sci</u>; Pelletier.

BASS, EDGAR WALES (1843-1918). Astronomy; Mathematics. OCCUP: Army officer; Engineer; Professor. WORKS ABOUT: <u>Am Men Sci</u> 1-3 d4; Wallace; <u>Who Was Who Am</u> 1.

BASTIN, EDSON SEWELL (1843-1897). Botany. OCCUP: Professor. WORKS ABOUT: Adams O; Allibone <u>Supp</u>; <u>Am Journ Pharm</u> 69 (1897):385-91; <u>Natl Cycl Am Biog</u> 5; <u>Sci</u> 5(1897):617; Wallace.

BATCHELDER, JOHN MONTGOMERY (1811-1892). Engineering; Physics; Instrumentation. OCCUP: Civil engineer; U.S. government scientist. MSS: <u>Ind NUCMC</u> (?) (81-421). WORKS ABOUT: Elliott.

BATCHELDER, LOREN HARRISON (1846-   ). Chemistry. OCCUP: Professor; Academic administrator. WORKS ABOUT: <u>Who Was Who Am</u> 4.

BATES, EDWARD PAYSON (1844?-1919). Engineering. OCCUP: Steam engineer. WORKS ABOUT: <u>Sci</u> 50(1919):302.

BATES, ORIC (1883-1918). Archeology. OCCUP: Explorer; Museum curator. WORKS ABOUT: <u>Am Lit Yearbook</u>; Wallace; <u>Who Was Who Am</u> 1.

BATTEY, ROBERT (1828-1895). Botany (pharmaceutical); Surgery. OCCUP: Surgeon. Scientific work: see Roy Soc <u>Cat</u>. MSS: <u>Ind NUCMC</u> (61-3147, 79-54). WORKS ABOUT: <u>Dict Am Biog</u>; <u>Dict Am Med Biog</u>; <u>Natl Cycl Am Biog</u> 9; <u>Who Was Who Am</u> h. See also: <u>Biog Geneal Master Ind</u>.

BATTY, JOSEPH H. (-1906). Taxidermy. OCCUP: Natural history collector; Taxidermist. MSS: <u>Ind NUCMC</u> (?). WORKS ABOUT: Allibone <u>Supp</u>; <u>Sci</u> 24(1906):190; Wallace.

BAUMGARTEN, GUSTAV (1837-1910). Medicine. OCCUP: Physician; Professor. MSS: <u>Ind NUCMC</u>. WORKS ABOUT: <u>Am Men Sci</u> 1-2 d3; Groce and Wallace; <u>Natl Cycl Am Biog</u> 12; <u>Who Was Who Am</u> 1.

BAUR, GEORG HERMANN CARL LUDWIG (1859-1898). Paleontology.
OCCUP: Professor. WORKS ABOUT: Am Nat 32(1898):717-18,
33(1899):15-30; Auk 15(1898):286-87; Geol Mag 5(1898):379-
81; Leopoldina 34(1898):126; Nature 58(1898):350; Ornith
Monatsberichte 6(1898):167-68; Sci 8(1898):20, 68-71;
Zoologist 3(1899):95-96.

BAUSCH, HENRY (1859?-1909). Optics. OCCUP: Industrial
officer. WORKS ABOUT: Sci 29(1909):451.

BAYMA, JOSEPH (1816-1892). Mathematics; Physics. OCCUP:
Clergyman; Professor; College president. WORKS ABOUT: Dict
Am Biog; Elliott; Natl Cycl Am Biog 17; Who Was Who Am h.
See also: Biog Geneal Master Ind.

BEACH, ALFRED ELY (1826-1896). Invention. OCCUP: Editor;
Inventor. WORKS ABOUT: Dict Am Biog; Natl Cycl Am Biog 8;
Who Was Who Am h. See also: Barr; Biog Geneal Master Ind.

BEACH, HENRY HARRIS AUBREY (1843-1910). Anatomy. OCCUP:
Surgeon; Academic instructor (medicine). WORKS ABOUT:
Appleton's; Claghorn; Natl Cycl Am Biog 15; Sci
32(1910):54; Who Was Who Am 1.

BEACHLER, CHARLES S. (1870-1894). Geology. OCCUP: Not
determined. MSS: Ind NUCMC (?). WORKS ABOUT: Geol Mag
1(1894):287-88.

BEAL, FOSTER ELLENBOROUGH LASCELLES (1840-1916). Natural
history; Ornithology. OCCUP: Professor (engineering); U.S.
government scientist. WORKS ABOUT: Am Men Sci 1-2 d3;
Appleton's; Sci 44(1916):528; Who Was Who Am 1.

BEAN, TARLETON HOFFMAN (1846-1916). Zoology; Ichthyology.
OCCUP: Museum associate; U.S. and state government
scientist. MSS: Ind NUCMC. WORKS ABOUT: Am Men Sci 1-2 d3;
Dict Am Biog; Elliott (s); Natl Cycl Am Biog 24; Who Was
Who Am 1. See also: Barr; Biog Geneal Master Ind.

BEARDSLEY, ARTHUR (1843-1920). Engineering. OCCUP:
Professor. MSS: Ind NUCMC (?). WORKS ABOUT: Am Men Sci 1-
2 d3.

BEAUMONT, WILLIAM (1785-1853). Physiology; Surgery. OCCUP:
Army surgeon; Physician. MSS: Ind NUCMC (60-1967, 64-39,
68-2114). WORKS ABOUT: Dict Am Biog; Dict Am Med Biog; Dict
Sci Biog; Elliott; Natl Cycl Am Biog 18; Who Was Who Am h.
See also: Biog Geneal Master Ind; Ireland Ind Sci;
Pelletier.

BEBB, MICHAEL SCHUCK (1833-1895). Botany. OCCUP: Farmer.
MSS: Ind NUCMC. WORKS ABOUT: Am Journ Sci 151(1896):78; Bot
Gaz 21(1896):53-66; Erythea 4(1896):29-31; Gard and Forest
8(1895):510; Sci 2(1895):890.

BECK, CARL (1856-1911). Surgery. OCCUP: Surgeon; Professor. WORKS ABOUT: <u>Am Men Sci</u> 2 d3; <u>Dict Am Biog</u>; <u>Natl Cycl Am Biog</u> 10, 30; <u>Who Was Who Am</u> 1. See also: Barr; <u>Biog Geneal Master Ind</u>.

BECK, LEWIS CALEB (1798-1853). Chemistry; Mineralogy; Natural history. OCCUP: Physician; Professor. MSS: <u>Ind NUCMC</u>. WORKS ABOUT: <u>Dict Am Biog</u>; Elliott; <u>Natl Cycl Am Biog</u> 5; <u>Who Was Who Am</u> h. See also: Barr; <u>Biog Geneal Master Ind</u>; Ireland <u>Ind Sci</u>; Pelletier.

BECK, THEODRIC ROMEYN (1791-1855). Medicine; Medical jurisprudence; Meteorology. OCCUP: Physician; Educator; Professor. MSS: <u>Ind NUCMC</u> (68-1054). WORKS ABOUT: <u>Dict Am Biog</u>; <u>Dict Am Med Biog</u>; <u>Who Was Who Am</u> h. See also: Barr; <u>Biog Geneal Master Ind</u>.

BECKER, GEORGE FERDINAND (1847-1919). Geology; Mathematics; Physics. OCCUP: U.S. government scientist. MSS: <u>Ind NUCMC</u> (60-24). WORKS ABOUT: <u>Am Men Sci</u> 1 d3; <u>Dict Am Biog</u>; <u>Dict Sci Biog</u>; Elliott (s); Natl Ac Scis <u>Biog Mem</u> [<u>Mem</u> 21(2)]; <u>Natl Cycl Am Biog</u> 20; <u>Who Was Who Am</u> 1. See also: Barr; <u>Biog Geneal Master Ind</u>; Pelletier.

BECKHAM, CHARLES WICKLIFFE (1856-1888). Ornithology. OCCUP: Lawyer; U.S. government employee; Museum associate. WORKS ABOUT: Allibone <u>Supp</u>; <u>Auk</u> 5(1888):445-46.

BEEBE, WILLIAM (1851-1917). Astronomy; Mathematics. OCCUP: Professor. WORKS ABOUT: <u>Am Journ Sci</u> 193(1917):342; <u>Am Men Sci</u> 1-2 d3; <u>Sci</u> 45(1917):289; <u>Who Was Who Am</u> 1.

BEECHER, CHARLES EMERSON (1856-1904). Geology; Paleontology. OCCUP: Professor; Museum curator. MSS: <u>Ind NUCMC</u>. WORKS ABOUT: <u>Dict Am Biog</u>; Elliott; Natl Ac Scis <u>Biog Mem</u> 6; <u>Natl Cycl Am Biog</u> 13; <u>Who Was Who Am</u> 1. See also: Barr; <u>Biog Geneal Master Ind</u>.

BEHR, HANS HERMANN (1818-1904). Botany; Entomology. OCCUP: Physician; Professor; Curator. MSS: <u>Ind NUCMC</u>. WORKS ABOUT: Elliott. See also: Barr; <u>Biog Geneal Master Ind</u>; Ireland <u>Ind Sci</u>; Pelletier.

BELDING, LYMAN (1829-1917). Ornithology. OCCUP: Not determined. WORKS ABOUT: <u>Am Men Sci</u> 1-2 d3.

BELKNAP, GEORGE EUGENE (1832-1903). Hydrography. OCCUP: Naval officer; Observatory director. MSS: <u>Ind NUCMC</u> (68-2004). WORKS ABOUT: <u>Dict Am Biog</u>; <u>Natl Cycl Am Biog</u> 4, 42; <u>Who Was Who Am</u> 1. See also: Barr; <u>Biog Geneal Master Ind</u>.

BELL, AGRIPPA NELSON (1820-1911). Medicine. OCCUP: Physician; Editor. WORKS ABOUT: Adams O; <u>Am Men Sci</u> 1-2 d3; Knight; <u>Natl Cycl Am Biog</u> 8; <u>Pop Sci Mo</u> 65(1904):282-83; <u>Twentieth Cent Biog Dict</u>; Wallace; <u>Who Was Who Am</u> 1.

BELL, CHARLES J. (1854-1903). Chemistry. OCCUP: Professor. WORKS ABOUT: <u>Am Journ Sci</u> 165(1903):242; <u>Sci</u> 17(1903):79, 119.

BELL, JOHN G. (1812-1889). Natural history; Ornithology; Taxidermy. OCCUP: Natural history explorer; Taxidermist; Businessman. WORKS ABOUT: <u>Auk</u> 7(1890):98-99; Gruson.

BENDIRE, CHARLES EMIL (1836-1897). Ornithology. OCCUP: Army officer; Museum associate. MSS: <u>Ind NUCMC</u> (71-1333). WORKS ABOUT: Adams O; <u>Auk</u> 14(1897):253, 327-29, 15(1898):1-6; Gruson; <u>Sci</u> 5(1897):261-62, 304, 7(1898):241; Wallace.

BENNER, HENRY (-1901). Mathematics. OCCUP: Professor. WORKS ABOUT: <u>Sci</u> 14(1901):339.

BENT, SILAS (1820-1887). Oceanography. OCCUP: Naval officer; Administrator. WORKS ABOUT: <u>Dict Am Biog</u>; Elliott; <u>Who Was Who Am</u> h. See also: <u>Biog Geneal Master Ind</u>.

BENTON, FRANK (1852-1919). Entomology. OCCUP: Teacher; U.S. government scientist. MSS: <u>Ind NUCMC</u> (70-593). WORKS ABOUT: Adams O; <u>Am Men Sci</u> 1-2 d3; Wallace; <u>Who Was Who Am</u> 4.

BERCKMANS, PROSPER J. A. (1829?-1910). Horticulture. OCCUP: Horticulturist. WORKS ABOUT: <u>Sci</u> 32(1910):711.

BERG, WALTER GILMAN (1858-1908). Civil engineering. OCCUP: Engineer. WORKS ABOUT: <u>Am Men Sci</u> 1 d3; Wallace; <u>Who Was Who Am</u> 1.

BERGEN, JOSEPH YOUNG (1851-1917). Botany. OCCUP: School teacher. WORKS ABOUT: Adams O; Allibone <u>Supp</u>; <u>Am Men Sci</u> 2 d3; Appleton's; <u>Ohio Authors</u>; <u>Sci</u> 46(1917):379-80, 47(1918):14-15; Wallace; <u>Who Was Who Am</u> 4.

BERGSTROM, JOHN ANDREW (1867-1910). Psychology. OCCUP: Professor; Laboratory director. WORKS ABOUT: <u>Am Men Sci</u> 1 d3; <u>Sci</u> 31(1910):410.

BERNADOU, JOHN BAPTISTE (1858-1908). Chemistry. OCCUP: Naval officer. WORKS ABOUT: Adams O; <u>Natl Cycl Am Biog</u> 9; <u>Sci</u> 21(1905):879; Wallace; <u>Who Was Who Am</u> 1.

BESSELS, EMIL (1847-1888). Natural history; Exploration. OCCUP: Physician; Explorer. WORKS ABOUT: Allibone <u>Supp</u>; Appleton's; <u>Geog Jahrbuch</u> 14(1891):201-02; <u>Leopoldina</u> 24(1888):110; Phil Soc Washington <u>Bull</u> 11(1892):465-66; <u>Sci</u> 1st ser 11(1888):169, 219-20.

BESSEY, CHARLES EDWIN (1845-1915). Botany; Education. OCCUP: Professor; Academic administrator; State government scientist. MSS: <u>Ind NUCMC</u> (70-316). WORKS ABOUT: <u>Am Men Sci</u> 1-2, d3; <u>Dict Am Biog</u>; <u>Dict Sci Biog</u>; Elliott (s); <u>Natl Cycl Am Biog</u> 8; <u>Who Was Who Am</u> 1. See also: Barr; <u>Biog Geneal Master Ind</u>; Ireland <u>Ind Sci</u>.

BESSEY, EDWARD A. (-1910). Engineering. OCCUP:
Industrial engineer; Professor. WORKS ABOUT: Sci
32(1910):151.

BEYER, HENRY GUSTAV (1850-1918). Hygiene; Physiology.
OCCUP: Naval medical officer. WORKS ABOUT: Am Men Sci 1-2
d3; Elliott (s); Natl Cycl Am Biog 14; Sci 48(1918):616;
Who Was Who Am 1.

BICKMORE, ALBERT SMITH (1839-1914). Natural history;
Geology. OCCUP: Collector; Museum curator. MSS: Ind
NUCMC (?). WORKS ABOUT: Am Men Sci 1-2 d3; Dict Am Biog;
Natl Cycl Am Biog 8; Who Was Who Am 1. See also: Barr;
Biog Geneal Master Ind.

BIDDLE, OWEN (1737-1799). Astronomy. OCCUP: Businessman;
Public official; Surveyor. MSS: Ind NUCMC. WORKS ABOUT:
Elliott. See also: Ireland Ind Sci.

BIEN, JULIUS (1826-1909). Map making. OCCUP: Printer; Map-
maker. MSS: Ind NUCMC (?). WORKS ABOUT: Dict Am Biog; Natl
Cycl Am Biog 28; Who Was Who Am 1. See also: Biog Geneal
Master Ind.

BIGELOW, ARTEMAS (1818-1901). Botany; Chemistry; Geology.
OCCUP: Natural history explorer; Teacher; Industrial
scientist. WORKS ABOUT: Allibone; Wesleyan Alumni Rec.

BIGELOW, JACOB (1786-1879). Botany; Technology. OCCUP:
Physician; Professor. MSS: Ind NUCMC (65-1225). WORKS
ABOUT [birth dates may differ]: Dict Am Biog; Dict Am Med
Biog; Elliott; Natl Cycl Am Biog 4; Who Was Who Am h. See
also: Barr; Biog Geneal Master Ind; Ireland Ind Sci;
Pelletier.

BILLINGS, FRANK SEAVER (1845-1912). Medicine. OCCUP:
Pathologist; Academic researcher. WORKS ABOUT: Allibone
Supp; Am Men Sci 1; Sci 36(1912):552; Wallace; Who Was Who
Am 4.

BILLINGS, JOHN SHAW (1838-1913). Medicine. OCCUP: Surgeon;
Librarian. MSS: Ind NUCMC (74-525). WORKS ABOUT: Am Men
Sci 1-2 d3; Dict Am Biog; Dict Am Med Biog; Natl Ac Scis
Biog Mem 8; Natl Cycl Am Biog 4; Who Was Who Am 1. See
also: Barr; Biog Geneal Master Ind; Ireland Ind Sci;
Pelletier.

BINNEY, AMOS (1803-1847). Natural history; Zoology;
Conchology. OCCUP: Businessman; Legislator; Engaged in
independent study. MSS: Ind NUCMC. WORKS ABOUT: Dict Am
Biog; Elliott; Natl Cycl Am Biog 7:510; Who Was Who Am h.
See also: Barr; Biog Geneal Master Ind.

BINNEY, WILLIAM GREENE (1833-1909). Conchology. OCCUP:
Inherited wealth; Businessman; Engaged in independent
study. MSS: Ind NUCMC (?). WORKS ABOUT: Elliott.

BIRD, C. (1843?-1910). Geography. OCCUP: Educator. WORKS ABOUT: <u>Sci</u> 31(1910):663.

BIRKINBINE, JOHN (1844-1915). Engineering. OCCUP: Engineer; U.S. government scientist. WORKS ABOUT: <u>Am Men Sci</u> 1-2 d3; <u>Natl Cycl Am Biog</u> 12; <u>Who Was Who Am</u> 1.

BIRTWELL, FRANCIS J. (-1901). Ornithology. OCCUP: Not determined. WORKS ABOUT: <u>Sci</u> 14(1901):77.

BISHOP, IRVING PRESCOTT (1849-1912). Geology. OCCUP: Teacher. WORKS ABOUT: <u>Am Men Sci</u> 1-2 d3; <u>Who Was Who Am</u> 4.

BISHOP, SERENO EDWARDS (1827-1909). Geology. OCCUP: Clergyman; Educator; Editor. MSS: <u>Ind NUCMC</u> (61-175). WORKS ABOUT: <u>Sci</u> 29(1909):654; <u>Who Was Who Am</u> 1.

BISSELL, WILLIAM GROSVENOR (1870-1919). Bacteriology. OCCUP: Physician; City government scientist. WORKS ABOUT: <u>Sci</u> 50(1919):480; <u>Who Was Who Am</u> 1.

BISSETT, JOHN J. (1836?-1915). Engineering. OCCUP: Naval engineer. WORKS ABOUT: <u>Sci</u> 42(1915):157.

BLACK, CHARLES W. M. (-1902). Mathematics. OCCUP: Professor. WORKS ABOUT: <u>Sci</u> 16(1902):318.

BLACK, GREENE VARDIMAN. (1836-1915). Chemistry; Dentistry. OCCUP: Dentist; Professor; Academic administrator. MSS: <u>Ind NUCMC</u> (64-683). WORKS ABOUT: <u>Dict Am Biog</u>; <u>Dict Am Med Biog</u>; <u>Natl Cycl Am Biog</u> 13; <u>Who Was Who Am</u> 1. See also: Barr; <u>Biog Geneal Master Ind</u>; Pelletier.

BLACKBURN, ISAAC W. (1851?-1911). Pathology. OCCUP: Professor; Pathologist. WORKS ABOUT: <u>Sci</u> 33(1911):991.

BLACKFORD, EUGENE GILBERT (1839-1904). Ichthyology. OCCUP: Merchant. WORKS ABOUT: Appleton's; <u>Natl Cycl Am Biog</u> 3; <u>Sci</u> 21(1905):38, 232-33; <u>Twentieth Cent Biog Dict</u>; <u>Who Was Who Am</u> 1.

BLACKWELL, ELIZABETH (1821-1910). Medicine. OCCUP: Physician; Professor. MSS: <u>Ind NUCMC</u> (61-2920). WORKS ABOUT: <u>Dict Am Biog</u>; <u>Dict Am Med Biog</u>; <u>Natl Cycl Am Biog</u> 9; <u>Notable Am Wom</u>; Ogilvie; <u>Who Was Who Am</u> 1. See also: <u>Biog Geneal Master Ind</u>; Ireland <u>Ind Sci</u>; Pelletier.

BLAIR, H. W. (-1884). Geography. OCCUP: U.S. government scientist. WORKS ABOUT: <u>Sci</u> 1st ser 5(1885):264.

BLAIR, R. A. (-1902). Geology. OCCUP: Not determined. WORKS ABOUT: <u>Sci</u> 17(1903):37.

BLAKE, CLARENCE JOHN (1843-1919). Otology. OCCUP: Professor. WORKS ABOUT: <u>Am Men Sci</u> 1-2 d3; Appleton's; <u>Sci</u> 49(1919):167; <u>Twentieth Cent Biog Dict</u>; Wallace; <u>Who Was Who Am</u> 1.

BLAKE, ELI WHITNEY (1795-1886). Aerodynamics; Invention;
Physics. OCCUP: Manufacturer. MSS: Ind NUCMC. WORKS ABOUT:
Dict Am Biog; Natl Cycl Am Biog 9; Who Was Who Am h. See
also: Biog Geneal Master Ind.

BLAKE, ELI WHITNEY, JR. (1836-1895). Physics. OCCUP:
Professor. MSS: Ind NUCMC. WORKS ABOUT: Am Journ Sci
150(1895):434; Appleton's; Sci 2(1895):483; Twentieth Cent
Biog Dict.

BLAKE, FRANCIS (1850-1913). Physics. OCCUP: U.S. government
scientist; Inventor; Industrial employee. MSS: Ind NUCMC
(84-1991). WORKS ABOUT: Am Men Sci 1-2 d3; Dict Am Biog;
Elliott (s); Natl Cycl Am Biog 13, 22; Who Was Who Am 1.
See also: Barr; Biog Geneal Master Ind.

BLAKE, JAMES (1815-1893). Chemistry; Physics. OCCUP:
Physician; Professor. WORKS ABOUT: Am Chem and Chem Eng;
Dict Am Med Biog.

BLAKE, JOHN H. (1808-1899). Chemistry; Geology. OCCUP:
Industrial scientist; Civil engineer; Mining manager. WORKS
ABOUT: Assoc Am Eng Soc Journ 25(1900):302-03; Blake p9-10,
40-41, 44-46, 62.

BLAKE, JOHN MARCUS (1831?-1920). Crystallography. OCCUP:
Not determined. WORKS ABOUT: Am Journ Sci 200(1920):316.

BLAKE, LUCIEN IRA (1854-1916). Physics. OCCUP: Professor;
Industrial engineer; Inventor. WORKS ABOUT [birth dates may
differ]: Am Men Sci 1-2 d3; Natl Cycl Am Biog 18; Sci
43(1916):684; Twentieth Cent Biog Dict; Wallace; Who Was
Who Am 1.

BLAKE, WILLIAM PHIPPS (1825-1910). Geology; Metallurgy;
Mineralogy. OCCUP: State government scientist; Professor.
MSS: Ind NUCMC (62-2173). WORKS ABOUT [birth dates may
differ]: Am Men Sci 1 d3; Dict Am Biog; Elliott (s); Natl
Cycl Am Biog 10, 25; Who Was Who Am 1. See also: Barr; Biog
Geneal Master Ind; Ireland Ind Sci.

BLANCHARD, FREDERICK (1843-   ). Entomology. OCCUP:
Banker. WORKS ABOUT: Am Men Sci 1-2 d3.

BLAND, JAMES H. B. (ca.1832-1911). Entomology. OCCUP: Not
determined. Scientific work: see Roy Soc Cat. WORKS ABOUT:
Ent News 23(1911):47.

BLAND, THOMAS (1809-1885). Natural history; Conchology.
OCCUP: Lawyer; Mining manager; Engaged in independent
study. WORKS ABOUT: Dict Am Biog; Elliott; Who Was Who Am
h. See also: Barr.

BLASIUS, WILLIAM (1818-1899). Meteorology. OCCUP:
Businessman. WORKS ABOUT: Meteorol Zeitschrift
16(1899):215, 555-57.

BLODGET, LORIN (1823-1901). Physics; Climatology; Statistics. OCCUP: U.S. government employee. WORKS ABOUT: <u>Dict Am Biog</u>; Elliott; <u>Natl Cycl Am Biog</u> 4; <u>Who Was Who Am</u> 1. See also: <u>Biog Geneal Master Ind</u>.

BLUNT, EDMUND (1799-1866). Geodesy. OCCUP: U.S. government scientist; Publisher. WORKS ABOUT: Allibone; <u>Am Journ Sci</u> 92(1866):433-34; Appleton's; Drake; <u>Twentieth Cent Biog Dict</u>.

BLUNT, EDMUND MARCH (1770-1862). Hydrography. OCCUP: Editor and publisher. WORKS ABOUT: <u>Dict Am Biog</u>; <u>Who Was Who Am</u> h. See also: <u>Biog Geneal Master Ind</u>.

BLUNT, GEORGE WILLIAM (1802-1878). Hydrography. OCCUP: Publisher; U.S. government scientist. WORKS ABOUT: <u>Dict Am Biog</u>; <u>Natl Cycl Am Biog</u> 21; <u>Who Was Who Am</u> h. See also: <u>Biog Geneal Master Ind</u>.

BOCHER, MAXIME (1867-1918). Mathematics. OCCUP: Professor. MSS: <u>Ind NUCMC</u> (65-1226). WORKS ABOUT: <u>Am Men Sci</u> 1 d3; <u>Dict Am Biog</u>; <u>Dict Sci Biog</u>; <u>Natl Cycl Am Biog</u> 18; <u>Who Was Who Am</u> 1. See also: Barr; <u>Biog Geneal Master Ind</u>; Ireland <u>Ind Sci</u>.

BODINE, DONALDSON (1866-1916?). Geology; Zoology. OCCUP: Professor. WORKS ABOUT: <u>Am Men Sci</u> 1-2 d3; <u>Sci</u> 43(1916):131.

BODLEY, RACHEL LITTLER (1831-1888). Botany; Chemistry. OCCUP: Professor; Academic administrator (medicine). WORKS ABOUT: Elliott; <u>Notable Am Wom</u>; Ogilvie; Siegel and Finley. See also: <u>Biog Geneal Master Ind</u>; Pelletier.

BOGUE, ERNEST EVERETT (1864-1907). Forestry. OCCUP: Professor. WORKS ABOUT: <u>Am Men Sci</u> 1 d3; <u>Sci</u> 26(1907):295.

BOGUE, VIRGIL GAY (1846-1916). Engineering. OCCUP: Engineer. WORKS ABOUT: <u>Dict Am Biog</u>; <u>Natl Cycl Am Biog</u> 13; <u>Who Was Who Am</u> 1. See also: Barr; <u>Biog Geneal Master Ind</u>.

BOLANDER, HENRY NICHOLAS (ca.1831-1897). Botany; Botanical exploration. OCCUP: State government scientist; Educator. MSS: <u>Ind NUCMC</u>. WORKS ABOUT: <u>Erythea</u> 6(1898):100-07.

BOLL, JACOB (1828-1880). Natural history; Geology. OCCUP: Pharmacist; Museum affiliate; Natural history explorer/collector. WORKS ABOUT: <u>Dict Am Biog</u>; Elliott; <u>Who Was Who Am</u> h. See also: <u>Biog Geneal Master Ind</u>; Ireland <u>Ind Sci</u>.

BOLLES, FRANK (1856-1894). Nature writing. OCCUP: Academic administrator. WORKS ABOUT: <u>Dict Am Biog</u>; <u>Who Was Who Am</u> h. See also: <u>Biog Geneal Master Ind</u>.

BOLLES, WILLIAM PALMER (-1916). Botany. OCCUP: Professor; Surgeon. WORKS ABOUT: <u>Sci</u> 43(1916):566.

BOLLMAN, CHARLES HARVEY (1868-1889). Zoology. OCCUP: U.S. government scientist. WORKS ABOUT: U S Natl Mus <u>Bull</u> 46(1893):7.

BOLTER, ANDREW (1820-1900). Entomology. OCCUP: Manufactuer; Collector. MSS: <u>Ind NUCMC</u> (?). WORKS ABOUT: <u>Natl Cycl Am Biog</u> 24; <u>Sci</u> 11(1900):518-19.

BOLTON, HENRY CARRINGTON (1843-1903). Chemistry. OCCUP: Laboratory administrator; Professor. MSS: <u>Ind NUCMC</u> (?). WORKS ABOUT: <u>Dict Am Biog</u>; Elliott; <u>Natl Cycl Am Biog</u> 10; <u>Who Was Who Am</u> 1. See also: Barr; <u>Biog Geneal Master Ind</u>; Pelletier.

BONAPARTE, CHARLES LUCIEN (Prince of Canino and Musignano) (1803-1857). Zoology; Ornithology. OCCUP: Engaged in independent study. MSS: <u>Ind</u> <u>NUCMC</u>. WORKS ABOUT [form of name may differ]: <u>Dict Sci Biog</u>; Elliott; <u>Who Was Who Am</u> h. See also: Barr; <u>Biog Geneal Master Ind</u>.

BONAPARTE, LUCIEN. See BONAPARTE, CHARLES LUCIEN.

BOND, GEORGE PHILLIPS (1825-1865). Astronomy. OCCUP: Observatory director. MSS: <u>Ind NUCMC</u> (62-3730, 65-1227). WORKS ABOUT: <u>Dict Am Biog</u>; <u>Dict Sci Biog</u>; Elliott; <u>Natl Cycl Am Biog</u> 5; <u>Who Was Who Am</u> h. See also: Barr; <u>Biog Geneal Master Ind</u>; Ireland <u>Ind Sci</u>; Pelletier.

BOND, WILLIAM CRANCH (1789-1859). Astronomy. OCCUP: Craftsman; Observatory director. MSS: <u>Ind NUCMC</u> (65-1228). WORKS ABOUT: <u>Dict Am Biog</u>; <u>Dict Sci Biog</u>; Elliott; <u>Natl Cycl Am Biog</u> 8; <u>Who Was Who Am</u> h. See also: Barr; <u>Biog Geneal Master Ind</u>; Ireland <u>Ind Sci</u>; Pelletier.

BONNER, JOHN (ca.1643-1725/26). Map making. OCCUP: Businessman; Seaman. WORKS ABOUT: <u>Dict Am Biog</u>; <u>Who Was Who Am</u> h.

BONNYCASTLE, CHARLES (1792-1840). Mathematics. OCCUP: Professor. MSS: <u>Ind NUCMC</u> (?). WORKS ABOUT: Allibone; Appleton's; Drake; <u>Natl Cycl Am Biog</u> 13; Wallace.

BOORAEM, ROBERT ELMER (1856-1918). Mining engineering. OCCUP: Engineer; Mining manager. WORKS ABOUT: <u>Am Men Sci</u> 1-3 d4; <u>Who Was Who Am</u> 1.

BOOTH, BRADFORD ALLEN (1878-1919). Medicine. OCCUP: Physician. WORKS ABOUT: <u>Am Men Sci</u> 3 d3.

BOOTH, EDWARD (-1917). Chemistry. OCCUP: Professor. WORKS ABOUT: <u>Sci</u> 46(1917):337.

BOOTH, JAMES CURTIS (1810-1888). Chemistry. OCCUP: Educator; U.S. government scientist. WORKS ABOUT: <u>Dict Am Biog</u>; Elliott; <u>Natl Cycl Am Biog</u> 13; <u>Who Was Who Am</u> h. See also: Barr; <u>Biog Geneal Master Ind</u>; Ireland <u>Ind Sci</u>; Pelletier.

BOOTH, SAMUEL COLTON. (1812-1895). Mineralogy. OCCUP: Farmer; Collector; Engaged in independent study. WORKS ABOUT: Natl Cycl Am Biog 15; Sci 2(1895):452.

BOOTT, FRANCIS (1792-1863). Botany. OCCUP: Physician. MSS: Ind NUCMC (74-346). Allibone Supp; Am Journ Sci 87(1864):288-92.

BOOTT, WILLIAM (1805-1887). Botany. OCCUP: Businessman. WORKS ABOUT: Am Journ Sci 134(1887):160, 135(1888):262.

BORDEN, GAIL (1801-1874). Invention; Surveying. OCCUP: Surveyor; Public official; Manufacturer. MSS: Ind NUCMC (62-404, 62-4508, 67-1996). WORKS ABOUT: Dict Am Biog; Natl Cycl Am Biog 7:306; Who Was Who Am h. See also: Biog Geneal Master Ind; Pelletier.

BORDEN, SIMEON (1798-1856). Civil engineering; Surveying. OCCUP: Industrial manager; Engineer; State government surveyor. WORKS ABOUT: Dict Am Biog; Natl Cycl Am Biog 24; Who Was Who Am h. See also: Biog Geneal Master Ind.

BORDLEY, JOHN BEAL (1727-1804). Agriculture. OCCUP: Lawyer; Farmer. MSS: Ind NUCMC. WORKS ABOUT: Dict Am Biog; Who Was Who Am h. See also: Biog Geneal Master Ind; Ireland Ind Sci.

BORLAND, JOHN NELSON (1828-1890). Medicine; Zoological studies. OCCUP: Physician. WORKS ABOUT: Atkinson; Yale Obit Rec (1891):39.

BOSS, LEWIS (1846-1912). Astronomy; Positional astronomy. OCCUP: Observatory director. MSS: Ind NUCMC (?). WORKS ABOUT: Am Men Sci 1-2 d3; Dict Am Biog; Dict Sci Biog; Elliott (s); Natl Ac Scis Biog Mem 9; Natl Cycl Am Biog 13; Who Was Who Am 1. See also: Barr; Ireland Ind Sci.

BOURKE, JOHN GREGORY (1846-1896). Anthropology; Ethnology. OCCUP: Army officer. MSS: Ind NUCMC (65-1423, 70-1322, 79-1392). WORKS ABOUT: Dict Am Biog; Natl Cycl Am Biog 13; Who Was Who Am h. See also: Barr; Biog Geneal Master Ind.

BOUVE, THOMAS TRACY (1815-1896). Natural history. OCCUP: Merchant. WORKS ABOUT: Allibone Supp; Am Ac Arts Scis Proc 32(1897):340-44; Appleton's; Boston Soc Nat Hist Proc 27(1897):219-41; Natl Cycl Am Biog 7:506; Shaw; Wallace.

BOWDITCH, HENRY PICKERING (1840-1911). Physiology. OCCUP: Professor; Academic administrator (medicine). MSS: Ind NUCMC (62-3533). WORKS ABOUT: Am Men Sci 1-2 d3; Dict Am Biog; Dict Am Med Biog; Dict Sci Biog; Elliott (s); Natl Ac Scis Biog Mem [Mem 17]; Natl Cycl Am Biog 12; Who Was Who Am 1. See also: Barr; Biog Geneal Master Ind; Ireland Ind Sci; Pelletier.

9303327

BOWDITCH, JONATHAN INGERSOLL (1806-1889). Astronomy; Mathematics; Meteorology. OCCUP: Businessman. MSS: <u>Ind NUCMC</u> (66-1668). WORKS ABOUT: Am Ac Arts Scis <u>Proc</u> 24(1888-1889):435-37; Hall.

BOWDITCH, NATHANIEL (1773-1838). Astronomy; Mathematics. OCCUP: Seaman; Businessman; Actuary. MSS: <u>Ind NUCMC</u>. WORKS ABOUT: <u>Dict Am Biog</u>; <u>Dict Sci Biog</u>; Elliott; <u>Natl Cycl Am Biog</u> 6; <u>Who Was Who Am</u> h. See also: Barr; <u>Biog Geneal Master Ind</u>; Ireland <u>Ind Sci</u>; Pelletier.

BOWDOIN, JAMES (1726-1790). Astronomy; Physics. OCCUP: Merchant; Legislator; Governor. MSS: <u>Ind NUCMC</u>. WORKS ABOUT: <u>Dict Am Biog</u>; Elliott; <u>Natl Cycl Am Biog</u> 2; <u>Who Was Who Am</u> h. See also: <u>Biog Geneal Master Ind</u>.

BOWEN, GEORGE THOMAS (1803-1828). Chemistry; Medicine. OCCUP: Physician; Professor. WORKS ABOUT: <u>Am Journ Sci</u> 15(1829):403-04; Appleton's; <u>Natl Cycl Am Biog</u> 12; Smith E p222; <u>Twentieth Cent Biog Dict</u>.

BOWER, HENRY (1833-1896). Chemistry. OCCUP: Manufacturer. WORKS ABOUT: <u>Am Chem and Chem Eng</u>.

BOWSER, EDWARD ALBERT (1837-1910). Engineering; Mathematics. OCCUP: Professor; U.S. government scientist. WORKS ABOUT [birth dates may differ]: Elliott; <u>Natl Cycl Am Biog</u> 19; <u>Who Was Who Am</u> 1. See also: Barr; <u>Biog Geneal Master Ind</u>.

BOYDEN, SETH (1788-1870). Invention. OCCUP: Inventor; Manufacturer. WORKS ABOUT: <u>Dict Am Biog</u>; <u>Natl Cycl Am Biog</u> 11; <u>Who Was Who Am</u> h. See also: <u>Biog Geneal Master Ind</u>; Pelletier.

BOYDEN, URIAH ATHERTON (1804-1879). Engineering; Invention. OCCUP: Industrial engineer. MSS: <u>Ind NUCMC</u> (80-220). WORKS ABOUT: <u>Dict Am Biog</u>; <u>Natl Cycl Am Biog</u> 11; <u>Who Was Who Am</u> h. See also: <u>Biog Geneal Master Ind</u>; Pelletier.

BOYE, MARTIN HANS (1812-1909). Chemistry; Geology; Medicine; Physics. OCCUP: State government scientist; Teacher. WORKS ABOUT: <u>Am Men Sci</u> 1 d3; <u>Dict Am Biog</u>; Elliott (s); <u>Who Was Who Am</u> 1. See also: Barr; <u>Biog Geneal Master Ind</u>; Pelletier.

BOYLSTON, ZABDIEL (1680-1766). Medicine. OCCUP: Physician. WORKS ABOUT: <u>Dict Am Biog</u>; <u>Dict Am Med Biog</u>; Elliott; <u>Natl Cycl Am Biog</u> 7:270; <u>Who Was Who Am</u> h. See also: <u>Biog Geneal Master Ind</u>; Ireland <u>Ind Sci</u>; Pelletier.

BRACE, DeWITT BRISTOL (1859-1905). Physics; Optics. OCCUP: Professor. WORKS ABOUT: <u>Dict Am Biog</u>; <u>Dict Sci Biog</u>; Elliott; <u>Who Was Who Am</u> 1. See also: Barr; <u>Biog Geneal Master Ind</u>.

BRACE, JOHN PIERCE (1793-1872). Natural history. OCCUP: Educator; Editor. Scientific work: see Roy Soc <u>Cat</u>. MSS: <u>Ind NUCMC</u>. WORKS ABOUT: <u>Dict Am Biog</u>; <u>Who Was Who Am</u> h. See also: <u>Biog Geneal Master Ind</u>.

BRACKENRIDGE, WILLIAM DUNLOP (1810-1893). Botany. OCCUP: Horticulturist. MSS: <u>Ind NUCMC</u> (67-1336, 78-882). WORKS ABOUT: <u>Dict Am Biog</u>; Elliott; <u>Who Was Who Am</u> h.

BRACKETT, CYRUS FOGG (1833-1915). Physics. OCCUP: Professor. WORKS ABOUT: <u>Am Journ Sci</u> 189(1915):326; <u>Am Men Sci</u> 1-2 d3; <u>Dict Am Med Biog</u>; Elliott (s); <u>Sci</u> 41(1915):204, 523-25; <u>Who Was Who Am</u> 1.

BRACKETT, FOSTER H. (ca.1863-1900). Ornithology. OCCUP: Banker. WORKS ABOUT: <u>Auk</u> 17(1900):197.

BRADFORD, JOSHUA TAYLOR (1818-1871). Medicine; Surgery. OCCUP: Physician. WORKS ABOUT: <u>Dict Sci Biog</u>.

BRADFORD, ROYAL BIRD (1844-1914). Naval science. OCCUP: Naval officer. WORKS ABOUT: <u>Am Men Sci</u> 1-2 d3; Appleton's <u>Supp</u> 1901; <u>Who Was Who Am</u> 1.

BRADLEY, FRANCIS (mid-19th century). Astronomy. OCCUP: Not determined. Scientific work: see Roy Soc <u>Cat</u>. WORKS ABOUT: Bradley; Yale <u>Cat</u>.

BRADLEY, FRANK HOWE (1838-1879). Geology. OCCUP: Professor; U.S. and state government scientist. WORKS ABOUT: <u>Dict Am Biog</u>; Elliott; <u>Who Was Who Am</u> h. See also: Barr.

BRADLEY, GUY M. (1870-1905). Ornithology. OCCUP: Public official. WORKS ABOUT: <u>Sci</u> 22(1905):190; <u>Who Was Who Am</u> h.

BRAID, ANDREW (1846-1919). Hydrography. OCCUP: U.S. government scientist. WORKS ABOUT: <u>Sci</u> 49(1919):167; <u>Who Was Who Am</u> 4.

BRANDEGEE, MARY KATHARINE LAYNE CURRAN (1844-1920). Botany. OCCUP: Physician; Herbarium curator. MSS: <u>Ind NUCMC</u>. WORKS ABOUT: Elliott; <u>Notable Am Wom</u>; Ogilvie; Siegel and Finley.

BRASHEAR, JOHN ALFRED (1840-1920). Astronomy; Physics; Mechanics; Lens making; Astrophysical instrumentation. OCCUP: Instrument maker. MSS: <u>Ind NUCMC</u>. WORKS ABOUT: <u>Am Men Sci</u> 1-2 d3; <u>Dict Am Biog</u>; <u>Dict Sci Biog</u>; Elliott (s); <u>Natl Cycl Am Biog</u> 4; <u>Who Was Who Am</u> h (add), 1, 4. See also: Barr; <u>Biog Geneal Master Ind</u>; Ireland <u>Ind Sci</u>.

BRATTLE, THOMAS (1658-1713). Astronomy. OCCUP: Businessman; Academic administrator. MSS: <u>Ind NUCMC</u> (?). WORKS ABOUT: <u>Dict Am Biog</u>; Elliott; <u>Who Was Who Am</u> h. See also: <u>Biog Geneal Master Ind</u>; Ireland <u>Ind Sci</u>.

BREED, DANIEL (ca.1825?-   ). Chemistry; Physiological chemistry. OCCUP: Army physician; U.S. government employee. Scientific work: see Roy Soc <u>Cat</u>. WORKS ABOUT: New York Univ <u>Gen Cat</u>.

BREWER, THOMAS MAYO (1814-1880). Ornithology; Oology. OCCUP: Physician; Publisher. MSS: <u>Ind NUCMC</u>. WORKS ABOUT: <u>Dict Am Biog</u>; Elliott; <u>Natl Cycl Am Biog</u> 22; <u>Who Was Who Am</u> h. See also: <u>Biog Geneal Master Ind</u>; Pelletier.

BREWER, WILLIAM HENRY (1828-1910). Agriculture; Botany; Chemistry; Geology. OCCUP: Professor. MSS: <u>Ind NUCMC</u> (71-2011). WORKS ABOUT: <u>Am Men Sci</u> 1-2 d3; <u>Dict Am Biog</u>; Elliott (s); Natl Ac Scis <u>Biog Mem</u> 12; <u>Natl Cycl Am Biog</u> 13; <u>Who Was Who Am</u> 1. See also: Barr; <u>Biog Geneal Master Ind</u>.

BREWSTER, JONATHAN (1593-1659). Alchemy. OCCUP: Businessman. WORKS ABOUT: <u>Am Chem and Chem Eng</u>.

BREWSTER, WILLIAM (1851-1919). Ornithology. OCCUP: Museum curator. MSS: <u>Ind NUCMC</u>. WORKS ABOUT: <u>Am Men Sci</u> 1-2 d3; <u>Dict Am Biog</u>; Elliott (s); <u>Natl Cycl Am Biog</u> 12, 22; <u>Who Was Who Am</u> 1. See also: <u>Biog Geneal Master Ind</u>; Pelletier.

BRICKNER, SAMUEL MAX (1867-1916). Gynecology. OCCUP: Physician. WORKS ABOUT: <u>Am Men Sci</u> 1-2 d3; <u>Sci</u> 43(1916):710.

BRIDGES, ROBERT (1806-1882). Botany; Chemistry. OCCUP: Physician; Professor. WORKS ABOUT: <u>Dict Am Biog</u>; Elliott; <u>Natl Cycl Am Biog</u> 5; <u>Who Was Who Am</u> h. See also: <u>Biog Geneal Master Ind</u>; Pelletier.

BRIGGS, ROBERT (1822-1882). Civil engineering. OCCUP: Engineer; Editor. WORKS ABOUT: Am Soc Civil Eng <u>Trans</u> 36(1896):542-45.

BRINCKERHOFF, WALTER REMSEN (1874-1911). Medicine; Pathology. OCCUP: U.S. government scientist; Professor. WORKS ABOUT: <u>Am Men Sci</u> 1-2 d3; <u>Sci</u> 33(1911):368.

BRINLEY, CHARLES A. (1847-1919). Chemistry; Metallurgy. OCCUP: Industrial manager; Educator. WORKS ABOUT: Appleton's; Wallace; <u>Who Was Who Am</u> 4.

BRINTON, DANIEL GARRISON (1837-1899). Anthropology. OCCUP: Physician; Professor. MSS: <u>Ind NUCMC</u> (68-446). WORKS ABOUT: <u>Dict Am Biog</u>; <u>Natl Cycl Am Biog</u> 9; <u>Who Was Who Am</u> 1. See also: Barr; <u>Biog Geneal Master Ind</u>; Ireland <u>Ind Sci</u>.

BROADHEAD, GARLAND CARR (1827-1912). Geology. OCCUP: Professor. MSS: <u>Ind NUCMC</u> (64-266). WORKS ABOUT: <u>Am Men Sci</u> 1-2 d3; <u>Dict Am Biog</u>; Elliott (s); <u>Natl Cycl Am Biog</u> 13; <u>Who Was Who Am</u> 1. See also: <u>Biog Geneal Master Ind</u>.

BROCKLESBY, JOHN (1811-1889). Meteorology; Microscopy.
OCCUP: Professor. WORKS ABOUT: Elliott; <u>Natl Cycl Am Biog</u>
12; <u>Who Was Who Am</u> h. See also: <u>Biog Geneal Master Ind</u>.

BROCKWAY, FRED JOHN (1860-1901). Anatomy. OCCUP:
Physician (?); Academic instructor. WORKS ABOUT: <u>Sci</u>
13(1901):676; Wallace; <u>Who Was Who Am</u> 1.

BROOKE, JOHN MERCER (1826-1906). Astronomy; Physics;
Invention. OCCUP: Naval officer; Professor. MSS: <u>Ind NUCMC</u>
(78-1855). WORKS ABOUT: <u>Dict Am Biog</u>; Elliott; <u>Natl Cycl Am</u>
<u>Biog</u> 22; <u>Who Was Who Am</u> 1. See also: Barr; <u>Biog Geneal</u>
<u>Master Ind</u>.

BROOKS, THOMAS BENTON (1836-1900). Geology. OCCUP: Mining
engineer; State government scientist; Farmer. WORKS ABOUT:
<u>Dict Am Biog</u>; <u>Natl Cycl Am Biog</u> 3, 45; <u>Who Was Who Am</u> 1.
See also: Barr; <u>Biog Geneal Master Ind</u>; Ireland <u>Ind Sci</u>.

BROOKS, WILLIAM KEITH (1848-1908). Natural history;
Zoology; Embryology. OCCUP: Professor. WORKS ABOUT: <u>Am Men</u>
<u>Sci</u> 1 d3; <u>Dict Am Biog</u>; <u>Dict Sci Biog</u>; Elliott (s); Natl Ac
Scis <u>Biog Mem</u> 7; <u>Natl Cycl Am Biog</u> 23; <u>Who Was Who Am</u> 1.
See also: Barr; <u>Biog Geneal Master Ind</u>; Ireland <u>Ind Sci</u>.

BROTHWICK, JOHN LIVINGSTON DINWIDDIE (1840?-1904).
Engineering. OCCUP: Naval engineer. WORKS ABOUT: <u>Sci</u>
20(1904):573.

BROUN, WILLIAM LeROY (1827-1902). Mathematics; Physics.
OCCUP: Professor; College president. MSS: <u>Ind</u> NUCMC (60-
15). WORKS ABOUT: <u>Natl Cycl Am Biog</u> 19; <u>Sci</u> 15(1902):198,
316-17; <u>Twentieth Cent Biog Dict</u>; <u>Who Was Who Am</u> 1.

BROWER, JACOB VRADENBERG (1844-1905). Archeology. OCCUP:
Public official. MSS: <u>Ind NUCMC</u> (82-1252). WORKS ABOUT: <u>Am</u>
<u>Men Sci</u> 1 d3; <u>Dict Am Biog</u>; <u>Who Was Who Am</u> 1. See also:
<u>Biog Geneal Master Ind</u>.

BROWN, ADDISON (1830-1913). Botany. OCCUP: Lawyer. WORKS
ABOUT: <u>Dict Am Biog</u>; <u>Who Was Who Am</u> 1. See also: Barr; <u>Biog</u>
<u>Geneal Master Ind</u>.

BROWN, AMOS PEASLEE (1864-1917). Geology; Mineralogy.
OCCUP: Professor. WORKS ABOUT: <u>Am Men Sci</u> 1-2 d3; <u>Sci</u>
47(1918):37; <u>Who Was Who Am</u> 1.

BROWN, ARTHUR ERWIN (1850-1910). Natural history; Zoology.
OCCUP: Curator. MSS: <u>Ind NUCMC</u> (66-17). WORKS ABOUT: <u>Am Men</u>
<u>Sci</u> 1-2 d3; <u>Who Was Who Am</u> 1.

BROWN, BENJAMIN G. (1837-1903). Mathematics. OCCUP:
Professor. WORKS ABOUT: <u>Sci</u> 18(1903):477.

BROWN, C. N. (1858?-1902). Engineering. OCCUP: Professor;
Academic administrator. WORKS ABOUT: <u>Sci</u> 15(1902):476.

BROWN, CHARLES HENRY (1847?-1901). Medicine. OCCUP: Physician; Editor. WORKS ABOUT: <u>Sci</u> 14(1901):659-60.

BROWN, FRANCIS C. (1830-1900). Ornithology. OCCUP: Not determined. WORKS ABOUT: <u>Sci</u> 11(1900):635.

BROWN, JOSEPH (1733-1785). Astronomy; Physics. OCCUP: Businessman; Professor. WORKS ABOUT: <u>Dict Am Biog</u>; Elliott; <u>Natl Cycl Am Biog</u> 8; <u>Who Was Who Am</u> h. See also: <u>Biog Geneal Master Ind</u>.

BROWN, SAMUEL (1769-1830). Medicine (vaccination). OCCUP: Physician; Professor. MSS: <u>Ind NUCMC</u>. WORKS ABOUT: <u>Dict Am Biog</u>; <u>Dict Am Med Biog</u>; <u>Natl Cycl Am Biog</u> 4; <u>Who Was Who Am</u> h. See also: <u>Biog Geneal Master Ind</u>.

BROWN, WILLIAM GEORGE (1853-1920). Chemistry. OCCUP: Professor; Laboratory administrator. MSS: <u>Ind NUCMC</u> (?). WORKS ABOUT: <u>Am Men Sci</u> 1-2 d3; <u>Who Was Who Am</u> 1.

BROWNE, DANIEL JAY (1804-1867). Agriculture. OCCUP: Farmer; Editor; U.S. government employee. WORKS ABOUT: <u>Dict Am Biog</u>; Elliott; <u>Who Was Who Am</u> h. See also: <u>Biog Geneal Master Ind</u>.

BROWNE, DAVID HENRY (1864-1917). Engineering; Metallurgy. OCCUP: Industrial metallurgist. WORKS ABOUT: <u>Natl Cycl Am Biog</u> 23; <u>Sci</u> 45(1917):334.

BROWNE, FRANCIS C. (ca.1830-1900). Ornithology. OCCUP: Farmer. WORKS ABOUT: <u>Auk</u> 17(1900):194-96.

BROWNE, PETER ARRELL (1782-1860). Geology; Mineralogy. OCCUP: Lawyer; Professor. Scientific work: see Roy Soc <u>Cat</u>. MSS: <u>Ind NUCMC</u> (61-387). WORKS ABOUT: Allibone; Coffin p113 (mentioned); Martin p85, 90, 91, 253 (mentioned); Sinclair.

BROWNELL, JOHN T. (1836-1886). Microscopy. OCCUP: Clergyman. WORKS ABOUT: Am Soc Micr <u>Proc</u> (1886):202-03.

BROWNELL, WALTER ABNER (1838-1904). Geology. OCCUP: Teacher; Educator. WORKS ABOUT: <u>Natl Cycl Am Biog</u> 2; <u>Twentieth Cent Biog Dict</u>; <u>Who Was Who Am</u> 1.

BROWN-SEQUARD, CHARLES-EDOUARD (1817-1894). Physiology. OCCUP: Physician; Professor. MSS: <u>Ind NUCMC</u>. WORKS ABOUT: <u>Dict Sci Biog</u>; Natl Ac Scis <u>Biog Mem</u> 4. See also: Barr; <u>Biog Geneal Master Ind</u>; Pelletier.

BROWN-SEQUARD, C. E. (1866?-1896). Biology. OCCUP: Physician (?). WORKS ABOUT: <u>Sci</u> 4(1896):653.

BRUCE, ADAM TODD (-1887). Zoology; Morphology. OCCUP: Academic instructor. WORKS ABOUT: Johns Hopkins <u>Circ</u> [6](1886-1887):87.

BRUCE, ARCHIBALD (1777-1818). Mineralogy. OCCUP: Physician; Professor; Editor. WORKS ABOUT: Dict Am Biog; Elliott; Natl Cycl Am Biog 9; Who Was Who Am h. See also: Barr; Biog Geneal Master Ind; Ireland Ind Sci.

BRUCE, CATHERINE WOLFE (1816-1900). Astronomy. OCCUP: Benefactor. WORKS ABOUT: Notable Am Wom. See also: Barr.

BRUMBACK, ARTHUR MARION (1869-1916). Chemistry. OCCUP: Professor; College president. WORKS ABOUT: Am Men Sci 2 d3; Sci 44(1916):306.

BRUMBY, RICHARD TRAPIER (1804-1875). Chemistry; Mineralogy. OCCUP: Lawyer; Professor. MSS: Ind NUCMC. WORKS ABOUT: Dict Am Biog; Who Was Who Am h.

BRUNNOW, FRANZ FRIEDRICH ERNST (1821-1891). Astronomy. OCCUP: Professor; Observatory director; Government scientist (foreign). MSS: Ind NUCMC (?). WORKS ABOUT: Elliott; Natl Cycl Am Biog 13.

BRUSH, CHARLES BENJAMIN (1848-1897). Engineering. OCCUP: Industrial engineer; Professor; Academic administrator. WORKS ABOUT: Natl Cycl Am Biog 9; Sci 5(1897):916; Twentieth Cent Biog Dict.

BRUSH, GEORGE JARVIS (1831-1912). Mineralogy. OCCUP: Professor; Academic administrator. MSS: Ind NUCMC. WORKS ABOUT: Am Men Sci 1-2 d3; Dict Am Biog; Elliott (s); Natl Ac Scis Biog Mem [Mem 17]; Natl Cycl Am Biog 10, 28; Who Was Who Am 1. See also: Barr; Biog Geneal Master Ind.

BRYAN, ALBERT HUGH (1874-1920). Chemistry. OCCUP: Industrial and U.S. government scientist. WORKS ABOUT: Am Men Sci 2 d3.

BRYAN, ELIZABETH LETSON (MRS. WILLIAM ALANSON) (1874-1919). Conchology. OCCUP: Museum administrator. MSS: Ind NUCMC. WORKS ABOUT: Sci 49(1919):305.

BRYANT, HENRY (1820-1867). Ornithology. OCCUP: Physician. MSS: Ind NUCMC. WORKS ABOUT: Am Ac Arts Scis Proc 7(1865-1868):304-05; Boston Soc Nat Hist Proc 11(1868):205-15.

BRYANT, JOSEPH DECATUR (1845-1914). Surgery. OCCUP: Surgeon; Professor. MSS: Ind NUCMC (?). WORKS ABOUT: Dict Am Biog; Natl Cycl Am Biog 14, 23; Who Was Who Am 1. See also: Barr; Biog Geneal Master Ind.

BUCHANAN, ROBERDEAU (1839-1916). Astronomy. OCCUP: Engineer; U.S. government scientist. MSS: Ind NUCMC (60-3313, as collector of mss). WORKS ABOUT: Allibone Supp; Am Men Sci 1-2 d3; Sci 45(1917):112; Wallace; Who Was Who Am 1.

BUCK, LEFFERT LEFFERTS (1837-1909). Engineering. OCCUP: Civil engineer. WORKS ABOUT: Dict Am Biog; Natl Cycl Am Biog 10; Who Was Who Am 1. See also: Barr.

BUCK, SAMUEL JAY (1835-1918). Mathematics. OCCUP:
Clergyman; Professor. WORKS ABOUT: Who Was Who Am 1.

BUCKHOUT, WILLIAM A. (1846-1912). Botany. OCCUP:
Professor. WORKS ABOUT: Am Journ Sci 185(1913):120; Am Men
Sci 1-2 d3; Sci 36(1912):822.

BUCKLEY, ERNEST ROBERTSON (1872-1912). Geology; Mining
geology. OCCUP: Industrial and state government scientist.
WORKS ABOUT: Am Men Sci 1-2 d3; Natl Cycl Am Biog 19; Who
Was Who Am 1.

BUCKLEY, SAMUEL BOTSFORD (1809-1884). Natural history;
Botany. OCCUP: Teacher; Farmer; State geologist. MSS: Ind
NUCMC. WORKS ABOUT [death dates may differ]: Dict Am Biog;
Elliott; Natl Cycl Am Biog 5; Who Was Who Am h. See also:
Barr; Biog Geneal Master Ind.

BUDD, JOSEPH LANCASTER (1835-1904). Horticulture. OCCUP:
Horticulturist; Professor. WORKS ABOUT: Dict Am Biog. See
also: Barr; Biog Geneal Master Ind.

BUEL, JESSE (1778-1839). Agriculture. OCCUP: Publisher;
Farmer; Legislator. MSS: Ind NUCMC (?). WORKS ABOUT: Dict
Am Biog; Natl Cycl Am Biog 11; Who Was Who Am h. See also:
Biog Geneal Master Ind.

BULKELEY, GERSHOM (1636-1713). Alchemy; Chemistry. OCCUP:
Clergyman; Physician; Public official. MSS: Ind NUCMC (60-
2822). WORKS ABOUT: Am Chem and Chem Eng.

BULL, CHARLES STEDMAN (1845-1911). Ophthalmology. OCCUP:
Surgeon; Professor. WORKS ABOUT: Adams O; Am Men Sci 1-2
d3; Appleton's Supp 1918-31; Natl Cycl Am Biog 9; Sci
33(1911):686; Who Was Who Am 1.

BULL, EPHRAIM WALES (1806-1895). Horticulture. OCCUP:
Tradesman; Horticulturist. MSS: Ind NUCMC (66-1617). WORKS
ABOUT: Dict Am Biog; Who Was Who Am h. See also: Barr.

BULL, STORM (1856-1907). Steam engineering. OCCUP:
Professor. MSS: Ind NUCMC (82-1382). WORKS ABOUT: Am Men
Sci 1 d3; Sci 26(1907):767; Who Was Who Am 1.

BULL, WILLIAM TILLINGHAST (1849-1909). Surgery. OCCUP:
Surgeon; Professor. WORKS ABOUT: Dict Am Biog; Natl Cycl Am
Biog 9; Who Was Who Am 1. See also: Barr; Biog Geneal
Master Ind; Ireland Ind Sci.

BULLOCK, CHARLES (1826-1900). Chemistry; Pharmacy. OCCUP:
Pharmacist; Businessman; Educator. WORKS ABOUT: Am Journ
Pharm 72(1900):411-22; Natl Cycl Am Biog 5.

BULMER, THOMAS SANDERSON (-1898). Archeology. OCCUP:
Physician (?). WORKS ABOUT: Sci 8(1898):866.

BUMSTEAD, HENRY ANDREWS (1870-1920). Physics. OCCUP: Professor; Laboratory director. WORKS ABOUT: <u>Am Men Sci</u> 1-3 d4; <u>Dict Am Biog</u>; Natl Ac Scis <u>Biog Mem</u> 13; <u>Natl Cycl Am Biog</u> 15, 21; <u>Who Was Who Am</u> 1.

BURBANK, JOHN EMERSON (1872-1919). Geology; Physics. OCCUP: U.S. government scientist. WORKS ABOUT: <u>Am Men Sci</u> 2 d3.

BURBECK, ALLAN BEAL (1882-1920). Astronomy. OCCUP: Not determined. WORKS ABOUT: <u>Am Men Sci</u> 3 d4.

BURGESS, EDWARD (1848-1891). Entomology. OCCUP: Academic instructor; Craftsman. WORKS ABOUT: <u>Dict Am Biog</u>; Elliott; <u>Natl Cycl Am Biog</u> 1; <u>Who Was Who Am</u> h. See also: Barr; <u>Biog Geneal Master Ind</u>; Ireland <u>Ind Sci</u>.

BURNETT, CHARLES HENRY (1842-1902). Medicine. OCCUP: Physician; Professor. WORKS ABOUT: <u>Dict Am Biog</u>; <u>Natl Cycl Am Biog</u> 25; <u>Who Was Who Am</u> h. See also: Barr; <u>Biog Geneal Master Ind</u>.

BURNETT, JOSEPH (1820-1894). Chemistry (manufacturing). OCCUP: Manufacturer. WORKS ABOUT: <u>Dict Am Biog</u>; <u>Natl Cycl Am Biog</u> 26; <u>Who Was Who Am</u> h.

BURNETT, WALDO IRVING (1827-1854). Biology; Histology. OCCUP: Physician (?). WORKS ABOUT [birth dates may differ]: Elliott. See also: Barr; <u>Biog Geneal Master Ind</u>.

BURR, THEODORE (1771-ca.1822). Engineering. OCCUP: Engineer. WORKS ABOUT: Gies.

BURRILL, THOMAS JONATHAN (1839-1916). Botany. OCCUP: Professor; Academic administrator. MSS: <u>Ind NUCMC</u> (65-1306). WORKS ABOUT: <u>Am Men Sci</u> 1 d3; <u>Dict Am Biog</u>; Elliott (s); <u>Natl Cycl Am Biog</u> 12, 18; <u>Who Was Who Am</u> 1. See also: <u>Biog Geneal Master Ind</u>.

BURRITT, ELIJAH HINSDALE (1794-1838). Astronomy. OCCUP: Civil engineer; Publisher; Teacher. WORKS ABOUT: Elliott. See also: <u>Biog Geneal Master Ind</u>.

BURROUGHS, WILLIAM SEWARD (1855-1898). Invention. OCCUP: Inventor; Manufacturer. WORKS ABOUT: <u>Dict Am Biog</u> supp 1; <u>Natl Cycl Am Biog</u> 27; <u>Who Was Who Am</u> h. See also: <u>Biog Geneal Master Ind</u>; Pelletier.

BURTON, BEVERLY S. (-1904). Chemistry. OCCUP: Not determined. WORKS ABOUT: <u>Sci</u> 19(1904):117.

BURTON, GEORGE DEXTER (1855-1918). Electrical engineering. OCCUP: Inventor; Industrialist. WORKS ABOUT: <u>Natl Cycl Am Biog</u> 12; <u>Who Was Who Am</u> 1.

BUSCH, FREDERICK CARL (1873-1914). Physiology. OCCUP: Professor. WORKS ABOUT: <u>Am Men Sci</u> 1-2 d3; Wallace.

BUSHNELL, DAVID (ca.1742-1824). Invention. OCCUP: Inventor; Teacher; Physician. WORKS ABOUT: <u>Dict Am Biog</u>; <u>Natl Cycl Am Biog</u> 9; <u>Who Was Who Am</u> h. See also: <u>Biog Geneal Master Ind</u>; Ireland <u>Ind Sci</u>; Pelletier.

BUTLER, THOMAS BELDEN (1806-1873). Meteorology. OCCUP: Physician; Lawyer; Public official. WORKS ABOUT: <u>Dict Am Biog</u>; <u>Natl Cycl Am Biog</u> 12; <u>Who Was Who Am</u> h. See also: Barr; <u>Biog Geneal Master Ind</u>.

BYFORD, WILLIAM HEATH (1817-1890). Medicine; Gynecology. OCCUP: Physician; Professor. WORKS ABOUT: <u>Dict Am Biog</u>; <u>Dict Am Med Biog</u>; <u>Natl Cycl Am Biog</u> 2; <u>Who Was Who Am</u> h. See also: <u>Biog Geneal Master Ind</u>.

BYRD, WILLIAM, II (1674-1744). Natural history; Botany. OCCUP: Planter; Legislator. MSS: <u>Ind NUCMC</u> (62-4347, 79-722). WORKS ABOUT: <u>Dict Am Biog</u>; <u>Natl Cycl Am Biog</u> 7:247; Stearns; <u>Who Was Who Am</u> h. See also: <u>Biog Geneal Master Ind</u>.

BYRNE, JOHN (1825-1902). Gynecology. OCCUP: Physician; Professor. WORKS ABOUT: <u>Dict Am Biog</u>; <u>Natl Cycl Am Biog</u> 9; <u>Who Was Who Am</u> h. See also: Barr.

BYRON, JOHN W. (1861-1895). Bacteriology. OCCUP: Physician; Laboratory administrator. WORKS ABOUT: <u>Sci</u> 1(1895):585-86; <u>Twentieth Cent Biog Dict</u>.

C

CABELL, JAMES LAWRENCE (1813-1889). Anatomy. OCCUP: Physician; Professor. MSS: Ind NUCMC (61-195). WORKS ABOUT: Dict Am Biog; Dict Am Med Biog; Natl Cycl Am Biog 12; Who Was Who Am h. See also: Barr; Biog Geneal Master Ind.

CABOT, SAMUEL (1850-1906). Chemistry. OCCUP: Manufacturer. MSS: Ind NUCMC. WORKS ABOUT: Natl Cycl Am Biog 23.

CABOT, SAMUEL, JR. (1815-1885). Ornithology. OCCUP: Physician. MSS: Ind NUCMC. WORKS ABOUT: Elliott; Natl Cycl Am Biog 25. See also: Barr.

CADWALADER, THOMAS (1707 or 1708-1779). Medicine. OCCUP: Physician; Public official. MSS: Ind NUCMC. WORKS ABOUT [death dates may differ]: Dict Am Biog; Dict Am Med Biog; Who Was Who Am h. See also: Biog Geneal Master Ind; Ireland Ind Sci; Pelletier.

CAHOON, JOHN C. (1863-1891). Ornithology. OCCUP: Natural history collector. WORKS ABOUT: Auk 8(1891):320-21; Ornith Ool 16(1891):73-75.

CALDWELL, CHARLES (1772-1853). Medicine. OCCUP: Physician; Professor. MSS: Ind NUCMC. WORKS ABOUT: Dict Am Biog; Dict Am Med Biog; Natl Cycl Am Biog 7:276; Who Was Who Am h. See also: Biog Geneal Master Ind; Ireland Ind Sci.

CALDWELL, EUGENE WILSON (1870-1918). Medicine; Roentgenology. OCCUP: Professor; Laboratory administrator. WORKS ABOUT: Am Men Sci 1-2 d3; Dict Am Biog; Dict Am Med Biog; Natl Cycl Am Biog 18; Who Was Who Am 1. See also: Barr.

CALDWELL, GEORGE CHAPMAN (1834-1907). Chemistry. OCCUP: Professor. MSS: Ind NUCMC (62-3892, 66-766). WORKS ABOUT: Adams O; Allibone Supp; Am Chem and Chem Eng; Am Men Sci 1 d3; Appleton's; Browne and Weeks p475; Elliott (s); Natl Cycl Am Biog 4, 26; Twentieth Cent Biog Dict; Wallace; Who Was Who Am 1.

CALDWELL, JOSEPH (1773-1835). Mathematics. OCCUP: Professor; College president. MSS: Ind NUCMC (78-1872). WORKS ABOUT: Dict Am Biog; Natl Cycl Am Biog 13; Who Was Who Am h. See also: Biog Geneal Master Ind.

CALL, RICHARD ELLSWORTH (1856-1917). Zoology. OCCUP: Teacher; Educator; Curator. MSS: Ind NUCMC (?). WORKS ABOUT: Am Men Sci 1-2 d3; Who Was Who Am 4.

CALVIN, SAMUEL (1840-1911). Geology. OCCUP: Professor; State geologist. WORKS ABOUT: Am Men Sci 1-2 d3; Dict Am Biog; Elliott (s); Natl Cycl Am Biog 13; Who Was Who Am 1. See also: Barr; Biog Geneal Master Ind.

CAMPBELL, JOHN H. (1847-1897). Conchology. OCCUP: Lawyer. WORKS ABOUT: Nautilus 10(1896-1897):116-17.

CAMPBELL, JOHN LYLE (1818-1886). Chemistry; Geology. OCCUP: Professor. WORKS ABOUT: Adams O; Allibone Supp; Am Journ Sci 131(1886):240; Appleton's; Knight; Twentieth Cent Biog Dict; Wallace.

CAMPBELL, JOHN LYLE (1827-1904). Astronomy; Physics. OCCUP: Professor. MSS: Ind NUCMC (?). WORKS ABOUT: Indiana Authors 1917; Wallace; Who Was Who Am 1.

CAMPBELL, JOHN PENDLETON (1863-1918). Biology; Physiology. OCCUP: Professor. MSS: Ind NUCMC (?). WORKS ABOUT: Am Men Sci 1-2 d3; Sci 49(1919):21; Who Was Who Am 4.

CAMPBELL, JOHN TenBROOK (1833-1911). Civil engineering; Geology. OCCUP: Public official; Surveyor; Engineer. WORKS ABOUT: Adams O; Natl Cycl Am Biog 12; Wallace; Who Was Who Am 1.

CAPSHAW, WALTER L. (-1915). Anatomy. OCCUP: Professor. Sci 43(1916):64.

CARBUTT, JOHN (1832-1905). Chemistry. OCCUP: Photographer; Manufacturer. WORKS ABOUT: Dict Am Biog; Natl Cycl Am Biog 22.

CAREY, JOHN (1797?-1880). Botany. OCCUP: Natural history explorer. MSS: Ind NUCMC. WORKS ABOUT: Am Journ Sci 119(1880):421-23, 127(1884):242.

CARHART, HENRY SMITH (1844-1920). Physics. OCCUP: Professor. MSS: Ind NUCMC. WORKS ABOUT: Adams O; Am Chem and Chem Eng; Am Men Sci 1-2 d3; Appleton's; Elliott (s); Natl Cycl Am Biog 4; Twentieth Cent Biog Dict; Wallace; Who Was Who Am 1.

CARLL, JOHN FRANKLIN (1828-1904). Geology. OCCUP: Civil engineer; Surveyor; State government scientist. WORKS ABOUT: Dict Am Biog; Elliott; Natl Cycl Am Biog 12; Who Was Who Am 1. See also: Biog Geneal Master Ind.

CARMALT, CHURCHILL (1866-1905). Anatomy. OCCUP: Academic instructor. WORKS ABOUT: <u>Am Men Sci</u> 1 d3.

CARNELL, FREDERICK JAMES (-1902). Physics. OCCUP: Academic affiliate. WORKS ABOUT: <u>Sci</u> 16(1902):877.

CARPENTER, FRANKLIN REUBEN (1848-1910). Geology. OCCUP: Mining engineer. WORKS ABOUT: <u>Am Men Sci</u> 1 d3; <u>Dict Am Biog</u>; <u>Who Was Who Am</u> 4.

CARPENTER, GEORGE WASHINGTON (1802-1860). Chemistry; Mineralogy; Pharmacology. OCCUP: Pharmacist; Businessman. MSS: <u>Ind NUCMC</u> (?). WORKS ABOUT: Elliott; <u>Natl Cycl Am Biog</u> 10. See also: <u>Biog Geneal Master Ind</u>.

CARPENTER, ROLLA CLINTON (1852-1919). Mechanical engineering. OCCUP: Professor. MSS: <u>Ind NUCMC</u> (?). WORKS ABOUT: Adams O; <u>Am Men Sci</u> 1-2 d3; Appleton's; <u>Sci</u> 49(1919):118; <u>Twentieth Cent Biog Dict</u>; Wallace; <u>Who Was Who Am</u> 1.

CARPENTER, WILLIAM MARBURY (1811-1848). Botany; Geology. OCCUP: Physician; Professor. WORKS ABOUT: Johnson; <u>Tulane Graduates Mag</u> (January 1914):122-27.

CARR, EZRA SLOCUM (1819-1894). Chemistry; Geology. OCCUP: Professor; Farmer. MSS: <u>Ind NUCMC</u>. WORKS ABOUT: Allibone <u>Supp</u>; <u>Am Chem and Chem Eng</u>; Appleton's <u>Supp</u> 1901.

CARROLL, CHARLES GEIGER (1875-1916). Chemistry. OCCUP: Professor. WORKS ABOUT: <u>Am Men Sci</u> 2 d3.

CARROLL, JAMES (1854-1907). Bacteriology; Pathology. OCCUP: Army physician; Professor. WORKS ABOUT: <u>Am Men Sci</u> 1 d3; <u>Dict Am Biog</u>; <u>Dict Am Med Biog</u>; <u>Dict Sci Biog</u>; <u>Who Was Who Am</u> 1. See also: Ireland <u>Ind Sci</u>.

CARRUTH, JAMES HARRISON (1807-1896). Botany. OCCUP: Clergyman; Professor; State government scientist. MSS: <u>Ind NUCMC</u>. WORKS ABOUT: Appleton's; Kansas Ac Sci <u>Trans</u> 15(1898):135-36; <u>Twentieth Cent Biog Dict</u>.

CARSON, JOSEPH (1808-1876). Medical botany; Pharmacy. OCCUP: Physician; Professor. MSS: <u>Ind NUCMC</u> (76-897). WORKS ABOUT: <u>Dict Am Biog</u>; <u>Natl Cycl Am Biog</u> 5; <u>Who Was Who Am</u> h. See also: Barr; <u>Biog Geneal Master Ind</u>.

CARTER, JAMES MADISON GORE (1843-1919). Medicine. OCCUP: Physician; Professor. WORKS ABOUT: <u>Am Men Sci</u> 1-2 d3; Wallace; <u>Who Was Who Am</u> 1.

CARTER, OSCAR CHARLES SUMNER (1857-1917). Chemistry; Geology; Mineralogy. OCCUP: Teacher. WORKS ABOUT: <u>Am Men Sci</u> 1-2 d3.

CARTWRIGHT, SAMUEL ADOLPHUS (1793-1863). Medicine; Study of biological functions. OCCUP: Physician. Scientific work: see Roy Soc <u>Cat</u>. MSS: <u>Ind NUCMC</u> (72-1548). WORKS ABOUT: Appleton's; <u>Dict Am Med Biog</u>; <u>New Orleans Med Surg Journ</u> 19(1866-1867):432-36; <u>Twentieth Cent Biog Dict</u>.

CARUS, PAUL (1852-1919). Philosphy of science. OCCUP: Editor and publisher. WORKS ABOUT: <u>Am Men Sci</u> 1; <u>Dict Am Biog</u>; <u>Dict Sci Biog</u>; <u>Natl Cycl Am Biog</u> 14; <u>Who Was Who Am</u> 1. See also: <u>Biog Geneal Master Ind</u>.

CARY, PHILLIP H. (-1916). Geology. OCCUP: Industrial scientist (?). WORKS ABOUT: <u>Sci</u> 44(1916):672.

CASAMAJOR, PAUL (1831-1887). Chemistry. OCCUP: Consultant; Industrial scientist. WORKS ABOUT: <u>Am Chem and Chem Eng</u>; Am Chem Soc <u>Journ</u> 9(1887):206-08; Browne and Weeks p495.

CASE, WILLARD ERASTUS (1857-1918). Electricity; Electrochemistry. OCCUP: Businessman. WORKS ABOUT: <u>Am Men Sci</u> 1-2 d3; <u>Natl Cycl Am Biog</u> 19; <u>Sci</u> 48(1918):467.

CASEY, THOMAS LINCOLN (1831-1896). Engineering. OCCUP: Army engineer. MSS: <u>Ind NUCMC</u>. WORKS ABOUT: <u>Dict Am Biog</u> supp 1; Natl Ac Scis <u>Biog Mem</u> 4; <u>Natl Cycl Am Biog</u> 4. See also: Barr; <u>Biog Geneal Master Ind</u>.

CASSEDAY, S. A. (-1860). Paleontology. OCCUP: Not determined. WORKS ABOUT: <u>Am Journ Sci</u> 81(1861):155.

CASSIN, JOHN (1813-1869). Ornithology. OCCUP: Public official; Printer. MSS: <u>Ind NUCMC</u>. WORKS ABOUT: <u>Dict Am Biog</u>; Elliott; <u>Natl Cycl Am Biog</u> 22; <u>Who Was Who Am</u> h. See also: Barr; <u>Biog Geneal Master Ind</u>; Ireland <u>Ind Sci</u>; Pelletier.

CASTNER, HAMILTON YOUNG (1859-1899). Chemistry. OCCUP: Consultant (analytical chemistry); Industrialist. WORKS ABOUT: <u>Am Chem and Chem Eng</u>; Howard; <u>Sci</u> 10(1899):620, 21(1905):881; Williams.

CASWELL, ALEXIS (1799-1877). Mathematics; Natural philosophy; Astronomy. OCCUP: Professor; College president. MSS: <u>Ind NUCMC</u> (?). WORKS ABOUT: <u>Dict Am Biog</u>; Elliott; Natl Ac Scis <u>Biog Mem</u> 6; <u>Natl Cycl Am Biog</u> 1, 8; <u>Who Was Who Am</u> h. See also: Barr; <u>Biog Geneal Master Ind</u>.

CASWELL, JOHN HENRY (1846-1909). Mineralogy. OCCUP: Academic affiliate; Businessman. WORKS ABOUT: Elliott.

CATESBY, MARK (1683-1749). Natural history; Botany. OCCUP: Natural history explorer/ collector. MSS: <u>Ind NUCMC</u>. WORKS ABOUT [birth dates may differ]: <u>Dict Am Biog</u>; <u>Dict Sci Biog</u>; Elliott; <u>Who Was Who Am</u> h. See also: <u>Biog Geneal Master Ind</u>; Ireland <u>Ind Sci</u>; Pelletier.

CATLIN, CHARLES ALBERT (1849-1916). Chemistry. OCCUP:
Industrial scientist. WORKS ABOUT: Am Men Sci 1-2 d3; Sci
43(1916):639; Who Was Who Am 1.

CATLIN, GEORGE (1796-1872). Anthropology; Ethnology. OCCUP:
Artist. MSS: Ind NUCMC (67-122, 77-310). WORKS ABOUT: Dict
Am Biog; Natl Cycl Am Biog 3; Who Was Who Am h. See also:
Barr; Biog Geneal Master Ind; Ireland Ind Sci.

CATON, JOHN DEAN (1812-1895). Natural history. OCCUP:
Lawyer; Businessman. MSS: Ind NUCMC (62-4603). WORKS ABOUT:
Dict Am Biog; Natl Cycl Am Biog 4; Who Was Who Am h. See
also: Biog Geneal Master Ind.

CATTELL, WILLIAM ASHBURNER (1863-1920). Civil engineering.
OCCUP: Engineer. WORKS ABOUT: Am Men Sci 3 d4; Who Was Who
Am 1.

CAVERLY, CHARLES SOLOMON (1856-1918). Hygiene. OCCUP:
Professor; Public official. WORKS ABOUT: Dict Am Med Biog;
Sci 48(1918):467.

CHAILLE, STANFORD EMERSON (1830-1911). Physiology. OCCUP:
Physician; Professor; Academic administrator (medicine).
MSS: Ind NUCMC. WORKS ABOUT: Adams O; Allibone Supp; Am Men
Sci 1-2 d3; Appleton's; Dict Am Med Biog; Knight; Natl Cycl
Am Biog 9; Sci 27(1908):517, 33(1911):339; Twentieth Cent
Biog Dict; Wallace; Who Was Who Am 1.

CHALMERS, LIONEL (1715-1777). Medicine; Meteorology. OCCUP:
Physician. MSS: Ind NUCMC (?). WORKS ABOUT: Elliott. See
also: Biog Geneal Master Ind.

CHALMOT, GUILLAUME LOUIS JACQUES de. See DeCHALMOT,
GUILLAUME LOUIS JACQUES.

CHAMBERLAIN, ALEXANDER FRANCIS (1865-1914). Anthropology.
OCCUP: Professor. WORKS ABOUT: Am Men Sci 1-2 d3; Dict Am
Biog; Who Was Who Am 1. See also: Barr; Biog Geneal Master
Ind.

CHAMBERLIN, BENJAMIN B. (1831-1888). Mineralogy. OCCUP:
Businessman. WORKS ABOUT: Am Journ Sci 136(1888):396; New
York Ac Scis Trans 8(1888-1889):46-47; Sci 1st ser
12(1888):215.

CHAMBERS, VACTOR TOUSEY (1830-1883). Entomology. OCCUP:
Lawyer. WORKS ABOUT: Canadian Ent 15(1883):178; Cincinnati
Soc Nat Hist Journ 6(1883):239-44; Ent Soc Ontario Rep
(1883):81; Sci 1st ser 2(1883):253-54.

CHANDLER, CHARLES HENRY (1840-1912). Mathematics. OCCUP:
Professor. WORKS ABOUT: Am Men Sci 1-2 d3; Appleton's; Sci
35(1912):689; Twentieth Cent Biog Dict; Wallace; Who Was
Who Am 1.

CHANDLER, SETH CARLO (1846-1913). Astronomy. OCCUP: Editor. MSS: Ind NUCMC (?). WORKS ABOUT: Am Men Sci 1 d3; Dict Am Biog; Dict Sci Biog; Elliott (s); Natl Cycl Am Biog 9; Who Was Who Am 1. See also: Barr; Ireland Ind Sci.

CHANDLER, THOMAS HENDERSON (1824?-1895). Dentistry. OCCUP: Academic administrator. WORKS ABOUT: Sci 2(1895):300.

CHANDLER, WILLIAM HENRY (1841-1906). Chemistry. OCCUP: Professor. WORKS ABOUT: Am Chem and Chem Eng; Am Men Sci 1 d3; Appleton's; Sci 24(1906):710; Twentieth Cent Biog Dict; Who Was Who Am 1.

CHANUTE, OCTAVE (1832-1910). Engineering. OCCUP: Engineer. MSS: Ind NUCMC (62-4604, 80-471). WORKS ABOUT: Am Men Sci 1-2 d3; Dict Am Biog; Natl Cycl Am Biog 10; Who Was Who Am 1. See also: Barr; Biog Geneal Master Ind; Ireland Ind Sci; Pelletier.

CHAPMAN, ALVAN WENTWORTH (1809-1899). Botany. OCCUP: Physician. MSS: Ind NUCMC. WORKS ABOUT: Dict Am Biog; Dict Sci Biog; Elliott; Natl Cycl Am Biog 13; Who Was Who Am h. See also: Biog Geneal Master Ind.

CHAPMAN, HENRY CADWALADER (1845-1909). Biology; Medicine. OCCUP: Physician; Professor (medicine). WORKS ABOUT: Am Men Sci 1 d3; Dict Am Biog; Natl Cycl Am Biog 22; Who Was Who Am 1. See also: Barr; Biog Geneal Master Ind.

CHAPMAN, NATHANIEL (1780-1853). Medicine; Materia medica. OCCUP: Physician; Professor. MSS: Ind NUCMC. WORKS ABOUT: Dict Am Biog; Dict Am Med Biog; Natl Cycl Am Biog 3; Who Was Who Am h. See also: Biog Geneal Master Ind; Pelletier.

CHAPMAN, ROBERT HOLLISTER (1868-1920). Topography; Topographical engineering. OCCUP: U.S. government scientist. WORKS ABOUT: Am Men Sci 2 d3; Natl Cycl Am Biog 18; Who Was Who Am 1.

CHARLES, FRED LEMAR (1872-1911). Zoology. OCCUP: Professor. WORKS ABOUT: Am Men Sci 2 d3.

CHARROPPIN, CHARLES M. (-1915). Astronomy. OCCUP: Clergyman; Academic instructor. WORKS ABOUT: Sci 42(1915):608.

CHASE, PLINY EARLE (1820-1886). Physical science. OCCUP: Businessman; Professor. MSS: Ind NUCMC. WORKS ABOUT: Dict Am Biog; Elliott; Natl Cycl Am Biog 6; Who Was Who Am h. See also: Biog Geneal Master Ind.

CHAUVENET, REGIS (1842-1920). Chemistry. OCCUP: Consultant; Mining engineer; Professor. WORKS ABOUT: Am Men Sci 3 d4; Appleton's; Natl Cycl Am Biog 7:446; Wallace; Who Was Who Am 1.

CHAUVENET, WILLIAM (1820-1870). Astronomy; Mathematics.
OCCUP: Professor; Academic administrator. MSS: Ind NUCMC.
WORKS ABOUT: Dict Am Biog; Elliott; Natl Ac Scis Biog Mem
1; Natl Cycl Am Biog 11; Who Was Who Am h. See also: Barr;
Biog Geneal Master Ind.

CHEESMAN, TIMOTHY MATLACK (1853-1919). Bacteriology;
Medicine. OCCUP: Physician; Academic instructor. WORKS
ABOUT: Am Men Sci 1-2 d3; Natl Cycl Am Biog 41; Sci
49(1919):233.

CHEEVER, BYRON WILLIAM (1841-1888). Metallurgy. OCCUP:
Professor. WORKS ABOUT: Am Inst Mining Eng Trans
16(1888):888-90.

CHEEVER, DAVID WILLIAMS (1831-1915). Medicine. OCCUP:
Professor. WORKS ABOUT: Allibone Supp; Am Men Sci 1-2 d3;
Dict Am Med Biog; Kagan p91; Leonardo p97-99; Natl Cycl Am
Biog 13; Sci 43(1916):18; Twentieth Cent Biog Dict;
Wallace; Who Was Who Am 1.

CHENEY, ALBERT NELSON (1849-1901). Ichthyology. OCCUP:
State government scientist; Editor. WORKS ABOUT: Sci
14(1901):380; Who Was Who Am 1.

CHESTER, ALBERT HUNTINGTON (1843-1903). Chemistry;
Mineralogy. OCCUP: Professor; Mining engineer. WORKS ABOUT:
Elliott; Natl Cycl Am Biog 11; Who Was Who Am 1. See also:
Barr; Biog Geneal Master Ind.

CHICKERING, JOHN WHITE (1831-1913). Botany. OCCUP:
Clergyman; Professor. WORKS ABOUT: Am Men Sci 1-2;
Twentieth Cent Biog Dict; Who Was Who Am 1.

CHILD, CHARLES T. (1867?-1902). Engineering. OCCUP:
Electrical engineer; Editor. WORKS ABOUT: Sci 16(1902):38.

CHILTON, GEORGE (ca.1767-ca.1836). Chemistry. OCCUP:
Teacher; Manufacturer. WORKS ABOUT: Am Journ Sci
31(1837):421-24.

CHILTON, JAMES RENWICK (1809?-1863). Chemistry. OCCUP:
Industrial scientist. WORKS ABOUT: Am Journ Sci
86(1863):314.

CHITTENDEN, HIRAM MARTIN (1858-1917). Civil engineering.
OCCUP: Army engineer. MSS: Ind NUCMC (70-1376). WORKS
ABOUT: Am Men Sci 2 d3; Dict Am Biog; Natl Cycl Am Biog 17;
Who Was Who Am 1. See also: Biog Geneal Master Ind.

CHOVET, ABRAHAM (1704-1790). Anatomy. OCCUP: Surgeon. WORKS
ABOUT: Dict Am Biog; Natl Cycl Am Biog 21; Who Was Who
Am h.

CHRISTIE, JAMES (1840-1911). Mechanical engineering. OCCUP:
Industrial engineer. WORKS ABOUT [death dates may differ]:
Am Men Sci 1-2 d3; Who Was Who Am 1.

CHRISTY, DAVID (1802-ca.1868?). Geology. OCCUP: Journalist; Industrial scientist. WORKS ABOUT: Dict Am Biog; Elliott; Who Was Who Am h. See also: Biog Geneal Master Ind; Ireland Ind Sci.

CHRISTY, SAMUEL BENEDICT (1853-1914). Metallurgy; Mining engineering. OCCUP: Professor; Academic administrator. WORKS ABOUT: Am Men Sci 1-2 d3; Natl Cycl Am Biog 20; Sci 41(1915):24; Who Was Who Am 1.

CHRYSOSTOM, REV. BROTHER (JOSEPH J. CONLAN) (1863-1917). Philosophy; Psychology. OCCUP: Professor. WORKS ABOUT: Am Men Sci 1; Sci 45(1917):112.

CHURCH, GEORGE EARL (1835-1910). Engineering; Geography. OCCUP: Civil engineer. MSS: Ind NUCMC. WORKS ABOUT: Dict Am Biog; Natl Cycl Am Biog 13. See also: Barr; Biog Geneal Master Ind.

CHURCH, JOHN ADAMS (1843-1917). Metallurgy. OCCUP: U.S. government scientist; Mining engineer; Consultant. WORKS ABOUT: Dict Am Biog; Who Was Who Am 1. See also: Biog Geneal Master Ind.

CHURCHILL, WILLIAM (1859-1920). Anthropology. OCCUP: Editor; Public official (diplomatic service); Research institution associate. WORKS ABOUT: Dict Am Biog; Who Was Who Am 1. See also: Biog Geneal Master Ind.

CHURCHILL, WILLIAM WILBERFORCE (1867-1910). Mechanical engineering. OCCUP: Industrial engineer and officer. WORKS ABOUT: Natl Cycl Am Biog 18.

CHURCHMAN, JOHN (1753-1805). Surveying. OCCUP: Surveyor. MSS: Ind NUCMC (60-1763). WORKS ABOUT: Appleton's; Drake; Natl Cycl Am Biog 9.

CIST, JACOB (1782-1825). Geology; Natural history. OCCUP: Postmaster; Businessman. MSS: Ind NUCMC (66-24). WORKS ABOUT: Dict Am Biog; Elliott; Who Was Who Am h. See also: Biog Geneal Master Ind.

CLAP, THOMAS (1703-1767). Astronomy. OCCUP: Clergyman; College president. MSS: Ind NUCMC. WORKS ABOUT: Dict Am Biog; Elliott; Natl Cycl Am Biog 1; Who Was Who Am h. See also: Biog Geneal Master Ind.

CLAPP, ASAHEL (1792-1862). Botany; Geology. OCCUP: Physician. WORKS ABOUT: Am Journ Sci 85(1863):306, 450-51; Dict Am Med Biog.

CLARK, ADMONT HALSEY (1888-1918). Pathology. OCCUP: Professor. WORKS ABOUT: Sci 48(1918):467.

CLARK, ALONZO HOWARD (1850-1918). History; Ichthyology.
OCCUP: U.S. government scientist; Museum curator; Editor.
WORKS ABOUT: Adams O; Allibone Supp; Appleton's; Sci
49(1919):234; Twentieth Cent Biog Dict; Wallace; Who Was
Who Am 1.

CLARK, ALVAN (1804-1887). Astronomy; Astronomical
instrumentation; Astronomical lens making. OCCUP: Artist;
Instrument maker. WORKS ABOUT: Dict Am Biog; Dict Sci Biog;
Elliott; Natl Cycl Am Biog 6; Who Was Who Am h. See also:
Barr; Biog Geneal Master Ind; Ireland Ind Sci.

CLARK, ALVAN GRAHAM (1832-1897). Astronomy; Astronomical
instrumentation; Astronomical lens making. OCCUP:
Instrument maker. WORKS ABOUT: Dict Am Biog; Dict Sci Biog;
Elliott; Natl Cycl Am Biog 5; Who Was Who Am h. See also:
Barr; Biog Geneal Master Ind; Pelletier.

CLARK, GAYLORD PARSONS (1856-1907). Physiology. OCCUP:
Professor. WORKS ABOUT: Am Men Sci 1 d3; Sci 26(1907):358;
Who Was Who Am 1.

CLARK, GEORGE ARCHIBALD (1864-1918). Administration;
Fisheries. OCCUP: Academic administrator. MSS: Ind NUCMC
(67-2066). WORKS ABOUT: Sci 48(1918):213-15; Who Was Who
Am 1.

CLARK, GEORGE BASSETT (1827-1891). Astronomical
instrumentation. OCCUP: Instrument maker. WORKS ABOUT: Dict
Sci Biog. See also: Biog Geneal Master Ind.

CLARK, HENRY JAMES (1826-1873). Botany; Zoology. OCCUP:
Professor. WORKS ABOUT: Dict Am Biog; Elliott; Natl Ac Scis
Biog Mem 1; Natl Cycl Am Biog 9; Who Was Who Am h. See
also: Barr; Biog Geneal Master Ind.

CLARK, JOHN NATHANIEL (1831?-1903). Ornithology. OCCUP: Not
determined. WORKS ABOUT: Sci 17(1903):159.

CLARK, WILLIAM BULLOCK (1860-1917). Geology. OCCUP:
Professor; U.S. and state government scientist. WORKS
ABOUT: Am Men Sci 1-2 d3; Dict Am Biog; Elliott (s); Natl
Ac Scis Biog Mem 9; Natl Cycl Am Biog 13, 18; Who Was Who
Am 1. See also: Barr; Biog Geneal Master Ind.

CLARK, WILLIAM SMITH (1826-1886). Botany; Chemistry. OCCUP:
Professor; College president. MSS: Ind NUCMC (84-1326).
WORKS ABOUT: Dict Am Biog; Natl Cycl Am Biog 5; Who Was Who
Am h. See also: Barr; Biog Geneal Master Ind; Pelletier.

CLARK, XENOS (1853-1889). Microscopy; Zoology. OCCUP:
Academic affiliate. WORKS ABOUT: Am Nat 23(1889):749-50.

CLARKE, BENJAMIN FRANKLIN (1831-1908). Mechanical
engineering. OCCUP: Professor. WORKS ABOUT: Am Men Sci 1-2
d3; Natl Cycl Am Biog 10; Twentieth Cent Biog Dict; Who Was
Who Am 1.

CLARKE, CORA HUIDEKOPER (1851-1916). Botany. OCCUP: Not
determined. MSS: Ind NUCMC. WORKS ABOUT: Am Men Sci 2 d3;
Sci 43(1916):815; Siegel and Finley.

CLARKE, THOMAS CURTIS (1827-1901). Engineering. OCCUP:
Civil engineer. WORKS ABOUT: Natl Cycl Am Biog 7:500; Sci
13(1901):1038-39; Twentieth Cent Biog Dict.

CLAYPOLE, EDITH JANE (1870-1915). Pathology; Physiology.
OCCUP: Academic instructor; Pathologist. WORKS ABOUT: Am
Men Sci 1-2 d3; Natl Cycl Am Biog 13; Ogilvie; Sci
41(1915):527, 754; Siegel and Finley; Who Was Who Am 5; Wom
Whos Who Am.

CLAYPOLE, EDWARD WALLER (1835-1901). Geology. OCCUP:
Professor. MSS: Ind NUCMC. WORKS ABOUT: Dict Am Biog;
Elliott; Natl Cycl Am Biog 13; Who Was Who Am 1. See also:
Barr; Biog Geneal Master Ind.

CLAYTON, JOHN (1694-1773). Botany. OCCUP: Public official.
MSS: Ind NUCMC. WORKS ABOUT [birth dates may differ]: Dict
Am Biog; Elliott; Natl Cycl Am Biog 19; Who Was Who Am h.
See also: Biog Geneal Master Ind; Ireland Ind Sci;
Pelletier.

CLEAVELAND, PARKER (1780-1858). Mineralogy. OCCUP:
Professor. MSS: Ind NUCMC (71-34, 72-1012). WORKS ABOUT:
Dict Am Biog; Dict Sci Biog; Elliott; Natl Cycl Am Biog 13;
Who Was Who Am h. See also: Biog Geneal Master Ind; Ireland
Ind Sci; Pelletier.

CLEGG, MOSES TRAN (1876-1918). Bacteriology. OCCUP:
Government scientist (foreign); U.S. government scientist.
WORKS ABOUT: Who Was Who Am 1.

CLEMENS, (JAMES) BRECKENRIDGE (1829 or 1830-1867).
Entomology. OCCUP: Not determined. WORKS ABOUT: Ent News
25(1914):289-92; Pop Sci Mo 76(1910):470-72.

CLEMSON, THOMAS GREEN (1807-1888). Chemistry; Geology;
Mining engineering. OCCUP: Mining engineer; Planter; Public
official. MSS: Ind NUCMC (61-2106). WORKS ABOUT: Dict Am
Biog; Elliott; Natl Cycl Am Biog 13; Who Was Who Am h. See
also: Biog Geneal Master Ind; Ireland Ind Sci; Pelletier.

CLEVENGER, SHOBAL VAIL (1843-1920). Neurology; Psychiatry;
Psychology. OCCUP: Physician; Administrator. MSS: Ind NUCMC
(66-1471). WORKS ABOUT: Dict Am Biog; Natl Cycl Am Biog 5;
Who Was Who Am 4. See also: Biog Geneal Master Ind.

CLIMENKO, HYMAN (1875-1920). Neurology. OCCUP:
Neurologist. WORKS ABOUT: Am Men Sci 3 d4.

CLINTON, DeWITT (1769-1828). Natural history. OCCUP:
Governor. MSS: Ind NUCMC (61-1573, 62-590, 66-2006, 68-
1096, 76-653, 82-1041). WORKS ABOUT: Dict Am Biog; Natl
Cycl Am Biog 3; Who Was Who Am h. See also: Biog Geneal
Master Ind.

CLINTON, GEORGE WILLIAM (1807-1885). Botany. OCCUP: Lawyer. MSS: <u>Ind NUCMC</u>. WORKS ABOUT: <u>Am Journ Sci</u> 131(1886):17-20; Torrey Bot Club <u>Bull</u> 12(1885):103-06; <u>Twentieth Cent Biog Dict</u>.

CLOUD, JOSEPH (1770-1845). Chemistry. OCCUP: U.S. government scientist. Scientific work: see Roy Soc <u>Cat</u>. WORKS ABOUT [mentioned in the following]: Baldwin C p75; <u>Chymia</u> 3(1950):110; Franklin Inst <u>Journ</u> 217(1934):122; Futhey and Cope p500; Poggendorff v1; Smith E p221; Stewart p95.

COAKLEY, GEORGE WASHINGTON (1814-1893). Astronomy; Mathematics. OCCUP: Professor. MSS: <u>Ind NUCMC</u> (65-1590). WORKS ABOUT: <u>Am Journ Sci</u> 146(1893):484; <u>Leopoldina</u> 29(1893):204; <u>Nature</u> 48(1893):398; <u>Twentieth Cent Biog Dict.</u>

COAN, TITUS (1801-1882). Geology. OCCUP: Clergyman. MSS: <u>Ind NUCMC</u> (60-1869, 76-152). WORKS ABOUT: <u>Dict Am Biog</u>; Elliott; <u>Natl Cycl Am Biog</u> 2; <u>Who Was Who Am</u> h. See also: Barr; <u>Biog Geneal Master Ind</u>.

COATES, REYNELL (1802-1886). Medicine; Zoology. OCCUP: Physician; Professor. WORKS ABOUT: Allibone; <u>Boston Med Surg Journ</u> 44(1851):135-37; Kelly and Burrage; Shaw (1972); <u>Twentieth Cent Biog Dict</u>.

COCHRANE, ALEXANDER (1802-1865). Chemistry. OCCUP: Manufacturer. WORKS ABOUT: <u>Am Chem and Chem Eng</u>; Haynes p57-73.

COCHRANE, ALEXANDER, JR. (1840-1919). Chemistry. OCCUP: Manufacturer. WORKS ABOUT: <u>Am Chem and Chem Eng</u>; Appleton's <u>Supp</u> 1918-31; <u>Natl Cycl Am Biog</u> 14, 27.

COE, HOWARD SHELDON (1888-1918). Agriculture. OCCUP: U.S. government scientist. WORKS ABOUT: <u>Natl Cycl Am Biog</u> 19; <u>Sci</u> 48(1918):467.

COFFIN, JAMES HENRY (1806-1873). Mathematics; Meteorology. OCCUP: Educator; Professor. MSS: <u>Ind NUCMC</u>. WORKS ABOUT: <u>Dict Am Biog</u>; Elliott; Natl Ac Scis <u>Biog Mem</u> 1; <u>Natl Cycl Am Biog</u> 8; <u>Who Was Who Am</u> h. See also: Barr; <u>Biog Geneal Master Ind</u>; Ireland <u>Ind Sci</u>.

COFFIN, JOHN HUNTINGTON CRANE (1815-1890). Mathematics. OCCUP: U.S. government scientist; Professor. MSS: <u>Ind NUCMC</u>. WORKS ABOUT: Elliott; Natl Ac Scis <u>Biog Mem</u> 8; <u>Natl Cycl Am Biog</u> 5; <u>Who Was Who Am</u> h. See also: Barr; <u>Biog Geneal Master Ind</u>.

COFFIN, SELDEN JENNINGS (1838-1915). Astronomy; Mathematics; Meteorology. OCCUP: Professor; Academic administrator. WORKS ABOUT: Adams O; Allibone <u>Supp</u>; <u>Am Men Sci</u> 1-2 d3; Appleton's; <u>Natl Cycl Am Biog</u> 11; <u>Twentieth Cent Biog Dict</u>; Wallace; <u>Who Was Who Am</u> 1.

COGSWELL, JOSEPH GREEN (1786-1871). Geology; Mineralogy.
OCCUP: Librarian; Professor; Educator. MSS: Ind NUCMC.
WORKS ABOUT: Dict Am Biog; Natl Cycl Am Biog 11; Who Was
Who Am h. See also: Biog Geneal Master Ind.

COLBURN, W. W. (-1899). Ornithology. OCCUP: Educator.
WORKS ABOUT: Sci 11(1900):159.

COLBURN, WARREN (1793-1833). Mathematics. OCCUP: Teacher;
Industrial manager. WORKS ABOUT: Dict Am Biog; Natl Cycl Am
Biog 10; Who Was Who Am h. See also: Biog Geneal Master
Ind.

COLBURN, ZERAH (1804-1839). Mathematics [prodigy]. OCCUP:
Clergyman; Professor (languages). WORKS ABOUT [death dates
may differ]: Dict Am Biog; Natl Cycl Am Biog 7:74; Who Was
Who Am h. See also: Biog Geneal Master Ind.

COLBY, CHARLES EDWARDS (1855-1897). Chemistry. OCCUP:
Professor. WORKS ABOUT: Sci 6(1897):628.

COLDEN, CADWALLADER (1688-1776). Botany; Medicine; Physics.
OCCUP: Surveyor; Public official. MSS: Ind NUCMC (61-546,
68-1129). WORKS ABOUT: Dict Am Biog; Dict Am Med Biog; Dict
Sci Biog; Elliott; Natl Cycl Am Biog 2; Who Was Who Am h.
See also: Barr; Biog Geneal Master Ind; Ireland Ind Sci;
Pelletier.

COLDEN, JANE (JANE COLDEN FARQUHAR) (1724-1766). Botany.
OCCUP: Engaged in independent study. WORKS ABOUT: Dict Am
Biog; Elliott; Notable Am Wom; Ogilvie; Siegel and Finley;
Who Was Who Am h. See also: Biog Geneal Master Ind;
Pelletier.

COLE, AARON HODGMAN (1856-1913). Biology. OCCUP: Teacher.
WORKS ABOUT: Am Men Sci 1-2 d3; Sci 39(1914):206; Who Was
Who Am 1.

COLEMAN, CLARENCE (-1918). Civil engineering. OCCUP:
U.S. government engineer. WORKS ABOUT: Am Men Sci 2 d3.

COLEMAN, W. L. (-1904). Medicine. OCCUP: Physician (?).
WORKS ABOUT: Sci 20(1904):814.

COLLES, CHRISTOPHER (1738-1816). Engineering; Physics
(pneumatics). OCCUP: Businessman; Inventor; Author. WORKS
ABOUT: Dict Am Biog; Natl Cycl Am Biog 9; Who Was Who Am h.
See also: Biog Geneal Master Ind.

COLLETT, JOHN (1828-1899). Geology. OCCUP: Farmer; State
government scientist. MSS: Ind NUCMC (?). WORKS ABOUT:
Elliott; Who Was Who Am 1. See also: Barr; Biog Geneal
Master Ind.

COLLIER, PETER (1835-1896). Chemistry; Agricultural chemistry. OCCUP: Professor; U.S. and state government scientist. MSS: Ind NUCMC (70-471). WORKS ABOUT: Dict Am Biog; Elliott; Natl Cycl Am Biog 8; Who Was Who Am h. See also: Barr; Biog Geneal Master Ind; Pelletier.

COLLIN, (HENRY) ALONZO (1837-1918). Physics. OCCUP: Professor. WORKS ABOUT: Am Men Sci 1-2 d3; Sci 47(1918):533; Who Was Who Am 1.

COLLINGWOOD, FRANCIS (1834-1911). Civil engineering. OCCUP: Civil engineer; Academic instructor. MSS:Ind NUCMC (?). WORKS ABOUT: Am Men Sci 1-2 d3; Natl Cycl Am Biog 13; Twentieth Cent Biog Dict; Who Was Who Am 1.

COLLINS, FRANK SHIPLEY (1848-1920). Botany. OCCUP: Industrial manager. MSS: Ind NUCMC (62-747). WORKS ABOUT: Am Men Sci 1-2 d3; Dict Am Biog; Elliott (s); Who Was Who Am 3. See also: Biog Geneal Master Ind.

COLLINS, JOSEPH WILLIAM (1839-1904). Ichthyology. OCCUP: Seaman; U.S. government scientist; Public official. WORKS ABOUT: Sci 41(1915):749-50; Twentieth Cent Biog Dict; Who Was Who Am 1.

COLLINS, ZACCHEUS (1764-1831). Botany. OCCUP: Benefactor; Collector. MSS: Ind NUCMC (61-383, 66-25). WORKS ABOUT: Am Journ Sci 23(1832-1833):398-99; Appleton's; Drake.

COLMAN, BENJAMIN (1673-1747). Medicine; Natural history. OCCUP: Clergyman. MSS: Ind NUCMC. WORKS ABOUT: Dict Am Biog; Elliott; Natl Cycl Am Biog 7:153; Who Was Who Am h. See also: Biog Geneal Master Ind.

COLSON, HAROLD ROY (1876-1913). Astronomy. OCCUP: Observatory affiliate; Publisher. WORKS ABOUT: Am Men Sci 1-2 d3.

COLTON, BUEL PRESTON (1852-1906). Biology. OCCUP: Professor. WORKS ABOUT: Allibone Supp; Am Men Sci 1-2 d3; Wallace.

COMPTON, ALFRED GEORGE (1835-1913). Astronomy; Mathematics; Physics. OCCUP: Professor. WORKS ABOUT: Am Men Sci 1-2 d3; Elliott (s); Sci 38(1913):879; Who Was Who Am 1.

COMSTOCK, CYRUS BALLOU (1831-1910). Engineering. OCCUP: Army officer; Engineer. MSS: Ind NUCMC (62-4637). WORKS ABOUT: Adams O: Allibone Supp; Am Men Sci 1 d3; Appleton's; Natl Ac Scis Biog Mem 7; Natl Cycl Am Biog 22; Sci 31(1910):901; Twentieth Cent Biog Dict; Wallace; Who Was Who Am 1.

COMSTOCK, THEODORE BRYANT (1849-1915). Geology; Mining
engineering. OCCUP: Academic administrator; Engineer;
Mining affiliate. MSS: Ind NUCMC. WORKS ABOUT: Adams O;
Allibone Supp; Am Men Sci 1-2 d3; Appleton's; Elliott (s);
Natl Cycl Am Biog 13; Twentieth Cent Biog Dict; Wallace;
Who Was Who Am 1.

CONANT, FRANKLIN STORY (1870?-1897). Biology. OCCUP:
Academic affiliate. WORKS ABOUT: Johns Hopkins Biol Lab Mem
4(1)(1898):ix-xi; John Hopkins Circ 17(1897-1898):19-21;
Sci 7(1898):277.

CONANT, LEVI LEONARD (1857-1916). Mathematics. OCCUP:
Professor. WORKS ABOUT: Adams O; Am Men Sci 1-2 d3; Sci
44(1916):564; Wallace; Who Was Who Am 1.

CONDON, THOMAS (1822-1907). Geology. OCCUP: Clergyman;
Professor. WORKS ABOUT: Dict Am Biog; Natl Cycl Am Biog 13.
See also: Barr; Biog Geneal Master Ind.

CONGDON, ERNEST ARNOLD (1866-1917). Chemistry. OCCUP:
Professor; Industrial scientist. WORKS ABOUT: Am Men Sci 1-
2 d3; Who Was Who Am 4.

CONKLIN, WILLIAM AUGUSTUS (1837-1913). Natural history;
Zoology. OCCUP: Veterinarian; City government scientist;
Editor. WORKS ABOUT: Appleton's; Natl Cycl Am Biog 2;
Twentieth Cent Biog Dict; Who Was Who Am 1.

CONN, HERBERT WILLIAM (1859-1917). Biology. OCCUP:
Professor; State government scientist; Laboratory
administrator. MSS: Ind NUCMC. WORKS ABOUT: Adams O;
Allibone Supp; Am Men Sci 1-2 d3; Bulloch p359; Natl Cycl
Am Biog 20; Sci 45(1917):407, 451-52, 429; Twentieth Cent
Biog Dict; Wallace; Who Was Who Am 1.

CONNEFFE, JAMES F. (-1910). Bacteriology. OCCUP:
Academic affiliate. WORKS ABOUT: Sci 31(1910):187.

CONRAD, SOLOMON WHITE (1779-1831). Natural history; Botany;
Mineralogy. OCCUP: Publisher; Professor. Scientific work:
see Roy Soc Cat. WORKS ABOUT: Harshberger; Pop Sci Mo
47(1895):257-59 (mentioned).

CONRAD, TIMOTHY ABBOTT (1803-1877). Conchology; Malacology;
Paleontology. OCCUP: Natural history explorer; State
government scientist. MSS: Ind NUCMC. WORKS ABOUT: Dict Sci
Biog; Elliott; Natl Cycl Am Biog 8; Who Was Who Am h. See
also: Barr; Biog Geneal Master Ind; Ireland Ind Sci;
Pelletier.

CONVERSE, HARRIET MAXWELL (1836-1903). Folklore. OCCUP:
Not determined; Author. MSS: Ind NUCMC (?). WORKS ABOUT:
Notable Am Wom; Who Was Who Am 1. See also: Biog Geneal
Master Ind.

COOK, A. GRACE (19th century). Astronomy. OCCUP: Not
determined. WORKS ABOUT: Ogilvie (app).

COOK, ALBERT JOHN (1842-1916). Natural history; Zoology; Entomology. OCCUP: Professor; State government scientist. WORKS ABOUT: Adams O; Allibone Supp; Am Men Sci 1-2 d3; Appleton's; Essig p578-81; Mallis; Osborn Herb p224-25, 348; Sci 44(1916):528; Twentieth Cent Biog Dict; Wallace; Who Was Who Am 1.

COOK, GEORGE HAMMELL (1818-1889). Geology. OCCUP: Professor; State government scientist. MSS: Ind NUCMC (65-1596). WORKS ABOUT: Dict Am Biog; Elliott; Natl Ac Scis Biog Mem 4; Natl Cycl Am Biog 6; Who Was Who Am h. See also: Barr; Biog Geneal Master Ind; Ireland Ind Sci; Pelletier.

COOK, JOSEPH (1838-1901). Biology; Science and religion. OCCUP: Clergyman; Lecturer. MSS: Ind NUCMC (60-1543). WORKS ABOUT: Dict Am Biog; Natl Cycl Am Biog 2; Who Was Who Am 1. See also: Barr; Biog Geneal Master Ind.

COOKE, JOSIAH PARSONS, JR. (1827-1894). Chemistry. OCCUP: Professor. MSS: Ind NUCMC (65-1230). WORKS ABOUT: Dict Am Biog; Dict Sci Biog; Elliott; Natl Ac Scis Biog Mem 4; Natl Cycl Am Biog 6; Who Was Who Am h. See also: Barr; Biog Geneal Master Ind; Ireland Ind Sci; Pelletier.

COOKE, WELLS WOODBRIDGE (1858-1916). Biology; Ornithology. OCCUP: Teacher; U.S. and state government scientist. MSS: Ind NUCMC (?). WORKS ABOUT: Natl Cycl Am Biog 21; Sci 43(1916):530.

COOLEY, LEROY CLARK (1833-1916). Physics. OCCUP: Professor. WORKS ABOUT: Adams O; Allibone Supp; Am Men Sci 1-2 d3; Appleton's; Biog Dict Am Educ; Elliott (s); Natl Cycl Am Biog 11; Preston; Sci 44(1916):461; Twentieth Cent Biog Dict; Wallace; Who Was Who Am 1.

COOPER, JAMES GRAHAM (1830-1902). Botany; Geology; Zoology. OCCUP: Surgeon; Natural history explorer/collector. MSS: Ind NUCMC. WORKS ABOUT: Dict Am Biog; Elliott; Who Was Who Am h. See also: Barr; Biog Geneal Master Ind; Ireland Ind Sci.

COOPER, PETER (1791-1883). Invention. OCCUP: Manufacturer; Inventor; Benefactor. MSS: Ind NUCMC. WORKS ABOUT: Dict Am Biog; Natl Cycl Am Biog 3; Who Was Who Am h. See also: Biog Geneal Master Ind; Ireland Ind Sci; Pelletier.

COOPER, THOMAS (1759-1839). Chemistry. OCCUP: Public official; Professor; College president. MSS: Ind NUCMC (66-524). WORKS ABOUT [death dates may differ]: Dict Am Biog; Dict Sci Biog; Elliott; Natl Cycl Am Biog 11; Who Was Who Am h. See also: Barr; Biog Geneal Master Ind; Ireland Ind Sci; Pelletier.

COOPER, WILLIAM (1797-1864). Natural history. OCCUP: Inherited wealth; Engaged in independent study. MSS: Ind NUCMC. WORKS ABOUT: Elliott. See also: Barr; Pelletier.

COPE, EDWARD DRINKER (1840-1897). Zoology; Vertebrate paleontology. OCCUP: Professor; Editor. MSS: Ind NUCMC (62-4336, 66-26). WORKS ABOUT: Dict Am Biog; Dict Sci Biog; Elliott; Natl Ac Scis Biog Mem 13; Natl Cycl Am Biog 7:474; Who Was Who Am h. See also: Barr; Biog Geneal Master Ind; Ireland Ind Sci; Pelletier.

COPPET, LOUIS CASIMIR de (1841-1911). Chemistry; Physics. OCCUP: Not determined. WORKS ABOUT: Sci 34(1911):343; World Whos Who Sci [as: DeCoppet].

COQUILLET, DANIEL WILLIAM (1856-1911). Entomology. OCCUP: U.S. government scientist; Museum curator. WORKS ABOUT: Am Men Sci 1-2 d3; Dict Am Biog; Who Was Who Am 1. See also: Barr; Ireland Ind Sci; Pelletier.

CORLISS, GEORGE HENRY (1817-1888). Invention. OCCUP: Inventor; Manufacturer. MSS: Ind NUCMC (60-2357). WORKS ABOUT: Dict Am Biog; Natl Cycl Am Biog 10; Who Was Who Am h. See also: Barr; Biog Geneal Master Ind; Ireland Ind Sci; Pelletier.

CORLISS, HARRY PERCIVAL (1886?-1918). Chemistry. OCCUP: Academic researcher. WORKS ABOUT: Sci 49(1919):42.

CORNWALL, HENRY BEDINGER (1844-1917). Chemistry. OCCUP: Professor. WORKS ABOUT: Adams O; Allibone Supp; Am Men Sci 1-2 d3; Appleton's; Elliott (s); Sci 45(1917):360; Twentieth Cent Biog Dict; Wallace; Who Was Who Am 1.

COTTIER, JOSEPH G. C. (1874?-1897). Mathematics; Physics. OCCUP: Academic affiliate. WORKS ABOUT: Sci 6(1897):363-64.

COTTING, JOHN RUGGLES (1778-1867). Chemistry; Geology. OCCUP: Clergyman; Teacher; State geologist. WORKS ABOUT [birth dates may differ]: Elliott; Natl Cycl Am Biog 5; Who Was Who Am h. See also: Barr; Biog Geneal Master Ind.

COUES, ELLIOTT (1842-1899). Natural history; Ornithology. OCCUP: Army surgeon; U.S. government scientist; Professor. MSS: Ind NUCMC (68-2159). WORKS ABOUT: Dict Am Biog; Dict Sci Biog; Elliott; Natl Ac Scis Biog Mem 6; Natl Cycl Am Biog 5; Who Was Who Am 1. See also: Barr; Biog Geneal Master Ind; Ireland Ind Sci; Pelletier.

COUPER, JAMES HAMILTON (1794-1866). Agriculture; Natural history. OCCUP: Planter; Manufacturer. MSS: Ind NUCMC (78-1919). WORKS ABOUT: Dict Am Biog; Elliott; Who Was Who Am h. See also: Biog Geneal Master Ind.

COURTENAY, EDWARD HENRY (1803-1853). Mathematics. OCCUP: Army engineer; Professor. MSS: Ind NUCMC. WORKS ABOUT: Elliott. See also: Biog Geneal Master Ind.

COUTHOUY, JOSEPH PITTY (1808-1864). Conchology. OCCUP: Seaman; U.S. government scientist. WORKS ABOUT: Biol Soc Washington <u>Proc</u> 4(1888):108-11; Dawdy; <u>Field and Lab</u> 25(1957):99-104; Groce and Wallace.

COWLES, EDWARD (1837-1919). Medicine. OCCUP: Physician; Professor (mental diseases); Administrator. MSS: <u>Ind</u> <u>NUCMC</u>. WORKS ABOUT: <u>Am Men Sci</u> 1-2 d3; <u>Natl Cycl Am Biog</u> 19; <u>Sci</u> 50(1919):109, 132-33; <u>Who Was Who Am</u> 1.

COWLES, EUGENE HUTCHINSON (1855-1892). Metallurgy. OCCUP: Industrialist. WORKS ABOUT: <u>Electrician</u> 29(1892):61; Franklin Inst <u>Journ</u> 133(1892):404-05; <u>Natl Cycl Am</u> <u>Biog</u> 23.

COX, CHARLES FINNEY (1846-1912). Biology. OCCUP: Museum (botanic garden) administrator. WORKS ABOUT: <u>Am Men Sci</u> 1-2 d3; <u>Sci</u> 35(1912):214; <u>Who Was Who Am</u> 1.

COX, JACOB DOLSON (1828-1900). Microscopy (diatoms). OCCUP: Governor; Public official; Academic administrator (law). MSS: <u>Ind NUCMC</u> (62-1419, 75-960). WORKS ABOUT: <u>Dict Am</u> <u>Biog</u>; <u>Natl Cycl Am Biog</u> 3, 4, 22; <u>Who Was Who Am</u> 1. See also: <u>Biog Geneal Master Ind</u>.

COX, ROWLAND, JR. (1871?-1916). Surgery. OCCUP: Physician (?); Academic instructor. WORKS ABOUT: <u>Sci</u> 44(1916):198.

COX, ULYSSES ORANGE (1864-1920). Biology. OCCUP: Professor. WORKS ABOUT: <u>Am Men Sci</u> 2 d3.

COXE, ECKLEY BRINTON (1839-1895). Engineering. OCCUP: Mining engineer; Industrial officer. MSS: <u>Ind NUCMC</u>. WORKS ABOUT: <u>Dict Am Biog</u>; <u>Natl Cycl Am Biog</u> 11; <u>Who Was Who Am</u> h. See also: Barr; <u>Biog Geneal Master Ind</u>.

COXE, JOHN REDMAN (1773-1864). Pharmacy. OCCUP: Physician; Professor. MSS: <u>Ind NUCMC</u>. WORKS ABOUT: <u>Dict Am Biog</u>; <u>Dict</u> <u>Am Med Biog</u>; <u>Natl Cycl Am Biog</u> 22; <u>Who Was Who Am</u> h. See also: Barr; <u>Biog Geneal Master Ind</u>; Ireland <u>Ind Sci</u>; Pelletier.

COXE, WILLIAM, JR. (1762-1831). Pomology. OCCUP: Horticulturist; Legislator. MSS: <u>Ind NUCMC</u>. WORKS ABOUT: <u>Dict Am Biog</u>; <u>Who Was Who Am</u> h. See also: <u>Biog Geneal</u> <u>Master Ind</u>.

COZZENS, ISSACHAR (1780-1865). Chemistry; Geology; Mineralogy. OCCUP: Consultant; Author. WORKS ABOUT [birth dates may differ]: Adams O; Allibone; <u>Am Geol</u> 24(1899):327-28; Appleton's; Drake; Wallace.

CRAFTS, JAMES MASON (1839-1917). Chemistry. OCCUP: Professor; College president. WORKS ABOUT: <u>Am Men Sci</u> 1-2 d3; <u>Dict Am Biog</u>; Elliott (s); Natl Ac Scis <u>Biog Mem</u> 9; <u>Natl Cycl Am Biog</u> 13; <u>Who Was Who Am</u> 1. See also: Barr; <u>Biog Geneal Master Ind</u>; Ireland <u>Ind Sci</u>; Pelletier.

CRAGIN, EDWIN BRADFORD (1859-1918). Obstetrics. OCCUP: Physician; Professor. WORKS ABOUT: Am Men Sci 1-2 d3; Appleton's Supp 1918-31; Leonardo p117-18; Natl Cycl Am Biog 15; Who Was Who Am 1.

CRAIG, JOHN (1864-1912). Horticulture. OCCUP: Government scientist (foreign); Professor. WORKS ABOUT: Am Men Sci 1-2 d3; Sci 36(1912):238; Who Was Who Am 1.

CRAIG, MOSES (-1913). Botany. OCCUP: Professor; Herbarium curator. MSS: Ind NUCMC (?). WORKS ABOUT: Sci 38(1913):401.

CRAIG, THOMAS (1855-1900). Mathematics. OCCUP: Professor. WORKS ABOUT: Dict Am Biog; Elliott; Who Was Who Am h. See also: Biog Geneal Master Ind.

CRAIGHILL, WILLIAM PRICE (1833-1909). Engineering. OCCUP: Army engineer; Professor. MSS: Ind NUCMC (?). WORKS ABOUT: Allibone Supp; Appleton's; Drake; Natl Cycl Am Biog 12; Sci 29(1909):178; Twentieth Cent Biog Dict; Wallace; Who Was Who Am 1.

CRAM, THOMAS JEFFERSON (ca.1807-1883). Engineering; Natural philosophy. OCCUP: Army engineer; Professor. MSS: Ind NUCMC (67-58). WORKS ABOUT: Appleton's; Cullum v1; Twentieth Cent Biog Dict.

CRAMPTON, CHARLES ALBERT (1858-1915). Chemistry. OCCUP: U.S. government scientist. WORKS ABOUT: Am Men Sci 1-2 d3; Who Was Who Am 1.

CRANDALL, CHARLES LEE (1850-1917). Civil engineering. OCCUP: Professor; Engineer. MSS: Ind NUCMC (?). WORKS ABOUT: Adams O; Am Men Sci 2 d3; Natl Cycl Am Biog 4; Sci 46(1917):310; Twentieth Cent Biog Dict; Wallace; Who Was Who Am 1.

CRAW, ALBERT (1850-1908). Entomology. OCCUP: State government scientist. WORKS ABOUT: Sci 28(1908):792.

CRAW, WILLIAM JARVIS (1830-1897). Chemistry. OCCUP: Businessman. WORKS ABOUT: Chamberlain v2:348-49.

CRAWE, ITHAMAR B. (1794-1847). Botany; Geology. OCCUP: Physician. WORKS ABOUT: Am Journ Sci 54(1847):300.

CREEVEY, CAROLINE ALATHEA STICKNEY (1843-1920). Botany. OCCUP: Teacher; Lecturer. WORKS ABOUT: Natl Cycl Am Biog 30; Wallace; Who Was Who Am 1.

CRITCHLEY, JOHN WILLIAM (-1910). Taxidermy. OCCUP: Taxidermist. WORKS ABOUT: Sci 32(1910):14.

CROCKER, LUCRETIA (1829-1886). Science education. OCCUP: Professor; Educator. WORKS ABOUT: Elliott; Notable Am Wom; Siegel and Finley. See also: Biog Geneal Master Ind.

CROES, JOHN JAMES ROBERTSON (1834-1906). Civil engineering. OCCUP: Engineer. WORKS ABOUT: Am Men Sci 1 d3; Natl Cycl Am Biog 6; Sci 23(1906):519; Who Was Who Am 1.

CROOM, HARDY BRYAN (1798-1837). Botany. OCCUP: Planter. MSS: Ind NUCMC (78-1926). WORKS ABOUT: Croom piii-x; Sargent v10:58.

CROTCH, GEORGE ROBERT (1842-1874). Entomology. OCCUP: Natural history explorer/ collector. WORKS ABOUT: Allibone Supp; California Ac Scis Proc 5(1873-1874):332-34; Essig p598-600.

CROUCH, H. CHESTER (1871-1903). Engineering. OCCUP: Professor. WORKS ABOUT: Sci 18(1903):607.

CROWELL, JAMES FOSTER (1848-1915). Engineering. OCCUP: Civil engineer. WORKS ABOUT: Sci 41(1915):527; Who Was Who Am 1.

CROZIER, ARTHUR ALGER (1856-1899). Botany. OCCUP: U.S. and state government scientist. WORKS ABOUT: Iowa Ac Scis Proc 7(1900):17-18; Wallace.

CUMING, (SIR) ALEXANDER (ca.1690-1775). Natural history. OCCUP: Explorer. WORKS ABOUT: Dict Am Biog; Dict Natl Biog; Who Was Who Am h. See also: Biog Geneal Master Ind.

CUMMINGS, CLARA EATON (1855-1906). Botany. OCCUP: Professor. WORKS ABOUT: Am Men Sci 1 d3; Ogilvie; Sci 25(1907):77-78; Siegel and Finley; Who Was Who Am 1.

CUMMINGS, J. B. (-1896). Science. OCCUP: Professor. WORKS ABOUT: Sci 3(1896):555.

CURLEY, JAMES (1796-1889). Astronomy. OCCUP: Clergyman; Academic instructor. MSS: Ind NUCMC (80-293). WORKS ABOUT: Elliott; Who Was Who Am h.

CURTIN, JEREMIAH (ca.1840-1906). Anthropology. OCCUP: Government employee; Researcher. MSS: Ind NUCMC (62-1972, 83-920). WORKS ABOUT: Am Men Sci 1 d3; Dict Am Biog; Who Was Who Am 1. See also: Barr; Biog Geneal Master Ind.

CURTIS, EDWARD (1838-1912). Medicine. OCCUP: Physician; Professor. WORKS ABOUT: Adams O; Allibone Supp; Appleton's; Natl Cycl Am Biog 9, 25; Oxford Canadian Hist (1973); Sci 36(1912):822; Twentieth Cent Biog Dict; Wallace; Who Was Who Am 1.

CURTIS, GEORGE EDWARD (1861-1895). Meteorology. OCCUP: U.S. government scientist. WORKS ABOUT: Am Meteorol Journ 12(1895-1896):1-4; Phil Soc Washington Bull 13(1900):384-87.

CURTIS, JOHN GREEN (1844-1913). Physiology. OCCUP: Physician; Professor. MSS: Ind NUCMC (62-2). WORKS ABOUT: Am Men Sci 1-2 d3; Dict Am Biog; Who Was Who Am 1. See also: Barr; Biog Geneal Master Ind.

CURTIS, MOSES ASHLEY (1808-1872). Botany. OCCUP: Clergyman; Teacher; State government scientist. MSS: Ind NUCMC (64-477, 78-1928, 83-1730). WORKS ABOUT: Dict Am Biog; Elliott; Natl Cycl Am Biog 5; Who Was Who Am h. See also: Barr; Biog Geneal Master Ind.

CURTISS, ALLEN HIRAM (1845-1907). Botany. OCCUP: U.S. government scientist. WORKS ABOUT: Am Men Sci 1 d3; Sci 26(1907):606.

CURTMAN, CHARLES OTTO (1829-1896). Chemistry. OCCUP: Professor. WORKS ABOUT: Am Journ Pharm 68(1896):351.

CURWEN, JOHN (1821-1901). Medicine. OCCUP: Physician; Medical administrator. MSS: Ind NUCMC. WORKS ABOUT: Sci 14(1900):77; Who Was Who Am 1.

CUSHING, EDWARD FITCH (1862?-1911). Medicine. OCCUP: Physician; Professor. WORKS ABOUT: Sci 33(1911):523.

CUSHING, FRANK HAMILTON (1857-1900). Anthropology; Ethnology. OCCUP: U.S. government scientist. MSS: Ind NUCMC. WORKS ABOUT: Dict Am Biog; Natl Cycl Am Biog 11; Who Was Who Am h. See also: Barr; Biog Geneal Master Ind.

CUSHING, JONATHAN PETER (1793-1835). Chemistry. OCCUP: Professor; College president. WORKS ABOUT: Am Chem and Chem Eng; Appleton's; Drake; Natl Cycl Am Biog 2; Twentieth Cent Biog Dict.

CUSHMAN, HOLBROOK (1857-1895). Physics. OCCUP: Academic administrator. WORKS ABOUT: Sci 2(1895):586, 757-61.

CUTBUSH, EDWARD (1772-1843). Chemistry. OCCUP: Physician; Professor; Academic administrator (medicine). WORKS ABOUT: Am Chem and Chem Eng; Natl Cycl Am Biog 18; Wallace.

CUTBUSH, JAMES (1788-1823). Chemistry. OCCUP: Manufacturer; Lecturer. MSS: Ind NUCMC. WORKS ABOUT: Dict Am Biog; Elliott; Who Was Who Am h. See also: Barr; Biog Geneal Master Ind; Ireland Ind Sci; Pelletier.

CUTLER, MANASSEH (1742-1823). Botany. OCCUP: Clergyman; Legislator; Businessman. MSS: Ind NUCMC (60-2228, 62-3547, 66-1892). WORKS ABOUT: Dict Am Biog; Elliott; Natl Cycl Am Biog 3; Who Was Who Am h. See also: Biog Geneal Master Ind; Ireland Ind Sci; Pelletier.

CUTTER, EPHRAIM (1832-1917). Medicine; Microscopy. OCCUP: Physician. MSS: Ind NUCMC (?). WORKS ABOUT: Dict Am Biog; Natl Cycl Am Biog 3, 27; Who Was Who Am 1. See also: Barr; Biog Geneal Master Ind.

CUTTING, HIRAM ADOLPHUS (1832-1892). Natural history; Geology. OCCUP: Businessman; Physician; State government scientist. WORKS ABOUT: Elliott; <u>Natl Cycl Am Biog</u> 10; <u>Who Was Who Am</u> h. See also: <u>Biog Geneal Master Ind</u>.

CUTTS, RICHARD DOMINICUS (1817-1883). Geodesy. OCCUP: U.S. government scientist. WORKS ABOUT: <u>Sci</u> 1st ser 2(1883):810; <u>Twentieth Cent Biog Dict</u>.

D

DABOLL, NATHAN (1750-1818). Almanac making; Mathematics; Navigation. OCCUP: Teacher; Editor. WORKS ABOUT: <u>Dict Am Biog</u>; Elliott; <u>Natl Cycl Am Biog</u> 23; <u>Who Was Who Am</u> h. See also: <u>Biog Geneal Master Ind</u>.

daCOSTA, JACOB MENDEZ (1833-1900). Medicine. OCCUP: Physician; Professor. WORKS ABOUT: <u>Dict Am Biog</u>; <u>Dict Am Med Biog</u>; <u>Natl Cycl Am Biog</u> 9; <u>Who Was Who Am</u> 1. See also: Barr; <u>Biog Geneal Master Ind</u>; Ireland <u>Ind Sci</u>; Pelletier.

DAGGETT, FRANK SLATER (1855-1920). Ornithology. OCCUP: Businessman; Museum administrator; Natural history collector. MSS: <u>Ind NUCMC</u>. WORKS ABOUT: <u>Natl Cycl Am Biog</u> 38.

DAKIN, JOHN A. (1852-1900). Entomology; Ornithology. OCCUP: Not determined. WORKS ABOUT: <u>Auk</u> 17(1900):196-97; <u>Ent News</u> 11(1900):451; <u>Sci</u> 11(1900):635.

DALTON, CHARLES X. (1840-1912). Instrumentation. OCCUP: Industrial affiliate. WORKS ABOUT: <u>Sci</u> 35(1912):444-46.

DALTON, JOHN CALL, JR. (1825-1889). Physiology; Medicine. OCCUP: Professor; Academic administrator. MSS: <u>Ind NUCMC</u>. WORKS ABOUT: <u>Dict Am Biog</u>; <u>Dict Am Med Biog</u>; <u>Dict Sci Biog</u> supp 1; Elliott; <u>Natl Ac Scis Biog Mem</u> 3; <u>Natl Cycl Am Biog</u> 10; <u>Who Was Who Am</u> h. See also: Barr; <u>Biog Geneal Master Ind</u>; Pelletier.

DALY, CHARLES PATRICK (1816-1899). Geography. OCCUP: Lawyer. MSS: <u>Ind NUCMC</u> (69-814). WORKS ABOUT: <u>Dict Am Biog</u>; <u>Natl Cycl Am Biog</u> 3; <u>Who Was Who Am</u> 1. See also: Barr; <u>Biog Geneal Master Ind</u>.

DALY, CLARENCE M. (-1904). Technology. OCCUP: Research institution associate. WORKS ABOUT: <u>Sci</u> 20(1904):509.

DAMON, WILLIAM EMERSON (1838-1911). Natural history. OCCUP: Businessman. WORKS ABOUT: Adams O; Allibone Supp; Natl Cycl Am Biog 15; Sci 35(1912):175-76; Wallace; Who Was Who Am 1.

DANA, JAMES DWIGHT (1813-1895). Geology; Zoology. OCCUP: Professor; U.S. government scientist; Editor. MSS: Ind NUCMC. WORKS ABOUT: Dict Am Biog; Dict Sci Biog; Elliott; Natl Ac Scis Biog Mem 9; Natl Cycl Am Biog 6, 30; Who Was Who Am h. See also: Barr; Biog Geneal Master Ind; Ireland Ind Sci; Pelletier.

DANA, JAMES FREEMAN [born JONATHAN] (1793-1827). Chemistry; Mineralogy. OCCUP: Physician; Professor. WORKS ABOUT: Dict Am Biog; Elliott; Natl Cycl Am Biog 10; Who Was Who Am h. See also: Biog Geneal Master Ind; Ireland Ind Sci; Pelletier.

DANA, SAMUEL LUTHER (1795-1868). Chemistry; Mineralogy. OCCUP: Physician; Industrial scientist. WORKS ABOUT: Dict Am Biog; Elliott; Natl Cycl Am Biog 8; Who Was Who Am h. See also: Barr; Biog Geneal Master Ind; Ireland Ind Sci.

DANFORTH, SAMUEL (1696-1777). Alchemy. OCCUP: Teacher; Public official; Researcher. WORKS ABOUT: Am Chem and Chem Eng; Appleton's.

DANFORTH, SAMUEL, JR. (1740-1827). Chemistry; Medicine. OCCUP: Physician. WORKS ABOUT: Am Chem and Chem Eng; Appleton's; Drake.

DANIELLS, WILLIAM WILLARD (1840-1912). Chemistry. OCCUP: Professor. WORKS ABOUT: Am Men Sci 1-2 d3; Sci 36(1912):552; Who Was Who Am 1.

DANIELS, ARCHIBALD LAMONT (1849-1918). Mathematics. OCCUP: Professor. WORKS ABOUT: Am Men Sci 1-2 d3; Sci 48(1918):137.

DANIELS, FRED HARRIS (1853-1913). Engineering; Metallurgy. OCCUP: Industrial engineer and officer. WORKS ABOUT: Dict Am Biog; Natl Cycl Am Biog 22; Who Was Who Am 1. See also: Biog Geneal Master Ind.

DARBY, JOHN (1804-1877). Botany; Mathematics. OCCUP: Teacher; Professor; College president. WORKS ABOUT: Dict Am Biog; Who Was Who Am h. See also: Barr; Biog Geneal Master Ind.

DARBY, WILLIAM (1775-1854). Geography; Surveying. OCCUP: Planter; Surveyor; Author. MSS: Ind NUCMC. WORKS ABOUT: Dict Am Biog; Who Was Who Am h. See also: Biog Geneal Master Ind.

DARLINGTON, WILLIAM (1782-1863). Botany. OCCUP: Physician; Legislator. MSS: Ind NUCMC (60-1883). WORKS ABOUT: Dict Am Biog; Dict Am Med Biog; Dict Sci Biog; Elliott; Natl Cycl Am Biog 10; Who Was Who Am h. See also: Barr; Biog Geneal Master Ind.

DAUGHERTY, LEWIS SYLVESTER (1857-1919). Zoology. OCCUP: Professor. WORKS ABOUT: Sci 49(1919):493-94; Who Was Who Am 1.

DAVENPORT, GEORGE EDWARD (1833-1907). Botany. OCCUP: Public official. WORKS ABOUT [death dates may differ]: Am Men Sci 1 d3; Elliott (s); Sci 27(1908):397; Who Was Who Am 1.

DAVENPORT, RUSSELL WHEELER (1849-1904). Metallurgy. OCCUP: Industrial manager. WORKS ABOUT: Natl Cycl Am Biog 25; Sci 19(1904):439; Shaw.

DAVENPORT, THOMAS (1802-1851). Invention. OCCUP: Inventor; Craftsman. WORKS ABOUT: Dict Am Biog; Natl Cycl Am Biog 3; Who Was Who Am h. See also: Biog Geneal Master Ind; Ireland Ind Sci; Pelletier.

DAVIDGE, JOHN BEALE (1768-1829). Anatomy. OCCUP: Physician; Professor. WORKS ABOUT: Dict Am Biog; Dict Am Med Biog; Natl Cycl Am Biog 22; Who Was Who Am h. See also: Biog Geneal Master Ind.

DAVIDSON, GEORGE (1825-1911). Astronomy; Geography; Geodesy. OCCUP: U.S. government scientist; Professor. MSS: Ind NUCMC (71-723). WORKS ABOUT: Am Men Sci 1-2 d3; Dict Am Biog; Elliott (s); Natl Ac Scis Biog Mem 18; Natl Cycl Am Biog 7:227; Who Was Who Am 1. See also: Barr; Biog Geneal Master Ind.

DAVIDSON, ROBERT JAMES (1862-1915). Chemistry. OCCUP: Professor; Academic administrator; State government scientist. WORKS ABOUT: Am Men Sci 2 d3; Sci 43(1916):64, 418-19; Who Was Who Am 1.

DAVIES, CHARLES (1798-1876). Mathematics; Natural philosophy. OCCUP: Professor. MSS: Ind NUCMC. WORKS ABOUT: Adams O; Allibone; Am Journ Sci 112(1876):320; Appleton's; Biog Dict Am Educ; Drake; Natl Cycl Am Biog 3; Twentieth Cent Biog Dict; Wallace; Who Was Who Am h.

DAVIES, JOHN EUGENE (1839-1900). Astronomy; Chemistry; Physics. OCCUP: Professor. WORKS ABOUT: Am Micr Soc Trans 21(1900):249-50; Twentieth Cent Biog Dict.

DAVIS, CHARLES ABBOTT (1869?-1908). Biology. OCCUP: Museum curator. WORKS ABOUT: Sci 27(1908):237.

DAVIS, CHARLES ALBERT (1861-1916). Botany; Geology. OCCUP: Professor; Curator; U.S. government scientist. MSS: Ind NUCMC. WORKS ABOUT: Am Men Sci 1-2 d3; Sci 43(1916):639; Who Was Who Am 1.

DAVIS, CHARLES HENRY (1807-1877). Astronomy; Hydrography. OCCUP: Naval officer; Observatory director. MSS: Ind NUCMC. WORKS ABOUT: Dict Am Biog; Elliott; Natl Ac Scis Biog Mem 4; Natl Cycl Am Biog 4; Who Was Who Am h. See also: Barr; Biog Geneal Master Ind; Pelletier.

DAVIS, ELLERY WILLIAMS (1857-1918). Mathematics. OCCUP: Professor; Academic administrator. WORKS ABOUT: Am Men Sci 1-2 d3; Natl Cycl Am Biog 8; Sci 47(1918):168; Wallace; Who Was Who Am 1.

DAVIS, FREDERIC W. (1823?-1854). Chemistry. OCCUP: Industrial manager. WORKS ABOUT: Am Journ Sci 69(1855):448-49.

DAVIS, GRANT TRAIN (1877-1920). Chemistry. OCCUP: Academic instructor. WORKS ABOUT: Am Men Sci 1-2 d3.

DAVIS, GWILYM GEORGE (1857-1918). Medicine. OCCUP: Surgeon; Professor. WORKS ABOUT: Am Men Sci 1-2 d3; Who Was Who Am 1.

DAVIS, JOHN E. (-1900). Mathematics; Physics. OCCUP: Professor. WORKS ABOUT: Sci 11(1900):198.

DAVIS, JOSEPH BAKER (1845-1920). Civil engineering. OCCUP: Professor. MSS: Ind NUCMC (65-215, 81-1799). WORKS ABOUT: Am Men Sci 1-2 d3; Natl Cycl Am Biog 18; Wallace; Who Was Who Am 1.

DAVIS, NATHAN SMITH (1817-1904). Medicine. OCCUP: Physician; Professor; Academic administrator. MSS: Ind NUCMC (65-1537). WORKS ABOUT: Dict Am Biog; Dict Am Med Biog; Natl Cycl Am Biog 10, 35; Who Was Who Am 1. See also: Barr; Biog Geneal Master Ind; Pelletier.

DAVIS, NATHAN SMITH, JR. (1858-1920). Medicine. OCCUP: Physician; Professor; Academic administrator. MSS: Ind NUCMC (65-1540). WORKS ABOUT: Adams O; Am Lit Yearbook; Am Men Sci 1-3 d4; Natl Cycl Am Biog 13, 16, 35; Twentieth Cent Biog Dict; Wallace; Who Was Who Am 1.

DAVIS, NOAH KNOWLES (1830-1910). Psychology. OCCUP: Professor; College president. MSS: Ind NUCMC. WORKS ABOUT: Am Men Sci 1 d3; Dict Am Biog; Natl Cycl Am Biog 4; Who Was Who Am 1. See also: Barr; Biog Geneal Master Ind.

DAVIS, WALTER GOULD (1851-1919). Meteorology. OCCUP: Government scientist (foreign). WORKS ABOUT: Sci 49(1919):493, 50(1919):11-13.

DAVISON, ALVIN (1868-1915). Biology; Zoology. OCCUP: Professor. WORKS ABOUT: Am Men Sci 1-2 d3; Natl Cycl Am Biog 18; Sci 43(1916):307; Wallace.

DAVISON, JOHN MASON (1840-1915). Chemistry. OCCUP: Banker; Researcher. WORKS ABOUT: Am Men Sci 1-2; Natl Cycl Am Biog 17.

DAWBARN, ROBERT HUGH MACKAY (1860-1915). Surgery. OCCUP: Surgeon; Professor. WORKS ABOUT: Allibone Supp; Am Men Sci 1-2 d3; Natl Cycl Am Biog 48; Sci 42(1915):157.

DAY, AUSTIN GOODYEAR (1824-1889). Chemistry. OCCUP: Not determined. WORKS ABOUT: Sci 21(1905):876.

DAY, DAVID FISHER (1829-1900). Botany. OCCUP: Lawyer. MSS: Ind NUCMC (?). WORKS ABOUT: Bot Gaz 30(1900):347-48.

DAY, FISK HOLBROOK (1826-1903). Natural history. OCCUP: Professor. WORKS ABOUT: Who Was Who Am 1.

DAY, JEREMIAH (1773-1867). Mathematics. OCCUP: Professor; College president. MSS: Ind NUCMC. WORKS ABOUT: Dict Am Biog; Natl Cycl Am Biog 1; Who Was Who Am h. See also: Barr; Biog Geneal Master Ind.

DAY, WILLIAM CATHCART (1857-1905). Chemistry. OCCUP: Professor. WORKS ABOUT: Elliott; Who Was Who Am 1. See also: Barr.

DEAN, GEORGE WASHINGTON (1825-1897). Astronomy. OCCUP: U.S. government scientist. WORKS ABOUT: Boston Evening Transcript (January 25, 1897):5; Poggendorff v3, 4; Twentieth Cent Biog Dict.

DEAN, JAMES (1776-1849). Natural philosophy; Astronomy; Mathematics. OCCUP: Professor. WORKS ABOUT: Appleton's; Drake; Twentieth Cent Biog Dict; Wallace.

DEANE, JAMES (1801-1858). Geology. OCCUP: Physician. WORKS ABOUT: Elliott. See also: Barr; Biog Geneal Master Ind; Ireland Ind Sci.

DEARBORN, HENRY ALEXANDER SCAMMELL (1783-1851). General science; Horticulture. OCCUP: Public official; Legislator; Author. Scientific work: see Roy Soc Cat. MSS: Ind NUCMC (61-2950, 65-9). WORKS ABOUT: Dict Am Biog; Natl Cycl Am Biog 9; Who Was Who Am h. See also: Biog Geneal Master Ind.

DeBRAHM, JOHN GERAR WILLIAM. See DeBRAHM, WILLIAM GERARD.

DeBRAHM, WILLIAM GERARD (1717-ca.1799). Geography; Hydrography; Natural history; Surveying. OCCUP: Surveyor; Engineer. MSS: Ind NUCMC. WORKS ABOUT: Dict Am Biog; Elliott; Natl Cycl Am Biog 24; Who Was Who Am h. See also: Biog Geneal Master Ind; Ireland Ind Sci.

DeCHALMOT, GUILLAUME LOUIS JACQUES (-1899). Chemistry; Electrochemistry; Metallurgy. OCCUP: Industrial affiliate. WORKS ABOUT: Am Chem and Chem Eng.

DeFOREST, ERASTUS LYMAN (1834-1888). Mathematics. OCCUP: Inherited wealth; Engaged in independent study. WORKS ABOUT: Dict Am Biog; Elliott; Who Was Who Am h.

DeGROOT, HENRY (1815-1893). Geology. OCCUP: Physician; Journalist; Researcher. Scientific work: see Roy Soc <u>Cat</u>. WORKS ABOUT: Hinkel; Wallace.

DeKAY, JAMES ELLSWORTH (1792-1851). Natural history; Zoology. OCCUP: Physician (?); State government scientist. MSS: <u>Ind NUCMC</u>. WORKS ABOUT: <u>Dict Am Biog</u>; Elliott; <u>Natl Cycl Am Biog</u> 9; <u>Who Was Who Am</u> h. See also: Barr; <u>Biog Geneal Master Ind</u>; Pelletier.

DELAFIELD, FRANCIS (1841-1915). Medicine. OCCUP: Physician; Professor. WORKS ABOUT: <u>Am Men Sci</u> 1-2 d3; <u>Dict Am Biog</u>; <u>Dict Am Med Biog</u>; <u>Natl Cycl Am Biog</u> 10; <u>Who Was Who Am</u> 1. See also: Barr; <u>Biog Geneal Master Ind</u>.

DELAFIELD, JOSEPH (1790-1875). Geodesy; Mineralogy. OCCUP: Lawyer; Public official; Collector. WORKS ABOUT: Elliott; <u>Natl Cycl Am Biog</u> 11. See also: <u>Biog Geneal Master Ind</u>.

DeMOTTE, HARVEY CLELLAND (1838-1904). Mathematics. OCCUP: Professor; College president; Administrator. WORKS ABOUT: <u>Twentieth Cent Biog Dict</u>; <u>Who Was Who Am</u> 1.

DENISON, CHARLES (1845-1909). Medicine. OCCUP: Physician; Professor. WORKS ABOUT: Adams O; Allibone <u>Supp</u>; <u>Dict Am Med Biog</u>; <u>Natl Cycl Am Biog</u> 17; <u>Sci</u> 29(1909):179; Wallace; <u>Who Was Who Am</u> 1.

DENISON, CHARLES SIMEON (1849-1913). Engineering; Stereotomy. OCCUP: Professor. MSS: <u>Ind NUCMC</u> (65-227). WORKS ABOUT: <u>Am Men Sci</u> 1-2 d3; <u>Sci</u> 38(1913):189, 43(1916):740; <u>Who Was Who Am</u> 1.

DENNIS, DAVID WORTH (1849-1916). Biology. OCCUP: Professor. WORKS ABOUT: <u>Am Men Sci</u> 1-2 d3; <u>Indiana Authors 1816</u>; <u>Natl Cycl Am Biog</u> 38; Wallace; <u>Who Was Who Am</u> 1.

DENNIS, MARTIN (1851-1916). Chemistry. OCCUP: Industrialist. MSS: <u>Ind NUCMC</u> (?). WORKS ABOUT: <u>Am Chem and Chem Eng</u>; <u>Chem Indus Contribution</u> p17-18; Haynes p197-208.

De POURTALES, LOUIS FRANCOIS. See POURTALES, LOUIS FRANCOIS de.

DERBY, ORVILLE ADELBERT (1851-1915). Geology. OCCUP: Government scientist (foreign). MSS: <u>Ind NUCMC</u> (?). WORKS ABOUT: <u>Am Journ Sci</u> 191(1916):152; <u>Am Men Sci</u> 1-2 d3; Appleton's <u>Supp</u> 1901; <u>Natl Cycl Am Biog</u> 10; <u>Sci</u> 42(1915):826, 43(1916):596; <u>Twentieth Cent Biog Dict</u>; <u>Who Was Who Am</u> 4.

DERNEHL, PAUL HERMAN (1878-1919). Zoology. OCCUP: Museum employee. WORKS ABOUT: <u>Am Men Sci</u> 2 d3.

DESOR, EDOUARD. See DESOR, PIERRE JEAN EDOUARD.

DESOR, PIERRE JEAN EDOUARD (1811-1882). Geology. OCCUP: Researcher; Public official (foreign). WORKS ABOUT: Dict Sci Biog. See also: Barr.

DEVARRE, SPENCER H. (-1899). Mathematics. OCCUP: Academic instructor. WORKS ABOUT: Sci 9(1899):629.

DEWEY, CHESTER (1784-1867). Botany. OCCUP: Professor; Educator. MSS: Ind NUCMC (62-4133). WORKS ABOUT: Dict Am Biog; Elliott; Natl Cycl Am Biog 6; Who Was Who Am h. See also: Barr; Biog Geneal Master Ind; Pelletier.

DeWITT, SIMEON (1756-1834). Surveying. OCCUP: Surveyor; Academic administrator; Public official. MSS: Ind NUCMC. WORKS ABOUT: Dict Am Biog; Natl Cycl Am Biog 3; Who Was Who Am h. See also: Barr; Biog Geneal Master Ind.

DEXTER, AARON (1750-1829). Chemistry. OCCUP: Physician; Professor. MSS: Ind NUCMC. WORKS ABOUT: Am Chem and Chem Eng; Dict Am Med Biog; Natl Cycl Am Biog 19; Smith E p148.

DEXTER, WILLIAM PRESCOTT (1820-1890). Chemistry. OCCUP: Physician; Researcher. WORKS ABOUT: Am Ac Arts Scis Proc ns19(1891-1892):363-67; Poggendorff v3.

DICKESON, MONTROVILLE WILSON (1810-1882). Archeology; Natural history. OCCUP: Physician (?); Explorer; Museum curator. WORKS ABOUT: Allibone Supp; Culin; Groce and Wallace; Samuels.

DICKSON, SAMUEL HENRY (1798-1872). Medicine. OCCUP: Physician; Professor. MSS: Ind NUCMC. WORKS ABOUT: Dict Am Biog; Dict Am Med Biog; Natl Cycl Am Biog 10; Who Was Who Am h. See also: Biog Geneal Master Ind.

DIXON, JEREMIAH (1733-1779). Astronomy. OCCUP: Surveyor. MSS: Ind NUCMC (68-1814). WORKS ABOUT: Dict Sci Biog. See also: Biog Geneal Master Ind; Pelletier.

DIXON, SAMUEL GIBSON (1851-1918). Bacteriology. OCCUP: Professor; Curator. MSS: Ind NUCMC (66-29, 68-440). WORKS ABOUT: Am Men Sci 1-2 d3; Natl Cycl Am Biog 13, 35; Sci 47(1918):236; Who Was Who Am 1. See also: Biog Geneal Master Ind.

DODGE, CHARLES KEENE (1844?-1918). Botany. OCCUP: Not determined. MSS: Ind NUCMC. WORKS ABOUT: Sci 47(1918):437-38.

DODGE, JACOB RICHARDS (1823-1902). Agriculture; Statistics. OCCUP: Editor; U.S. government employee. MSS: Ind NUCMC. WORKS ABOUT: Dict Am Biog; Who Was Who Am 1. See also: Barr; Biog Geneal Master Ind.

DODGE, RAYNAL (1844-   ). Botany. OCCUP: Craftsman; Lecturer; Author. WORKS ABOUT: Who Was Who Am 4.

DOLBEAR, AMOS EMERSON (1837-1910). Astronomy; Physics.
OCCUP: Professor. MSS: Ind NUCMC (70-2020). WORKS ABOUT:
Adams 0; Allibone Supp; Am Men Sci 1 d3; Appleton's; Biog
Dict Synopsis Books; Dunlap p74-76, 140; Elliott (s); Hawks
p114, 129-38; Lincoln Lib Language Arts; Lincoln Lib Social
Studies; Natl Cycl Am Biog 9; Pop Sci Mo 76(1910):415-16;
Sci 31(1910):341-42; Twentieth Cent Biog Dict; Wallace; Who
Was Who Am 1.

DOOLITTLE, CHARLES LEANDER (1843-1919). Astronomy;
Mathematics. OCCUP: Professor; Observatory director. WORKS
ABOUT: Am Men Sci 1-2 d3; Dict Am Biog; Elliott (s); Natl
Cycl Am Biog 20; Who Was Who Am 1. See also: Barr; Biog
Geneal Master Ind.

DOOLITTLE, ERIC (1869-1920). Astronomy. OCCUP: Professor.
MSS: Ind NUCMC. WORKS ABOUT: Am Men Sci 1-2 d3; Dict Am
Biog; Natl Cycl Am Biog 19; Who Was Who Am 1. See also:
Biog Geneal Master Ind.

DOREMUS, ROBERT OGDEN (1824-1906). Chemistry. OCCUP:
Professor. MSS: Ind NUCMC. WORKS ABOUT: Am Men Sci 1 d3;
Dict Am Biog; Elliott (s); Natl Cycl Am Biog 12, 28; Who
Was Who Am 1. See also: Barr; Biog Geneal Master Ind.

DORSEY, JAMES OWEN (1848-1895). Ethnology. OCCUP:
Clergyman; U.S. government scientist. WORKS ABOUT: Dict Am
Biog; Who Was Who Am h. See also: Barr; Biog Geneal Master
Ind.

DORSEY, JOHN SYNG (1783-1818). Anatomy; Surgery. OCCUP:
Surgeon; Professor. WORKS ABOUT: Dict Am Biog; Dict Am Med
Biog; Natl Cycl Am Biog 10; Who Was Who Am h. See also:
Biog Geneal Master Ind; Ireland Ind Sci.

DOUBLEDAY, NELTJE BLANCHAN deGRAFF (1865-1918). Natural
history; Nature writing; Botany; Ethnology; Ornithology.
OCCUP: Writer. WORKS ABOUT: Am Men Sci 1-2 d3; Dict Am
Biog; Natl Cycl Am Biog 13; Notable Am Wom; Siegel and
Finley; Who Was Who Am 1. See also: Biog Geneal Master
Ind.

DOUGLAS, JAMES (1837-1918). Chemistry; Metallurgy; Mining
engineering. OCCUP: Mining engineer; Industrial officer.
WORKS ABOUT: Am Men Sci 1-2 d3; Dict Am Biog; Natl Cycl Am
Biog 12, 23; Who Was Who Am 1. See also: Biog Geneal Master
Ind; Ireland Ind Sci; Pelletier.

DOUGLAS, ROBERT (1813-1897). Arboriculture. OCCUP:
Horticulturist. WORKS ABOUT: Gard and Forest 10(1897):230;
Sci 5(1897):916.

DOUGLAS, SILAS HAMILTON (1816-1890). Chemistry. OCCUP:
Physician; Professor. MSS: Ind NUCMC. WORKS ABOUT: Elliott;
Natl Cycl Am Biog 11; Who Was Who Am h. See also: Biog
Geneal Master Ind; Pelletier.

DOUGLASS, ANDREW ELLICOTT (1819-1901). Anthropology. OCCUP: Industrial officer; Collector. MSS: Ind NUCMC (75-1272). WORKS ABOUT: Sci 14(1901):570-71.

DOUGLASS, WILLIAM (ca.1691-1752). Medicine; Natural history. OCCUP: Physician. WORKS ABOUT: Dict Am Biog; Dict Am Med Biog; Elliott; Natl Cycl Am Biog 3; Who Was Who Am h. See also: Biog Geneal Master Ind; Ireland Ind Sci; Pelletier.

DOWNES, JOHN (1799-1882). Astronomy; Mathematics. OCCUP: Craftsman; U.S. government scientist. WORKS ABOUT: Adams O; Allibone; Appleton's; Wallace.

DRAKE, DANIEL (1785-1852). Medicine; Natural history. OCCUP: Physician; Professor (medicine). MSS: Ind NUCMC (71-1534). WORKS ABOUT: Dict Am Biog; Dict Am Med Biog; Elliott; Natl Cycl Am Biog 5; Who Was Who Am h. See also: Barr; Biog Geneal Master Ind; Ireland Ind Sci; Pelletier.

DRAPER, HENRY (1837-1882). Astronomy; Astronomical photography. OCCUP: Professor; Academic administrator (medicine). MSS: Ind NUCMC (68-1698, 80-294). WORKS ABOUT: Dict Am Biog; Dict Sci Biog; Elliott; Natl Ac Scis Biog Mem 3; Natl Cycl Am Biog 6; Who Was Who Am h. See also: Barr; Biog Geneal Master Ind; Ireland Ind Sci; Pelletier,

DRAPER, JOHN CHRISTOPHER (1835-1885). Anatomy; Chemistry; Physiology. OCCUP: Physician; Professor. MSS: Ind NUCMC (80-294). WORKS ABOUT: Adams O; Allibone Supp; Am Chem and Chem Eng; Am Journ Sci 131(1886):80; Appleton's; Duyckinck; Knight; Natl Cycl Am Biog 6; Nature 33(1886):254; New York Med Journ 42(1885):720; Preston; Twentieth Cent Biog Dict; Wallace.

DRAPER, JOHN WILLIAM (1811-1882). Chemistry; Physiology; Science and religion. OCCUP: Professor; Academic administrator (medicine). MSS: Ind NUCMC (78-1716, 80-294). WORKS ABOUT: Dict Am Biog; Dict Sci Biog; Elliott; Natl Ac Scis Biog Mem 2; Natl Cycl Am Biog 3; Who Was Who Am h. See also: Barr; Biog Geneal Master Ind; Ireland Ind Sci; Pelletier.

DRAPER, MARY ANNA PALMER (MRS. HENRY) (1839-1914). Astrophysics. OCCUP: Benefactor. WORKS ABOUT: Notable Am Wom; Ogilvie; Sci 41(1915):380-82.

DRAPER, WILLIAM HENRY (1830-1901). Medicine. OCCUP: Physician; Professor. WORKS ABOUT: Appleton's; Natl Cycl Am Biog 7:490; Sci 13(1901):716; Twentieth Cent Biog Dict; Who Was Who Am 1.

DROWN, THOMAS MESSINGER (1842-1904). Chemistry. OCCUP: Professor; State government scientist; College president. WORKS ABOUT: Dict Am Biog; Elliott; Natl Cycl Am Biog 7:112; Who Was Who Am 1. See also: Barr; Biog Geneal Master Ind; Pelletier.

DROWNE, SOLOMON (1753-1834). Botany; Medicine. OCCUP: Physician; Professor. MSS: Ind NUCMC. WORKS ABOUT: Appleton's; Natl Cycl Am Biog 8; Twentieth Cent Am Biog.

DuBOIS, AUGUSTUS JAY (1849-1915). Civil engineering. OCCUP: Professor. WORKS ABOUT: Am Men Sci 1-2 d3; Dict Am Biog; Natl Cycl Am Biog 15; Who Was Who Am 1. See also: Biog Geneal Master Ind

DUCATEL, JULIUS TIMOLEON (1796-1849). Chemistry; Geology. OCCUP: Professor; State government scientist. MSS: Ind NUCMC (?). WORKS ABOUT: Adams O; Am Journ Sci 58(1849):146-49; Appleton's; Natl Cycl Am Biog 4; Wallace.

DuCHAILLU, PAUL BELLONI (1835-1903). Exploration; Zoology. OCCUP: Explorer; Collector. MSS: Ind NUCMC (82-615). WORKS ABOUT: Dict Am Biog; Who Was Who Am 1. See also: Barr; Biog Geneal Master Ind; Pelletier.

DUCKWALL, EDWARD WILEY (1866-1916). Chemistry. OCCUP: Laboratory administrator; City government scientist. WORKS ABOUT: Am Men Sci 2 d3.

DUDLEY, CHARLES BENJAMIN (1842-1909). Chemistry. OCCUP: Industrial scientist. WORKS ABOUT: Am Men Sci 1 d3; Dict Am Biog; Elliott (s); Natl Cycl Am Biog 12; Who Was Who Am 1. See also: Barr; Biog Geneal Master Ind; Ireland Ind Sci; Pelletier.

DUDLEY, PAUL (1675-1751). Natural history. OCCUP: Public official; Judge. MSS: Ind NUCMC. WORKS ABOUT: Dict Am Biog; Elliott; Natl Cycl Am Biog 7:175; Who Was Who Am h. See also: Biog Geneal Master Ind; Ireland Ind Sci; Pelletier.

DUDLEY, WILLIAM HENRY (1811-1886). Medicine. OCCUP: Physician. WORKS ABOUT: Appleton's; Sci 1st ser 8(1886):364; Twentieth Cent Biog Dict.

DUDLEY, WILLIAM LOFLAND (1859-1914). Chemistry. OCCUP: Professor; Academic administrator (medicine). MSS: Ind NUCMC (71-233). WORKS ABOUT: Am Journ Sci 188(1914):490; Am Men Sci 1-2 d3; Appleton's; Natl Cycl Am Biog 8; Twentieth Cent Biog Dict; Who Was Who Am 1.

DUDLEY, WILLIAM RUSSEL (1849-1911). Botany. OCCUP: Professor. MSS: Ind NUCMC. WORKS ABOUT: Am Men Sci 1-2 d3; Dict Am Biog; Elliott (s); Natl Cycl Am Biog 22; Who Was Who Am 1. See also: Barr; Biog Geneal Master Ind.

DUFFIELD, JOHN THOMAS (1823-1901). Mathematics. OCCUP: Professor. WORKS ABOUT: Adams O; Allibone Supp; Appleton's; Sci 13(1901):636; Twentieth Cent Biog Dict; Wallace; Who Was Who Am 1.

DUFFIELD, SAMUEL PEARCE (1833-1916). Chemistry; Medicine. OCCUP: Physician; Public official. MSS: Ind NUCMC (80-1145). WORKS ABOUT: Am Chem and Chem Eng; Appleton's; Twentieth Cent Biog Dict; Who Was Who Am 4.

DUFFIELD, WILLIAM WARD (1823-1907). Geodesy. OCCUP: Civil engineer; U.S. government scientist. MSS: Ind NUCMC. WORKS ABOUT: Adams O; Allibone Supp; Appleton's; Sci 44(1916):48; Twentieth Cent Biog Dict; Wallace; Who Was Who Am 1.

DUFOUR, JOHN JAMES (ca.1763-1827). Botany. OCCUP: Horticulturist. WORKS ABOUT: Dict Am Biog; Who Was Who Am h. See also: Biog Geneal Master Ind.

DUN, WALTER ANGUS (1857-1887). Anthropology; Natural history. OCCUP: Physician; Academic instructor. WORKS ABOUT: Cincinnati Soc Nat Hist Journ 11([1889]):1-2, 55-61; Twentieth Cent Biog Dict.

DUNBAR, WILLIAM (1749-1810). Astronomy; Mathematics; Meteorology. OCCUP: Planter; Surveyor; Public official. MSS: Ind NUCMC (60-2578). WORKS ABOUT: Dict Am Biog; Elliott; Who Was Who Am h. See also: Biog Geneal Master Ind; Ireland Ind Sci.

DUNCAN, LOUIS (1862-1916). Electrical engineering. OCCUP: Professor. WORKS ABOUT: Am Men Sci 1-2 d3; Natl Cycl Am Biog 12, 14; Sci 43(1916):346; Twentieth Cent Biog Dict; Who Was Who Am 1.

DUNCAN, ROBERT KENNEDY (1868-1914). Chemistry. OCCUP: Professor. WORKS ABOUT: Am Men Sci 2 d3; Dict Am Biog; Natl Cycl Am Biog 21; Who Was Who Am 1. See also: Barr; Biog Geneal Master Ind.

DUNGLISON, ROBLEY (1798-1869). Medical education; Physiology. OCCUP: Physician; Professor. MSS: Ind NUCMC. WORKS ABOUT: Dict Am Biog; Dict Am Med Biog; Dict Sci Biog; Elliott; Natl Cycl Am Biog 10; Who Was Who Am h. See also: Biog Geneal Master Ind.

DUNN, LOUISE BRISBIN (-1902). Botany. OCCUP: Academic instructor. WORKS ABOUT: Sci 16(1902):1040.

DUNNE, JOHN AUGUSTINE (1871-1920). Astronomy. OCCUP: Observatory affiliate. WORKS ABOUT: Am Men Sci 1-2 d4.

DuPONT, ELEUTHERE IRENEE (1771-1834). Chemistry. OCCUP: Manufacturer. WORKS ABOUT: Dict Am Biog; Who Was Who Am h. See also: Biog Geneal Master Ind; Pelletier.

DuPONT, FRANCIS GURNEY (1850-1905). Chemistry. OCCUP: Industrial scientist and officer. MSS: Ind NUCMC (?). WORKS ABOUT: Am Men Sci 1 d3; Natl Cycl Am Biog 23.

DURAND, ELIAS (ELIE MAGLOIRE) (1794-1873). Botany; Pharmacy. OCCUP: Pharmacist. MSS: Ind NUCMC. WORKS ABOUT: Dict Am Biog; Elliott; Who Was Who Am h. See also: Barr; Biog Geneal Master Ind.

DURANT, CHARLES FERSON (1805-1873). Aeronautics; Botany. OCCUP: Aeronaut; Printer. WORKS ABOUT: Dict Am Biog. See also: Biog Geneal Master Ind; Ireland Ind Sci.

DURKEE, SILAS (1798-1878). Botany; Zoology; Entomology. OCCUP: Physician. WORKS ABOUT: Allibone Supp; Am Ac Arts Scis Proc 14(1878):343-44; Atkinson.

DuSIMITIERE, PIERRE EUGENE (ca.1736-1784). Natural history. OCCUP: Artist; Natural history collector; Museum curator. MSS: Ind NUCMC (61-2511, 79-1774). WORKS ABOUT: Dict Am Biog; Who Was Who Am h. See also: Biog Geneal Master Ind; Ireland Ind Sci.

DUSSAUCE, HYPPOLITE ETIENNE (1829-1869). Chemistry. OCCUP: Professor; Industrial employee. WORKS ABOUT: Am Chem and Chem Eng; Appleton's; Wallace.

DUTCHER, WILLIAM (1846-1920). Ornithology. OCCUP: Administrator. MSS: Ind NUCMC (?). WORKS ABOUT: Am Men Sci 1-2 d3; Who Was Who Am 1.

DUTTON, CLARENCE EDWARD (1841-1912). Geology. OCCUP: Army officer; U.S. government scientist. MSS: Ind NUCMC (69-2001). WORKS ABOUT: Am Men Sci 1-2 d3; Dict Am Biog; Dict Sci Biog; Elliott (s); Natl Ac Scis Biog Mem 32; Natl Cycl Am Biog 13; Who Was Who Am 1. See also: Barr; Biog Geneal Master Ind; Ireland Ind Sci; Pelletier.

DUTTON, W. T. (1852?-1914). Mathematics. OCCUP: Professor. WORKS ABOUT: Sci 39(1914):461.

DWIGHT, THOMAS (1843-1911). Anatomy. OCCUP: Professor. WORKS ABOUT: Am Men Sci 1-2 d3; Dict Am Biog; Elliott (s); Natl Cycl Am Biog 12; Who Was Who Am 1. See also: Barr; Biog Geneal Master Ind.

DWIGHT, WILLIAM BUCK (1833-1906). Geology. OCCUP: Professor; Museum curator. WORKS ABOUT: Adams O; Am Journ Sci 172(1906):352; Am Men Sci 1 d3; Appleton's; Natl Cycl Am Biog 10; Sci 24(1906):318; Twentieth Cent Biog Dict; Who Was Who Am 1.

DYCHE, LEWIS LINDSAY (1857-1915). Zoology. OCCUP: Professor; Curator. WORKS ABOUT [some as Louis]: Am Journ Sci 189(1915):486; Am Men Sci 1-2 d3; Natl Cycl Am Biog 21; Sci 41(1915):163, 280-82; Twentieth Cent Biog Dict; Who Was Who Am 1.

DYER, ISADORE (1865-1920). Dermatology. OCCUP: Professor. WORKS ABOUT: Dict Am Biog; Dict Am Med Biog; Who Was Who Am 1. See also: Biog Geneal Master Ind.

E

EADS, DARWIN D. (-1906). Botany. OCCUP: Physician.
WORKS ABOUT: <u>Sci</u> 23(1906):891.

EADS, JAMES BUCHANAN (1820-1887). Engineering. OCCUP:
Engineer; Businessman. MSS: <u>Ind NUCMC</u> (65-678). WORKS
ABOUT: <u>Dict Am Biog</u>; Natl Ac Scis <u>Biog Mem</u> 3; <u>Natl Cycl Am</u>
<u>Biog</u> 5; <u>Who Was Who Am</u> h. See also: Barr; <u>Biog Geneal</u>
<u>Master Ind</u>; Ireland <u>Ind Sci</u>; Pelletier.

EARLE, PLINY (1809-1892). Medicine (mental illness). OCCUP:
Physician; Medical administrator. MSS: <u>Ind NUCMC</u> (62-
3090). WORKS ABOUT: <u>Dict Am Biog</u>; <u>Dict Am Med Biog</u>; <u>Natl</u>
<u>Cycl Am Biog</u> 11; <u>Who Was Who Am</u> h. See also: <u>Biog Geneal</u>
<u>Master Ind</u>.

EARLL, ROBERT EDWARD (1853-1896). Ichthyology. OCCUP:
Museum curator; U.S. government scientist. WORKS ABOUT:
Phil Soc Washington <u>Bull</u> 13(1900):388-90; <u>Sci</u>
3(1896):471-72; U S Natl Mus <u>Rep</u> (1896):49.

EASTMAN, CHARLES ROCHESTER (1868-1918). Geology;
Paleontology. OCCUP: U.S. and state government scientist;
Museum curator; Professor. WORKS ABOUT: Adams O; <u>Am Journ</u>
<u>Sci</u> 196(1918):692; <u>Am Men Sci</u> 1-2 d3; <u>Sci</u> 48(1918):366,
49(1919):139-41; <u>Twentieth Cent Biog Dict</u>; <u>Who Was Who</u>
<u>Am</u> 1.

EASTMAN, GUY WARNER (1881-1907). Physics. OCCUP: Academic
researcher. WORKS ABOUT: <u>Sci</u> 25(1907):879.

EASTMAN, JOHN ROBIE (1836-1913). Astronomy. OCCUP:
Observatory affiliate. WORKS ABOUT: <u>Am Men Sci</u> 1-2 d3; <u>Dict</u>
<u>Am Biog</u>; Elliott (s); <u>Natl Cycl Am Biog</u> 13; <u>Who Was Who Am</u>
1. See also: Barr; <u>Biog Geneal Master Ind</u>.

EATON, ALVAH AUGUSTUS (1865-1908). Botany. OCCUP: Natural
history collector. MSS: <u>Ind NUCMC</u> (71-1790). WORKS ABOUT:
<u>Sci</u> 28(1908):752.

EATON, AMOS (1776-1842). Botany; Geology; Science education. OCCUP: Professor. MSS: <u>Ind NUCMC</u> (61-1417, 79-956). WORKS ABOUT: <u>Dict Am Biog</u>; <u>Dict Sci Biog</u>; Elliott; <u>Natl Cycl Am Biog</u> 5; <u>Who Was Who Am</u> h. See also: Barr; <u>Biog Geneal Master Ind</u>; Ireland <u>Ind Sci</u>; Pelletier.

EATON, DANIEL CADY (1834-1895). Botany. OCCUP: Professor. MSS: <u>Ind NUCMC</u> (71-2016). WORKS ABOUT: <u>Dict Am Biog</u>; Elliott; <u>Natl Cycl Am Biog</u> 11; <u>Who Was Who Am</u> h. See also: Barr; <u>Biog Geneal Master Ind</u>.

EATON, DARWIN G. (1823?-1895). Biology. OCCUP: Professor. WORKS ABOUT: <u>Sci</u> 1(1895):364.

EBERLE, JOHN (1787-1838). Botany. OCCUP: Physician; Professor. MSS: <u>Ind NUCMC</u>. WORKS ABOUT: <u>Dict Am Biog</u>; <u>Dict Am Med Biog</u>; <u>Natl Cycl Am Biog</u> 11; <u>Who Was Who Am</u> h. See also: Barr; <u>Biog Geneal Master Ind</u>.

EDDY, IMOGEN W. (-1904). Mathematics. OCCUP: Observatory affiliate. MSS: <u>Ind NUCMC</u> (?). WORKS ABOUT: <u>Sci</u> 20(1904):382.

EDDY, WILLIAM ABNER (1850-1909). Meteorology; Photography. OCCUP: Accountant; Researcher. WORKS ABOUT: <u>Sci</u> 31(1910):141; <u>Twentieth Cent Biog Dict</u>; <u>Who Was Who Am</u> 1.

EDGAR, ARTHUR (-1913). Chemistry. OCCUP: Academic instructor. WORKS ABOUT: <u>Sci</u> 38(1913):738.

EDGECOMB, DANIEL W. (1840?-1915). Instrumentation. OCCUP: Inventor; Instrument maker. WORKS ABOUT: <u>Sci</u> 41(1915):722.

EDMANDS, J. RAYNER (1850?-1910). Astronomy. OCCUP: Observatory affiliate. WORKS ABOUT: <u>Sci</u> 31(1910):535.

EDSON, ARTHUR WOODBURY (-1905). Botany. OCCUP: U.S. government scientist. WORKS ABOUT: <u>Sci</u> 22(1905):61-62.

EDSON, CYRUS (1857-1903). Sanitation. OCCUP: Physician; Public official. WORKS ABOUT: <u>Natl Cycl Am Biog</u> 3; <u>Sci</u> 18(1903):767; <u>Twentieth Cent Biog Dict</u>; <u>Who Was Who Am</u> 1.

EDSON, GEORGE C. (-1909). Geology. OCCUP: Not determined. WORKS ABOUT: <u>Sci</u> 29(1909):893.

EDWARDS, ARTHUR MEAD (1836-ca.1914). Chemistry; Geology; Physics. OCCUP: Physician (?); Professor. WORKS ABOUT: <u>Am Men Sci</u> 1-2 d3.

EDWARDS, HENRY (1830-1891). Entomology. OCCUP: Performer.
MSS: Ind NUCMC. WORKS ABOUT: California Ac Scis Proc
3(1893):367; Canadian Ent 23(1891):141-42, 259-67; Ent Mo
Mag 27(1891):226; Ent News 2(1891)129-30; Ent Rec
2(1891):143-44; Essig p611-12; Humming Bird 1(1891):74;
Insect Life 3(1891):489-90; Mallis; Osborn Herb p162, 347;
Ottawa Nat 5(1891):87-88; Pop Sci Mo 76(1910):470-72; Sci
1st ser 18(1891):18; Twentieth Cent Biog Dict; Victorian
Nat 8(1892):80; World Whos Who Sci.

EDWARDS, VINAL N. (-1919). Zoology. OCCUP: U.S.
government employee; Natural history collector. WORKS
ABOUT: Sci 50(1919):34-35.

EDWARDS, WILLIAM HENRY (1822-1909). Entomology. OCCUP:
Lawyer; Businessman. MSS: Ind NUCMC (62-2588, 66-34, 66-
652). WORKS ABOUT: Am Men Sci 1 d3; Dict Am Biog;
Elliott (s); Who Was Who Am 1. See also: Barr; Biog Geneal
Master Ind; Ireland Ind Sci; Pelletier.

EGLESTON, THOMAS (1832-1900). Metallurgy; Mineralogy;
Mining. OCCUP: Professor. MSS: Ind NUCMC (66-1580). WORKS
ABOUT: Dict Am Biog; Natl Cycl Am Biog 3; Who Was Who Am 1.
See also: Barr; Biog Geneal Master Ind.

EIGHTS, JAMES (1798-1882). Natural history; Geology. OCCUP:
Draftsman; Natural history explorer. WORKS ABOUT: Elliott.
See also: Biog Geneal Master Ind.

EILERS, FREDERIC ANTON (1839-1917). Metallurgy. OCCUP:
Industrial and U.S. government scientist; Manufacturer.
WORKS ABOUT: Dict Am Biog. See also: Biog Geneal Master
Ind.

EIMBECK, WILLIAM (1841-1909). Geodesy; Mathematics. OCCUP:
U.S. government engineer. WORKS ABOUT: Dict Am Biog;
Elliott; Who Was Who Am 4. See also: Barr.

ELD, HENRY (1814-1850). Exploration. OCCUP: Naval officer.
MSS: Ind NUCMC (63-344). WORKS ABOUT: Am Journ Sci
60(1850):137.

ELDER, WILLIAM (1843-1903). Chemistry. OCCUP: Professor.
WORKS ABOUT: Sci 18(1903):30.

ELDERHORST, WILLIAM (1828-1861). Chemistry. OCCUP:
Professor. WORKS ABOUT: Allibone Supp; Am Chem and Chem
Eng; Wallace.

ELDRIDGE, GEORGE (1828?-1900). Hydrography. OCCUP: Seaman.
WORKS ABOUT: Sci 12(1900):348.

ELDRIDGE, GEORGE HOMANS (1854-1905). Geology. OCCUP: U.S.
government scientist. WORKS ABOUT: Sci 22(1905):62;
Wallace; Who Was Who Am 1.

ELIOT, JARED (1685-1763). Agriculture; Chemistry. OCCUP: Clergyman; Physician. MSS: Ind NUCMC. WORKS ABOUT: Dict Am Biog; Who Was Who Am h. See also: Biog Geneal Master Ind.

ELLERY, JOHN GRAEM (-1855). Geology. OCCUP: Natural history explorer. WORKS ABOUT: Am Journ Sci 70(1855):296-97.

ELLET, CHARLES (1810-1862). Engineering. OCCUP: Civil engineer. MSS: Ind NUCMC (62-2071). WORKS ABOUT: Dict Am Biog; Natl Cycl Am Biog 4; Who Was Who Am h. See also: Biog Geneal Master Ind; Pelletier.

ELLET, WILLIAM H. (1806-1859). Chemistry. OCCUP: Professor; Industrial scientist. MSS: Ind NUCMC (?). WORKS ABOUT: Appleton's; Drake; Natl Cycl Am Biog 11; Twentieth Cent Biog Dict.

ELLICOTT, ANDREW (1754-1820). Mathematics; Surveying. OCCUP: Surveyor; Professor. MSS: Ind NUCMC (76-156, 80-295). WORKS ABOUT: Dict Am Biog; Elliott; Natl Cycl Am Biog 13; Who Was Who Am h. See also: Biog Geneal Master Ind; Ireland Ind Sci.

ELLIOT, DANIEL GIRAUD (1835-1915). Zoology. OCCUP: Museum curator. MSS: Ind NUCMC. WORKS ABOUT: Am Men Sci 1-2 d3; Dict Am Biog; Elliott (s); Natl Cycl Am Biog 5, 16; Who Was Who Am 1. See also: Barr; Biog Geneal Master Ind; Ireland Ind Sci; Pelletier.

ELLIOTT, ARTHUR H. (1848?-1918). Chemistry; Physics. OCCUP: Professor; Industrial scientist. WORKS ABOUT: Sci 47(1918):236; Wallace.

ELLIOTT, EZEKIEL BROWN (1823-1888). Electricity; Statistics. OCCUP: Actuary. WORKS ABOUT: Elliott; Natl Cycl Am Biog 2. See also: Barr; Biog Geneal Master Ind.

ELLIOTT, STEPHEN (1771-1830). Botany. OCCUP: Planter; Legislator; Professor. MSS: Ind NUCMC. WORKS ABOUT: Dict Am Biog; Elliott; Natl Cycl Am Biog 4; Who Was Who Am h. See also: Barr; Biog Geneal Master Ind.

ELLIS, ARVILLA (MRS. J. B.) (-1889). Botany. OCCUP: Not determined (personal scientific assistant). WORKS ABOUT: Sci 10(1899):191.

ELLIS, HENRY (1721-1806). Hydrography. OCCUP: Explorer; Governor. MSS: Ind NUCMC. WORKS ABOUT: Dict Am Biog; Natl Cycl Am Biog 1; Who Was Who Am h. See also: Biog Geneal Master Ind.

ELLIS, JOB BICKNELL (1829-1905). Botany; Mycology. OCCUP: Teacher; Collector. MSS: Ind NUCMC. WORKS ABOUT: Am Men Sci 1 d3; Dict Am Biog; Elliott (s); Who Was Who Am 1. See also: Biog Geneal Master Ind.

ELLIS, THEODORE GUNVILLE (1829-1883). Civil engineering. OCCUP: Engineer; State government surveyor. WORKS ABOUT: Allibone Supp; Appleton's; Twentieth Cent Biog Dict.

ELMER, JONATHAN (1745-1817). Anatomy. OCCUP: Physician; Legislator. MSS: Ind NUCMC. WORKS ABOUT: Dict Am Biog; Dict Am Med Biog; Natl Cycl Am Biog 11; Who Was Who Am h. See also: Biog Geneal Master Ind.

ELSBERG, LOUIS (1836-1885). Medicine (laryngology); Microscopy. OCCUP: Physician; Academic instructor. WORKS ABOUT: Dict Am Biog; Who Was Who Am h. See also: Biog Geneal Master Ind; Ireland Ind Sci.

ELSNER, HENRY LEOPOLD (1857-1916). Medicine. OCCUP: Physician; Professor. WORKS ABOUT [birth dates may differ]: Kagan p16; Kelly and Burrage; Sci 43(1916):309; Wallace; Who Was Who Am 1.

ELY, ACHSAH MOUNT (1848-1904). Mathematics. OCCUP: Professor. WORKS ABOUT: Sci 20(1904):895.

ELY, JOHN SLADE (1860-1906). Medicine. OCCUP: Professor. WORKS ABOUT: Sci 23(1906):279; Who Was Who Am 1.

ELY, THEODORE NEWEL (1846-1916). Engineering. OCCUP: Industrial engineer. WORKS ABOUT: Am Men Sci 1-2 d3; Natl Cycl Am Biog 32; Sci 44(1916):672; Who Was Who Am 1.

EMERSON, EDWIN (1823-1908). Optics; Photography. OCCUP: Clergyman; Editor. Scientific work: see Roy Soc Cat. MSS: Ind NUCMC (?). WORKS ABOUT: Princeton Theol Sem Biog Cat p169; Princeton Theol Sem Necrological Rep 3(1900-1909):586-87; Wallace.

EMERSON, FREDERICK VALENTINE (1871-1919). Geology. OCCUP: Academic instructor. WORKS ABOUT: Am Men Sci 2 d3.

EMERSON, GEORGE BARRELL (1797-1881). Botany. OCCUP: Educator. MSS: Ind NUCMC. WORKS ABOUT: Dict Am Biog; Elliott; Natl Cycl Am Biog 11; Who Was Who Am h. See also: Biog Geneal Master Ind.

EMERSON, GEORGE H. (1837-1864). Chemistry. OCCUP: Not determined. WORKS ABOUT: Am Journ Sci 89(1865):373-74.

EMERSON, RALPH (1787-1863). General science. OCCUP: Clergyman; Professor (ecclesiastical history). Scientific work: see Roy Soc Cat. WORKS ABOUT: Adams O; Appleton's; Appleton's Supp 1918-31; Dexter Biog Sketches 6(1912):380-84, 616; Drake; Natl Cycl Am Biog 10; Wallace; Who Was Who Am h.

EMERY, CHARLES EDWARD (1838-1898). Engineering. OCCUP: Naval engineer; Industrial engineer. WORKS ABOUT: Dict Am Biog; Natl Cycl Am Biog 9; Who Was Who Am h. See also: Barr; Biog Geneal Master Ind.

EMMET, JOHN PATTEN (1796-1842). Chemistry. OCCUP: Physician; Professor. MSS: Ind NUCMC. WORKS ABOUT: Elliott; Who Was Who Am h. See also: Barr; Biog Geneal Master Ind; Pelletier.

EMMONS, EBENEZER (1799-1863). Geology. OCCUP: Physician; Professor; State government scientist. MSS: Ind NUCMC. WORKS ABOUT: Dict Am Biog; Dict Sci Biog; Elliott; Natl Cycl Am Biog 8; Who Was Who Am h. See also: Barr; Biog Geneal Master Ind; Ireland Ind Sci.

EMMONS, SAMUEL FRANKLIN (1841-1911). Geology; Mining. OCCUP: U.S. government scientist. MSS: Ind NUCMC (?). WORKS ABOUT: Am Men Sci 1-2 d3; Dict Am Biog; Dict Sci Biog; Elliott (s); Natl Ac Scis Biog Mem 7; Natl Cycl Am Biog 10; Who Was Who Am 1. See also: Barr; Biog Geneal Master Ind.

EMORY, FREDERICK LINCOLN (1867-1919). Mechanics. OCCUP: Professor. WORKS ABOUT: Am Men Sci 2 d3; Who Was Who Am 1.

EMORY, WILLIAM HEMSLEY (1811-1887). Astronomy; Topographical engineering. OCCUP: Army officer; Engineer. MSS: Ind NUCMC (77-1408). WORKS ABOUT: Dict Am Biog; Elliott; Natl Cycl Am Biog 4; Who Was Who Am h. See also: Biog Geneal Master Ind.

ENDEMANN, SAMUEL THEODORE HERMANN CARL (1842-1909). Chemistry. OCCUP: Editor; Consultant. WORKS ABOUT: Am Men Sci 1-2 d3; Browne and Weeks p319, 496.

ENGELMANN, GEORGE (1809-1884). Botany; Meteorology. OCCUP: Physician. MSS: Ind NUCMC (77-1294, 84-231). WORKS ABOUT: Dict Am Biog; Dict Sci Biog supp 1; Elliott; Natl Ac Scis Biog Mem 4; Natl Cycl Am Biog 6; Who Was Who Am h. See also: Barr; Biog Geneal Master Ind.

ENGELMANN, GEORGE JULIUS (1847-1903). Gynecology. OCCUP: Physician; Professor. MSS: Ind NUCMC. WORKS ABOUT: Dict Am Biog; Natl Cycl Am Biog 11; Who Was Who Am 1. See also: Barr; Biog Geneal Master Ind.

ENGELMANN, HENRY (1831-1899). Geology. OCCUP: State government scientist; Industrial engineer and manager. WORKS ABOUT: Elliott.

ENGLER, EDMUND ARTHUR (1856-1918). Mathematics. OCCUP: Professor; College president. MSS: Ind NUCMC (?). WORKS ABOUT: Am Men Sci 1-2 d3; Sci 47(1918):217; Who Was Who Am 1.

ENNIS, JACOB (-ca.1890). Astronomy. OCCUP: Teacher; Educator. WORKS ABOUT: Sidereal Messenger 9(1890):185-86.

ERICSSON, JOHN (1803-1889). Engineering; Invention. OCCUP: Engineer. MSS: Ind NUCMC (60-2418, 62-4518, 72-848). WORKS ABOUT: Dict Am Biog; Natl Cycl Am Biog 4; Who Was Who Am h. See also: Barr; Biog Geneal Master Ind; Ireland Ind Sci; Pelletier.

ERNI, HENRI (1822-1885). Chemistry. OCCUP: Professor; U.S. government scientist; Public official (diplomatic service). WORKS ABOUT: <u>Am Chem and Chem Eng</u>.

ERSKINE, ROBERT (1735-1780). Geography. OCCUP: Businessman; Surveyor. MSS: <u>Ind NUCMC</u>. WORKS ABOUT: <u>Dict Am Biog</u>; <u>Natl Cycl Am Biog</u> 24; <u>Who Was Who Am</u> h. See also: <u>Biog Geneal Master Ind</u>.

ESPY, JAMES POLLARD (1785-1860). Meteorology. OCCUP: Teacher; U.S. government scientist. MSS: <u>Ind NUCMC</u>. WORKS ABOUT: <u>Dict Am Biog</u>; <u>Dict Sci Biog</u>; Elliott; <u>Natl Cycl Am Biog</u> 6; <u>Who Was Who Am</u> h. See also: Barr; <u>Biog Geneal Master Ind</u>; Ireland <u>Ind Sci</u>; Pelletier.

ESSIG, CHARLES J. (1827?-1901). Dentistry. OCCUP: Professor; Academic administrator. WORKS ABOUT: <u>Sci</u> 14(1901):981.

ESTY, WILLIAM COLE (1838-1916). Mathematics. OCCUP: Professor. WORKS ABOUT: <u>Am Men Sci</u> 1-2 d3; <u>Sci</u> 44(1916):165; <u>Who Was Who Am</u> 1.

EUSTIS, HENRY LAWRENCE (1819-1885). Engineering. OCCUP: Army engineer; Professor; Academic administrator. WORKS ABOUT: <u>Dict Am Biog</u>; Elliott; <u>Natl Cycl Am Biog</u> 22; <u>Who Was Who Am</u> h. See also: Barr; <u>Biog Geneal Master Ind</u>.

EVANS, EDGAR G. (1884-1914). Chemistry. OCCUP: Academic instructor. WORKS ABOUT: <u>Am Men Sci</u> 2 d3.

EVANS, JOHN (1812-1861). Geology. OCCUP: U.S. government employee and scientist. MSS: <u>Ind NUCMC</u> (78-884). WORKS ABOUT: Elliott. See also: Barr; <u>Biog Geneal Master Ind</u>.

EVANS, LEWIS (1700-1756). Cartography; Geography; Geology; Geomorphology. OCCUP: Surveyor; Map-maker. MSS: <u>Ind NUCMC</u>. WORKS ABOUT: <u>Dict Am Biog</u>; <u>Dict Sci Biog</u>; <u>Natl Cycl Am Biog</u> 11. See also: <u>Biog Geneal Master Ind</u>.

EVANS, OLIVER (1755-1819). Invention. OCCUP: Inventor; Manufacturer. MSS: <u>Ind NUCMC</u> (61-2096, 62-760). WORKS ABOUT: <u>Dict Am Biog</u>; <u>Natl Cycl Am Biog</u> 6; <u>Who Was Who Am</u> h. See also: Barr; <u>Biog Geneal Master Ind</u>; Ireland <u>Ind Sci</u>; Pelletier.

EVANS, THOMAS (1863-1907). Chemistry. OCCUP: Industrial scientist; Professor; Academic administrator. WORKS ABOUT: <u>Am Men Sci</u> 1 d3; <u>Sci</u> 26(1907):29-30.

EVERETT, HARRY DAY (1880?-1908). Forestry. OCCUP: Government scientist (foreign). WORKS ABOUT: <u>Sci</u> 28(1908):338.

EVERHART, BENJAMIN MATLACK (1818-1904). Botany. OCCUP: Merchant. WORKS ABOUT: Adams O; Appleton's; <u>Natl Cycl Am Biog</u> 10; Preston; <u>Sci</u> 20(1904):476; <u>Twentieth Cent Biog Dict</u>; Wallace; <u>Who Was Who Am</u> 1.

EVERHART, ISAIAH FAWKES (1840-1911). Natural history; Philanthropy. OCCUP: Physician; Museum founder. WORKS ABOUT: <u>Natl Cycl Am Biog</u> 5; <u>Sci</u> 33(1911):959.

EWBANK, THOMAS (1792-1870). Engineering; Science writing. OCCUP: Inventor; U.S. government scientist. WORKS ABOUT: <u>Dict Am Biog</u>; <u>Natl Cycl Am Biog</u> 7:559; <u>Who Was Who Am</u> h.

EWELL, ERVIN EDGAR (1867-1904). Chemistry. OCCUP: U.S. government scientist; Industrial manager. WORKS ABOUT: Elliott; <u>Who Was Who Am</u> 1. See also: Barr.

EWING, FAYETTE CLAY, JR. (1886?-1914). Engineering. OCCUP: Professor; Engineer. WORKS ABOUT: <u>Sci</u> 40(1914):930.

EWING, JOHN (1732-1802). Astronomy. OCCUP: Clergyman; Professor; Academic administrator. WORKS ABOUT: <u>Dict Am Biog</u>; <u>Natl Cycl Am Biog</u> 1; <u>Who Was Who Am</u> h. See also: <u>Biog Geneal Master Ind</u>.

F

FAHLBERG, CONSTANTIN (1850-1910). Chemistry. OCCUP: Industrial scientist; Consultant; Manufacturer. WORKS ABOUT: <u>Am Chem and Chem Eng</u>.

FAILYER, GEORGE HENRY (1849-   ). Chemistry. OCCUP: Professor; U.S. government scientist. WORKS ABOUT: <u>Am Men Sci</u> 2.

FAIRBANKS, HENRY (1830-1918). Mechanics. OCCUP: Professor; Manufacturer; Inventor. WORKS ABOUT: <u>Am Men Sci</u> 1-2 d3; <u>Dict Am Biog</u>; <u>Natl Cycl Am Biog</u> 10; <u>Who Was Who Am</u> 1. See also: <u>Biog Geneal Master Ind</u>.

FANNING, JOHN THOMAS (1837-1911). Hydraulics; Mechanics. OCCUP: Civil engineer. WORKS ABOUT: <u>Am Men Sci</u> 1-2 d3; <u>Dict Am Biog</u>; <u>Natl Cycl Am Biog</u> 9; <u>Who Was Who Am</u> 1. See also: <u>Biog Geneal Master Ind</u>; Ireland <u>Ind Sci</u>.

FARGIS, GEORGE ALOYSIUS (1854-1916). Astronomy. OCCUP: Clergyman; Academic instructor; Observatory affiliate. WORKS ABOUT: <u>Am Men Sci</u> 1-2 d3.

FARLOW, WILLIAM GILSON (1844-1919). Botany. OCCUP: Professor. MSS: <u>Ind NUCMC</u>. WORKS ABOUT: <u>Am Men Sci</u> 1-2 d3; <u>Dict Am Biog</u>; Elliott (s); Natl Ac Scis <u>Biog Mem</u> [<u>Mem</u> 21(4)]; <u>Natl Cycl Am Biog</u> 12, 22; <u>Who Was Who Am</u> 1. See also: Barr; <u>Biog Geneal Master Ind</u>.

FARMER, JOHN (1798-1859). Cartography. OCCUP: Teacher; Surveyor; Map-maker. MSS: <u>Ind NUCMC</u> (68-1234). WORKS ABOUT: <u>Dict Am Biog</u>; <u>Who Was Who Am</u> h. See also: <u>Biog Geneal Master Ind</u>.

FARQUHAR, JANE COLDEN. See COLDEN, JANE.

FARQUHARSON, ROBERT JAMES (1824-1884). Ethnography. OCCUP: Physician; Public official. WORKS ABOUT: Davenport Ac Nat Scis <u>Proc</u> 4(1886):201-06.

FARRAR, JOHN (1779-1853). Astronomy; Mathematics; Physics. OCCUP: Professor. MSS: Ind NUCMC (65-1237). WORKS ABOUT: Dict Am Biog; Dict Sci Biog; Elliott; Who Was Who Am h. See also: Biog Geneal Master Ind; Ireland Ind Sci.

FAUNCE, CHARLES M. (1867?-1899). Mathematics. OCCUP: Academic instructor. WORKS ABOUT: Sci 10(1899):126.

FAVA, FRANCIS R. (1860?-1896). Engineering. OCCUP: Professor. WORKS ABOUT: Sci 3(1896):510.

FAVILL, HENRY BAIRD (1860-1916). Medicine. OCCUP: Physician; Professor. WORKS ABOUT: Dict Am Biog; Dict Am Med Biog; Natl Cycl Am Biog 10; Who Was Who Am 1. See also: Barr; Biog Geneal Master Ind.

FAXON, CHARLES EDWARD (1846-1918). Botany. OCCUP: Academic instructor; Museum (arboretum) administrator; Artist. WORKS ABOUT: Am Men Sci 1-2 d3; Twentieth Cent Biog Dict; Who Was Who Am 1.

FAXON, WALTER (1848-1920). Zoology. OCCUP: Museum curator; Professor. WORKS ABOUT: Am Men Sci 1-2 d3; Wallace; Who Was Who Am 1.

FEATHERSTONHAUGH, GEORGE WILLIAM (1780-1866). Geology. OCCUP: Farmer; Natural history explorer; Public official (foreign). MSS: Ind NUCMC (82-1044). WORKS ABOUT: Dict Sci Biog; Natl Cycl Am Biog 8. See also: Barr; Biog Geneal Master Ind; Ireland Ind Sci.

FEAY, WILLIAM T. (1803?-1879). Botany. OCCUP: Physician (?); Teacher. WORKS ABOUT: Am Journ Sci 119(1880):76-77.

FENDLER, AUGUSTUS (1813-1883). Botany; Meteorology. OCCUP: Natural history collector; Farmer; Businessman. MSS: Ind NUCMC. WORKS ABOUT: Am Journ Sci 129(1885):169-71; Appleton's; Bot Gaz 9&10(1884-1885):111-12, 285-90, 301-04, 319-22; Gard Chronicle 22(1884):91; Pop Sci Mo 74(1909):240-43.

FERGUSON, JAMES (1797-1867). Astronomy. OCCUP: Civil engineer; U.S. government scientist. WORKS ABOUT: Elliott. See also: Biog Geneal Master Ind.

FERRARI, ANDREW (1839?-1915). Chemistry. OCCUP: Industrial scientist (?). WORKS ABOUT: Sci 42(1915):641.

FERREL, WILLIAM (1817-1891). Mathematical geophysics; Meteorology. OCCUP: Teacher; U.S. government scientist. MSS: Ind NUCMC. WORKS ABOUT: Dict Am Biog; Dict Sci Biog; Elliott; Natl Ac Scis Biog Mem 3; Natl Cycl Am Biog 9; Who Was Who Am h. See also: Barr; Biog Geneal Master Ind; Ireland Ind Sci.

FERRY, JOHN FARWELL (1877-1910). Ornithology. OCCUP: Museum associate; U.S. government scientist. WORKS ABOUT: Sci 31(1910):411.

FEUCHTWANGER, LEWIS (1805-1876). Chemistry; Mineralogy. OCCUP: Pharmacist; Physician; Manufacturer. MSS: Ind NUCMC. WORKS ABOUT: Adams O; Allibone Supp; Appleton's; Twentieth Cent Biog Dict; Wallace.

FIELD, MARTIN (1773-1833). Geology; Mineralogy. OCCUP: Lawyer; Public official. WORKS ABOUT: Allibone; Am Journ Sci 26(1834):204-05; Appleton's.

FIELDE, ADELE MARION (1839-1916). Biology. OCCUP: Missionary; Lecturer; Teacher. MSS: Ind NUCMC. WORKS ABOUT: Adams O; Am Men Sci 2 d3; Sci 43(1916):423; Wallace; Who Was Who Am 1; Wom Whos Who Am.

FILLEBROWN, THOMAS (1836-1908). Dentistry. OCCUP: Dentist; Professor. WORKS ABOUT: Dict Am Biog; Natl Cycl Am Biog 13. See also: Barr; Biog Geneal Master Ind.

FIRMSTONE, FRANK (1846-1917). Metallurgy. OCCUP: Industrial officer. WORKS ABOUT: Am Men Sci 1-2 d3.

FISHER, ALEXANDER METCALF (1794-1822). Mathematics; Physics. OCCUP: Professor. WORKS ABOUT: Allibone; Am Journ Sci 5(1822):367-76; Appleton's; Drake; Duyckinck; Natl Cycl Am Biog 19.

FISHER, GEORGE EGBERT (1863-1920). Mathematics. OCCUP: Professor. WORKS ABOUT: Am Men Sci 1-2 d3; Wallace; Who Was Who Am 1.

FISHER, WILLIAM REDWOOD (1808-1842). Chemistry; Pharmacy. OCCUP: Pharmacist; Professor. Scientific work: see Roy Soc Cat. WORKS ABOUT: Cordell; Philadelphia College Pharm First Cent; Simpson; Smith A p68-69 (mentioned).

FISKE, JOHN (1842-1901). Biology. OCCUP: Librarian; Lecturer; Author. MSS: Ind NUCMC (61-698, 65-906). WORKS ABOUT: Dict Am Biog; Natl Cycl Am Biog 3; Who Was Who Am 1. See also: Barr; Biog Geneal Master Ind; Ireland Ind Sci.

FITCH, ASA (1809-1879). Entomology; Economic entomology; Medicine. OCCUP: Physician; State government scientist. MSS: Ind NUCMC (62-4215, 74-975, 74-1190, 77-175). WORKS ABOUT: Dict Am Biog; Dict Sci Biog; Elliott; Natl Cycl Am Biog 7:252; Who Was Who Am h. See also: Barr; Biog Geneal Master Ind; Ireland Ind Sci; Pelletier.

FITZ, HENRY (1808-1863). Telescope construction. OCCUP: Craftsman; Instrument maker; Photographer. MSS: Ind NUCMC. WORKS ABOUT: Dict Am Biog; Elliott; Who Was Who Am h. See also: Barr; Biog Geneal Master Ind.

FITZ, REGINALD HEBER (1843-1913). Medicine. OCCUP: Physician; Professor. MSS: Ind NUCMC (62-3636). WORKS ABOUT: Am Men Sci 1-2 d3; Dict Am Biog; Dict Am Med Biog; Natl Cycl Am Biog 10; Who Was Who Am 1. See also: Barr; Biog Geneal Master Ind; Ireland Ind Sci; Pelletier.

FLAGG, THOMAS WILSON (1805-1884). Natural history. OCCUP: Clerk; Author. WORKS ABOUT: Dict Am Biog; Who Was Who Am h. See also: Biog Geneal Master Ind.

FLANERY, DAVID (1828-1900). Astronomy. OCCUP: Telegrapher. WORKS ABOUT: Pop Ast 8(1900):400, 484-86; Sci 12(1900):453.

FLEMING, WILLIAMINA PATON STEVENS (1857-1911). Astronomy. OCCUP: Curator; Observatory affiliate. MSS: Ind NUCMC. WORKS ABOUT: Am Men Sci 1-2 d3; Dict Am Biog; Dict Sci Biog; Elliott (s); Natl Cycl Am Biog 7:29; Notable Am Wom; Ogilvie; Siegel and Finley; Who Was Who Am 1. See also: Barr; Biog Geneal Master Ind; Ireland Ind Sci.

FLETCHER, HORACE (1849-1919). Dietetics. OCCUP: Businessman; Lecturer; Author. WORKS ABOUT: Am Men Sci 1-2 d3; Dict Am Biog; Natl Cycl Am Biog 14; Who Was Who Am 1. See also: Barr; Biog Geneal Master Ind.

FLETCHER, ROBERT (1823-1912). Medicine. OCCUP: Physician; Librarian. MSS: Ind NUCMC. WORKS ABOUT: Am Men Sci 1-2 d3; Dict Am Biog; Natl Cycl Am Biog 13. See also: Biog Geneal Master Ind.

FLINT, AUSTIN (1812-1886). Medicine. OCCUP: Physician; Professor. MSS: Ind NUCMC. WORKS ABOUT: Dict Am Biog; Dict Am Med Biog; Natl Cycl Am Biog 8; Who Was Who Am h. See also: Barr; Biog Geneal Master Ind; Ireland Ind Sci; Pelletier.

FLINT, AUSTIN, JR. (1836-1915). Physiology. OCCUP: Physician; Professor. MSS: Ind NUCMC. WORKS ABOUT: Am Men Sci 1-2 d3; Dict Am Biog; Dict Am Med Biog; Natl Cycl Am Biog 9; Who Was Who Am 1. See also: Barr; Biog Geneal Master Ind.

FLINT, JAMES MILTON (1838-1919). Medicine. OCCUP: Naval medical officer; Curator. WORKS ABOUT: Am Men Sci 1-2 d3; Who Was Who Am 1.

FOLGER, WALTER (1765-1849). Astronomy; Mathematics; Instrumentation; Invention. OCCUP: Craftsman; Lawyer; Legislator. WORKS ABOUT: Dict Am Biog; Elliott; Who Was Who Am h. See also: Barr; Biog Geneal Master Ind; Pelletier.

FOLSOM, CHARLES FOLLEN (1842-1907). Medicine. OCCUP: Physician; Professor. WORKS ABOUT: Adams O; Allibone Supp; Am Men Sci 1 d3; Appleton's; Natl Cycl Am Biog 19; Twentieth Cent Biog Dict; Wallace; Who Was Who Am 1.

FONTAINE, WILLIAM MORRIS (1835-1913). Paleobotany. OCCUP: Professor. WORKS ABOUT: Allibone Supp; Am Journ Sci 185(1913):642; Am Men Sci 1-2 d3; Elliott (s); Natl Cycl Am Biog 19; Pop Sci Mo 83(1913):207; Sci 37(1913):705; Wallace; Who Was Who Am 1.

FOOTE, ALBERT E. (1846-1895). Mineralogy. OCCUP: Professor; Collector; Businessman. MSS: Ind NUCMC (66-38). WORKS ABOUT: Am Journ Sci 150(1895):434; Geol Soc Am Bull 7(1896):481-85; Nature 53(1895-1896):12; Wallace.

FORBES, CHARLES (1844-1917). Physics. OCCUP: Professor; Physician; Curator. WORKS ABOUT: Am Men Sci 2-3 d4.

FORBES, FRANCIS BLACKWELL (1840?-1908). Botany. OCCUP: Not determined. WORKS ABOUT: Sci 27(1908):902.

FORBES, WILLIAM SMITH (1831-1905). Anatomy. OCCUP: Surgeon; Professor. WORKS ABOUT: Am Men Sci 1 d3.

FORCE, MANNING FERGUSON (1824-1899). Anthropology. OCCUP: Lawyer; Professor (law). MSS: Ind NUCMC (61-3296, 79-1782, 82-512, 83-934). WORKS ABOUT: Dict Am Biog; Who Was Who Am 1. See also: Barr; Biog Geneal Master Ind.

FORCHHEIMER, FREDERICK (1853-1913). Medicine. OCCUP: Physician; Professor. WORKS ABOUT: Am Men Sci 1-2 d3; Kagan p16; Natl Cycl Am Biog 16; Who Was Who Am 1.

FORD, CORYDON LA (1813-1894). Anatomy. OCCUP: Physician; Professor. MSS: Ind NUCMC (65-1331). WORKS ABOUT: Adams O; Allibone Supp; Appleton's; Sci 2(1895):162; Wallace.

FORSHEY, CALEB GOLDSMITH (1812-1881). Engineering. OCCUP: Engineer. MSS: Ind NUCMC (82-138). WORKS ABOUT: Appleton's; Knight; Twentieth Cent Biog Dict.

FOSDICK, NELLIE (-1917). Botany. OCCUP: Academic instructor. WORKS ABOUT: Sci 45(1917):562.

FOSTER, FRANK PIERCE (1841-1911). Medicine. OCCUP: Physician; Editor. WORKS ABOUT: Dict Am Biog; Natl Cycl Am Biog 22; Who Was Who Am 1. See also: Barr; Biog Geneal Master Ind.

FOSTER, HORATIO ALVAH (1858-1913). Electrical engineering. OCCUP: Industrial engineer; Consultant. WORKS ABOUT: Am Men Sci 2 d3; Wallace; Who Was Who Am 1.

FOSTER, JOHN (1814?-1897). Physics. OCCUP: Professor. WORKS ABOUT: Sci 6(1897):660.

FOSTER, JOHN WELLS (1815-1873). Geology. OCCUP: Lawyer; Civil engineer; U.S. and state government scientist. MSS: Ind NUCMC. WORKS ABOUT: Elliott; Natl Cycl Am Biog 10; Who Was Who Am h. See also: Barr; Biog Geneal Master Ind; Ireland Ind Sci.

FOSTER, WINTHROP D. (1880?-1918). Zoology. OCCUP: U.S. government scientist. WORKS ABOUT: <u>Sci</u> 48(1918):544.

FOTHERGILL, ANTHONY (1732 or 1735-1813). Medicine. OCCUP: Physician. MSS: <u>Ind NUCMC</u>. WORKS ABOUT: Allibone; <u>Dict Natl Biog</u>.

FOWLER, LORENZO NILES (1811-1896). Phrenology. OCCUP: Phrenologist. MSS: <u>Ind NUCMC</u>. WORKS ABOUT: Elliott. See also: <u>Biog Geneal Master Ind</u>.

FOWLER, LYDIA FOLGER (1822-1879). Medicine. OCCUP: Physician; Lecturer; Author. WORKS ABOUT: <u>Notable Am Wom</u>; Ogilvie. See also: <u>Biog Geneal Master Ind</u>; Ireland <u>Ind Sci</u>; Pelletier.

FOWLER, ORSON SQUIRE (1809-1887). Phrenology. OCCUP: Phrenologist; Editor. MSS: <u>Ind NUCMC</u>. WORKS ABOUT: <u>Dict Am Biog</u>; Elliott; <u>Natl Cycl Am Biog</u> 3; <u>Who Was Who Am</u> h. See also: <u>Biog Geneal Master Ind</u>.

FOX, CHARLES L. (-1898). Bacteriology. OCCUP: City government scientist (?). WORKS ABOUT: <u>Sci</u> 8(1898):509.

FOX, OSCAR CHAPMAN (1830-1902). Invention; Patent examination. OCCUP: Lawyer; U.S. government scientist. WORKS ABOUT: <u>Natl Cycl Am Biog</u> 1; <u>Sci</u> 15(1902):1037; <u>Twentieth Cent Biog Dict</u>; <u>Who Was Who Am</u> 1.

FRANCIS, JAMES BICHENO (1815-1892). Hydraulic engineering. OCCUP: Engineer; Industrial officer. MSS: <u>Ind NUCMC</u>. WORKS ABOUT: <u>Dict Am Biog</u>; <u>Natl Cycl Am Biog</u> 9; <u>Who Was Who Am</u> h. See also: <u>Biog Geneal Master Ind</u>.

FRANKLIN, BENJAMIN (1706-1790). Physics; Electricity; Meteorology; Oceanography. OCCUP: Printer; Public official; Researcher. MSS: <u>Ind NUCMC</u> (60-1814, 61-910, 68-451, 69-1593, 82-1808). WORKS ABOUT: <u>Dict Am Biog</u>; <u>Dict Sci Biog</u>; Elliott; <u>Natl Cycl Am Biog</u> 1; <u>Who Was Who Am</u> h. See also: Barr; <u>Biog Geneal Master Ind</u>; Ireland <u>Ind Sci</u>; Pelletier.

FRARY, HOBART DICKINSON (1887-1920). Mathematics; Mechanical engineering. OCCUP: Research institution scientist; Professor. WORKS ABOUT: <u>Natl Cycl Am Biog</u> 23.

FRASCH, HERMAN (1851-1914). Chemistry. OCCUP: Engineer; Inventor; Industrialist. WORKS ABOUT: <u>Dict Am Biog</u>; <u>Natl Cycl Am Biog</u> 19; <u>Who Was Who Am</u> h (add), 4. See also: Barr; <u>Biog Geneal Master Ind</u>; Ireland <u>Ind Sci</u>; Pelletier.

FRAZER, JOHN FRIES (1812-1872). Chemistry; Geophysics. OCCUP: Professor; Academic administrator. MSS: <u>Ind NUCMC</u> (60-2734). WORKS ABOUT: <u>Dict Am Biog</u>; Elliott; Natl Ac Scis <u>Biog Mem</u> 1; <u>Natl Cycl Am Biog</u> 1; <u>Who Was Who Am</u> h. See also: Barr; <u>Biog Geneal Master Ind</u>; Pelletier.

FRAZER, PERSIFOR (1844-1909). Chemistry; Geology. OCCUP: U.S. and state government scientist; Teacher; Consultant. MSS: Ind NUCMC (60-2982). WORKS ABOUT: Am Men Sci 1 d3; Dict Am Biog; Elliott (s); Natl Cycl Am Biog 4; Who Was Who Am 1. See also: Barr; Biog Geneal Master Ind.

FRAZIER, BENJAMIN WEST (1841-1905). Mineralogy. OCCUP: Professor. WORKS ABOUT: Am Journ Sci 169(1905):204; Sci 21(1905):78; Who Was Who Am 1.

FREEMAN, THOMAS (-1821). Astronomy; Civil engineering; Exploration; Surveying. OCCUP: Surveyor; Explorer; Public official. MSS: Ind NUCMC (80-2043). WORKS ABOUT: Dict Am Biog; Elliott; Natl Cycl Am Biog 24; Who Was Who Am h.

FREEMAN, THOMAS J. A. (1841-1907). Chemistry; Physics. OCCUP: Clergyman; Professor. WORKS ABOUT: Who Was Who Am 1.

FREER, PAUL CASPAR (1862-1912). Chemistry; Medicine. OCCUP: Professor; Government scientist (foreign). WORKS ABOUT: Adams 0; Am Chem and Chem Eng; Am Journ Sci 184(1912):404; Am Men Sci 1-2 d3; Natl Cycl Am Biog 19; Sci 36(1912):108-09, 140-41; Twentieth Cent Biog Dict; Wallace; Who Was Who Am 1.

FRIEDBURG, LOUIS HENRY (1846-   ). Chemistry. OCCUP: Professor. WORKS ABOUT: Am Men Sci 1-2.

FRITH, ARTHUR J. (-1913). Engineering. OCCUP: Professor. WORKS ABOUT: Sci 38(1913):738.

FRITZ, JOHN (1822-1913). Invention; Metallurgy. OCCUP: Industrial engineer and manager. MSS: Ind NUCMC (?). WORKS ABOUT: Dict Am Biog; Natl Cycl Am Biog 13; Who Was Who Am 1. See also: Barr; Biog Geneal Master Ind; Ireland Ind Sci.

FRIZELL, JOSEPH PALMER (1832-1910). Engineering. OCCUP: Civil engineer. WORKS ABOUT: Am Men Sci 1-2 d3; Dict Am Biog; Natl Cycl Am Biog 23; Who Was Who Am 1. See also: Biog Geneal Master Ind.

FROST, CARLETON PENNINGTON (1830?-1896). Medicine. OCCUP: Professor; Academic administrator. WORKS ABOUT: Sci 3(1896):839.

FROST, CHARLES CHRISTOPHER (1806-1880). Botany. OCCUP: Tradesman. WORKS ABOUT [birth dates may differ]: Am Journ Sci 119(1880):493, 127(1884):242; Appleton's; Journ Mycology 2(1886):114-18; Wallace.

FRY, JOSHUA (ca.1700-1754). Mathematics; Surveying. OCCUP: Professor; Surveyor; Public official. MSS: Ind NUCMC. WORKS ABOUT: Dict Am Biog; Elliott; Who Was Who Am h. See also: Biog Geneal Master Ind.

FUERTES, ESTEVAN ANTONIO (1838-1903). Civil engineering.
OCCUP: Engineer; Professor; Academic administrator. MSS:
Ind NUCMC. WORKS ABOUT: Dict Am Biog; Natl Cycl Am Biog 4;
Who Was Who Am 1. See also: Barr; Biog Geneal Master Ind.

FULLER, ANDREW S. (1828-1896). Botany. OCCUP:
Horticulturist; Editor. WORKS ABOUT: Dict Am Biog; Who Was
Who Am h. See also: Barr; Biog Geneal Master Ind.

FULLER, HOMER TAYLOR (1838-1908). Chemistry; Geology;
Psychology. OCCUP: Educator; College president. WORKS
ABOUT: Am Men Sci 1 d3; Natl Cycl Am Biog 13; Sci
28(1908):481; Twentieth Cent Biog Dict; Who Was Who Am 4.

FULMER, (HENRY) ELTON (1864-1916). Chemistry. OCCUP:
Professor; Academic administrator; State government
scientist. WORKS ABOUT: Am Men Sci 1-2 d3; Sci
43(1916):423, 44(1916):198; Who Was Who Am 1.

FULTON, ROBERT (1765-1815). Invention. OCCUP: Civil
engineer; Inventor; Businessman. MSS: Ind NUCMC (60-2170).
WORKS ABOUT: Dict Am Biog; Natl Cycl Am Biog 3; Who Was Who
Am h. See also: Barr; Biog Geneal Master Ind; Ireland Ind
Sci; Pelletier.

FULTON, ROBERT BURWELL (1849-1919). Astronomy; Physics.
OCCUP: Professor; Academic administrator; Educator. WORKS
ABOUT: Am Men Sci 1-2 d3; Dict Am Biog; Natl Cycl Am Biog
13; Who Was Who Am 1. See also: Biog Geneal Master Ind.

G

GABB, WILLIAM MORE (1839-1878). Geology; Paleontology. OCCUP: Natural history collector; State government scientist; Government scientist (foreign). MSS: <u>Ind NUCMC</u>. WORKS ABOUT: <u>Dict Am Biog</u>; <u>Dict Sci Biog</u>; Elliott; Natl Ac Scis <u>Biog Mem</u> 6; <u>Natl Cycl Am Biog</u> 4; <u>Who Was Who Am</u> h. See also: Barr; <u>Biog Geneal Master Ind</u>; Ireland <u>Ind Sci</u>.

GABLE, GEORGE DANIEL (1863-1911). Mathematics. OCCUP: Professor. WORKS ABOUT: <u>Am Men Sci</u> 1-2 d3; <u>Who Was Who Am</u> 1.

GAGE, SUSANNA PHELPS (1857-1915). Embryology; Comparative anatomy. OCCUP: Engaged in independent study (?). WORKS ABOUT: <u>Am Men Sci</u> 1-2 d3; Ogilvie (app); <u>Sci</u> 42(1915):523, 45(1917):82-83; Siegel and Finley; <u>Twentieth Cent Biog Dict</u>; <u>Who Was Who Am</u> 1; <u>Wom Whos Who Am</u>.

GALE, LEONARD DUNNELL (1800-1883). Chemistry; Geology. OCCUP: Professor; U.S. government scientist; Consultant. WORKS ABOUT: <u>Am Journ Sci</u> 126(1883):490; Smithsonian <u>Rep</u> (1883):47-48.

GAMBEL, WILLIAM (1823-1849). Ornithology. OCCUP: Natural history collector; Physician (?). WORKS ABOUT: Elliott; <u>Who Was Who Am</u> h. See also: Barr; Pelletier.

GANNETT, HENRY (1846-1914). Geography; Statistics. OCCUP: U.S. government scientist. WORKS ABOUT: <u>Am Men Sci</u> 1-2 d3; <u>Dict Am Biog</u>; Elliott (s); <u>Natl Cycl Am Biog</u> 19; <u>Who Was Who Am</u> 1. See also: Barr; <u>Biog Geneal Master Ind</u>.

GANZ, ALBERT FREDERICK (1872-1917). Electrical engineering. OCCUP: Professor; Consultant. WORKS ABOUT: <u>Am Men Sci</u> 2 d3; <u>Sci</u> 46(1917):135; <u>Who Was Who Am</u> 1.

GARBER, DAVIS (1829?-1896). Astronomy; Mathematics. OCCUP: Professor. WORKS ABOUT: <u>Sci</u> 4(1896):529.

GARDEN, ALEXANDER (1730-1791). Natural history; Botany. OCCUP: Physician. MSS: Ind NUCMC. WORKS ABOUT: Dict Am Biog; Dict Am Med Biog; Elliott; Natl Cycl Am Biog 23; Who Was Who Am h. See also: Biog Geneal Master Ind; Ireland Ind Sci; Pelletier.

GARDINER, EDWARD GARDINER (1854-1907). Biology; Zoology. OCCUP: Academic instructor; Researcher. WORKS ABOUT: Am Men Sci 1 d3; Natl Cycl Am Biog 14; Sci 26(1907):685, 27(1908):153.

GARDINER, FREDERIC (1858-1917). Embryology. OCCUP: Clergyman; Educator. WORKS ABOUT: Am Men Sci 1-2 d3; Who Was Who Am 1.

GARDINER, JAMES TERRY (1842-1912). Topographical engineering. OCCUP: Engineer; U.S. and state government scientist; Industrial officer. MSS: Ind NUCMC (62-937). WORKS ABOUT: Dict Am Biog; Elliott; Natl Cycl Am Biog 23; Who Was Who Am 1. See also: Barr; Biog Geneal Master Ind.

GARDINER, JOHN (1862?-1900). Biology. OCCUP: Professor. WORKS ABOUT: Sci 12(1900):934.

GARDNER, DANIEL PEREIRA (-1853). Chemistry; Natural history. OCCUP: Physician (?); Professor. WORKS ABOUT: Allibone; Barnhart; Hampden Sidney Gen Cat; Wallace.

GARLICK, THEODATUS (1805-1884). Pisciculture. OCCUP: Surgeon; Artist. MSS: Ind NUCMC (?). WORKS ABOUT: Dict Am Biog; Natl Cycl Am Biog 24; Who Was Who Am h. See also: Biog Geneal Master Ind.

GARNETT, JOHN (ca.1748-1820). Astronomy; Mathematics. OCCUP: Not determined; Farmer (?). Scientific work: see Roy Soc Cat. WORKS ABOUT: New Jersey Hist Soc Proc ns6(1921):18-20.

GARRETSON, JAMES EDMUND (pseud., JOHN DARBY) (1828-1895). Dentistry. OCCUP: Oral surgeon; Professor. WORKS ABOUT: Dict Am Biog; Dict Am Med Biog; Natl Cycl Am Biog 3; Who Was Who Am h. See also: Barr; Biog Geneal Master Ind.

GARRETT, ANDREW (1823-1887). Conchology. OCCUP: Museum associate; Natural history explorer/ collector. WORKS ABOUT: Appleton's Supp 1901; Conch Exchange 2(1887-1888):92-93; Journ Conch 5(1886-1888):317-18; Natl Cycl Am Biog 2; Sci 1st ser 11(1888):35-36.

GARRIGUES, HENRY JACQUES (1831-1913). Medicine. OCCUP: Physician; Professor. WORKS ABOUT: Adams O; Allibone Supp; Appleton's; Wallace; Who Was Who Am 1.

GARRIOTT, EDWARD BENNETT (1853-1910). Meteorology. OCCUP: U.S. government scientist. WORKS ABOUT: Am Men Sci 1-2 d3; Who Was Who Am 1.

GATSCHET, ALBERT SAMUEL (1832-1907). Anthropology. OCCUP: U.S. government scientist. MSS: Ind NUCMC (?). WORKS ABOUT: Am Men Sci 1 d3; Dict Am Biog; Natl Cycl Am Biog 21; Who Was Who Am 1. See also: Biog Geneal Master Ind; Ireland Ind Sci.

GATTINGER, AUGUSTIN (1825-1903). Botany. OCCUP: Physician; Librarian; State government scientist. MSS: Ind NUCMC (?). WORKS ABOUT: Elliott; Natl Cycl Am Biog 15. See also: Barr; Biog Geneal Master Ind.

GAY, MARTIN (1803-1850). Chemistry; Medicine. OCCUP: Physician. WORKS ABOUT: Am Journ Sci 59(1850):305-06; Appleton's; Drake.

GAYLEY, JAMES (1855-1920). Metallurgy. OCCUP: Industrial scientist and officer. WORKS ABOUT: Am Men Sci 2 d3; Dict Am Biog; Natl Cycl Am Biog 14; Who Was Who Am 1.

GAYLORD, WILLIS (1792-1844). Agriculture. OCCUP: Editor. WORKS ABOUT: Dict Am Biog; Natl Cycl Am Biog 23; Who Was Who Am h. See also: Biog Geneal Master Ind.

GEDDES, GEORGE (1809-1883). Agriculture. OCCUP: Civil engineer. WORKS ABOUT: Appleton's; Natl Cycl Am Biog 10.

GEHRMANN, ADOLPH (1868-1920). Bacteriology. OCCUP: City government scientist; Professor; Laboratory administrator. WORKS ABOUT: Am Men Sci 1-2 d3; Who Was Who Am 1.

GELSTON, LOUIS MERWIN (1879-1905). Bacteriology. OCCUP: Academic affiliate. WORKS ABOUT: Am Men Sci 1 d3.

GENOUD, ERNEST G. (1880?-1918). Bacteriology. OCCUP: Industrial scientist. WORKS ABOUT: Sci 48(1918):467.

GENTH, FREDERICK AUGUSTUS, SR. (1820-1893). Chemistry; Mineralogy. OCCUP: Consultant (analytical chemistry); Professor; State government scientist. MSS: Ind NUCMC (?). WORKS ABOUT: Dict Am Biog; Dict Sci Biog; Elliott; Natl Ac Scis Biog Mem 4; Natl Cycl Am Biog 7:493; Who Was Who Am h. See also: Barr; Biog Geneal Master Ind; Ireland Ind Sci; Pelletier.

GENTH, FREDERICK AUGUSTUS, JR. (1855-1910). Chemistry. OCCUP: Professor; State government scientist; Consultant. MSS: Ind NUCMC (?). WORKS ABOUT: Am Men Sci 1 d3; Sci 32(1910):339; Who Was Who Am 1.

GENTRY, THOMAS GEORGE (1843-1905). Natural history; Ornithology. OCCUP: Teacher; Educator. WORKS ABOUT: Adams O; Allibone Supp; Twentieth Cent Biog Dict; Wallace; Who Was Who Am 4.

GERRISH, FREDERIC HENRY (1845-1920). Anatomy. OCCUP: Surgeon; Professor. WORKS ABOUT: Am Men Sci 1-2 d3; Dict Am Biog; Dict Am Med Biog; Natl Cycl Am Biog 12; Who Was Who Am 1. See also: Biog Geneal Master Ind.

GIBBES, LEWIS REEVE (1810-1894). Astronomy; Chemistry; Natural history. OCCUP: Professor. MSS: Ind NUCMC (78-1722, 82-1081). WORKS ABOUT: Elliott. See also: Barr; Ireland Ind Sci; Pelletier.

GIBBES, ROBERT WILSON (1809-1866). Geology; Paleontology; Medicine. OCCUP: Professor; Physician; Businessman. MSS: Ind NUCMC. WORKS ABOUT: Dict Am Biog; Elliott; Natl Cycl Am Biog 11; Who Was Who Am h. See also: Barr; Biog Geneal Master Ind.

GIBBONS, WILLIAM PETERS (1812-1897). Botany; Zoology. OCCUP: Physician. WORKS ABOUT: Erythea 5(1897):74-76; Kelly and Burrage; Med Soc St California Trans 28(1898):296-97.

GIBBS, GEORGE (1776-1833). Mineralogy; Mineral collecting; Patronage of science. OCCUP: Inherited wealth. MSS: Ind NUCMC. WORKS ABOUT: Dict Am Biog; Elliott; Natl Cycl Am Biog 12; Who Was Who Am h. See also: Barr; Biog Geneal Master Ind.

GIBBS, GEORGE (1815-1873). Ethnology; Geology. OCCUP: Lawyer; U.S. government scientist; Author. MSS: Ind NUCMC. WORKS ABOUT: Dict Am Biog; Who Was Who Am h. See also: Biog Geneal Master Ind.

GIBBS, JOSIAH WILLARD (1839-1903). Mathematics; Theoretical physics. OCCUP: Professor. MSS: Ind NUCMC. WORKS ABOUT: Dict Am Biog; Dict Sci Biog; Elliott; Natl Ac Scis Biog Mem 6; Natl Cycl Am Biog 4; Who Was Who Am 1. See also: Barr; Biog Geneal Master Ind; Ireland Ind Sci; Pelletier.

GIBBS, MORRIS M. (-1908). Ornithology. OCCUP: Not determined; Physician (?). WORKS ABOUT: Sci 28(1908):441.

GIBBS, (OLIVER) WOLCOTT (1822-1908). Chemistry. OCCUP: Professor. MSS: Ind NUCMC (64-963). WORKS ABOUT: Am Men Sci 1 d3; Dict Am Biog; Dict Sci Biog; Elliott (s); Natl Ac Scis Biog Mem 7; Natl Cycl Am Biog 10; Who Was Who Am 1. See also: Barr; Biog Geneal Master Ind; Ireland Ind Sci; Pelletier.

GIBSON, JAMES A. (1867?-1917). Anatomy. OCCUP: Professor. WORKS ABOUT: Sci 46(1917):408.

GIBSON, WILLIAM HAMILTON (1850-1896). Natural history. OCCUP: Artist; Author. WORKS ABOUT: Dict Am Biog; Natl Cycl Am Biog 7:463; Who Was Who Am h. See also: Biog Geneal Master Ind.

GIHON, ALBERT LEARY (1833-1901). Medicine. OCCUP: Naval medical officer. WORKS ABOUT: Dict Am Biog; Dict Am Med Biog; Natl Cycl Am Biog 9; Who Was Who Am 1. See also: Barr; Biog Geneal Master Ind.

GILBERT, BENJAMIN DAVIS (1835-1907). Botany. OCCUP: Businessman; Editor; Administrator. MSS: Ind NUCMC. WORKS ABOUT: Wallace; Who Was Who Am 4.

GILBERT, GROVE KARL (1843-1918). Geography; Geology; Geomorphology. OCCUP: U.S. government scientist. MSS: Ind NUCMC. WORKS ABOUT: Am Men Sci 1-2 d3; Dict Am Biog; Dict Sci Biog; Elliott (s); Natl Ac Scis Biog Mem [Mem 21(5)]; Natl Cycl Am Biog 13; Who Was Who Am 1. See also: Barr; Biog Geneal Master Ind; Ireland Ind Sci; Pelletier.

GILDERSLEEVE, NATHANIEL (1871-1919). Bacteriology. OCCUP: Academic scientist. WORKS ABOUT: Am Men Sci 2 d3.

GILL, HARRY DOUGLAS (1861?-1918). Surgery. OCCUP: Veterinarian (?); Academic instructor. WORKS ABOUT: Sci 48(1918):616-17.

GILL, THEODORE NICHOLAS (1837-1914). Zoology; Ichthyology. OCCUP: Professor; Librarian; Museum associate. WORKS ABOUT: Am Men Sci 1-2 d3; Dict Am Biog; Dict Sci Biog; Elliott (s); Natl Ac Scis Biog Mem 8; Natl Cycl Am Biog 12; Who Was Who Am 1. See also: Barr; Biog Geneal Master Ind; Ireland Ind Sci; Pelletier.

GILLISS, JAMES MELVILLE (1811-1865). Astronomy. OCCUP: Naval officer; Observatory director. MSS: Ind NUCMC. WORKS ABOUT: Dict Am Biog; Elliott; Natl Ac Scis Biog Mem 1; Natl Cycl Am Biog 9; Who Was Who Am h. See also: Barr; Biog Geneal Master Ind; Pelletier.

GILLMAN, HENRY (1833-1915). Archeology. OCCUP: U.S. government scientist; Public official (diplomatic service); Librarian. MSS: Ind NUCMC. WORKS ABOUT: Am Men Sci 2 d3; Dict Am Biog; Natl Cycl Am Biog 7:359; Who Was Who Am 1. See also: Biog Geneal Master Ind.

GILMOR, ROBERT (1774-1848). Mineralogy. OCCUP: Merchant. MSS: Ind NUCMC (67-1462). WORKS ABOUT: Am Journ Sci 57(1847):142-43; Natl Cycl Am Biog 11.

GIRARD, CHARLES FREDERIC (1822-1895). Zoology. OCCUP: Museum employee; Physician. WORKS ABOUT: Dict Am Biog; Elliott; Natl Cycl Am Biog 16; Who Was Who Am h. See also: Biog Geneal Master Ind.

GIRAUD, JACOB P., JR. (1811?-1870). Ornithology. OCCUP: Not determined; Collector. WORKS ABOUT: Am Journ Sci 100(1870):293-94.

GLATFELTER, NOAH MILLER (1837-1911). Botany. OCCUP: Physician. MSS: Ind NUCMC (66-1702). WORKS ABOUT: Pop Sci Mo 74(1909):256-57.

GLENN, WILLIAM (1840-1907). Chemistry; Mining engineering. OCCUP: Mining engineer; Industrial manager. WORKS ABOUT: Am Men Sci 1 d3.

GLOVER, TOWNEND (1813-1883). Entomology. OCCUP: Inherited wealth; U.S. government scientist. MSS: Ind NUCMC (74-976). WORKS ABOUT: Dict Am Biog; Elliott; Natl Cycl Am Biog 23; Who Was Who Am h. See also: Barr; Biog Geneal Master Ind; Ireland Ind Sci; Pelletier.

GODDARD, PAUL BECK (1811-1866). Anatomy; Medicine. OCCUP: Physician; Professor. WORKS ABOUT: Dict Am Biog; Natl Cycl Am Biog 23; Who Was Who Am h. See also: Biog Geneal Master Ind.

GODFREY, THOMAS (1704-1749). Mathematics; Technology. OCCUP: Craftsman; Publisher (almanacs); Teacher. MSS: Ind NUCMC. WORKS ABOUT: Dict Am Biog; Dict Sci Biog; Elliott; Natl Cycl Am Biog 23; Who Was Who Am h. See also: Barr; Biog Geneal Master Ind; Pelletier.

GODMAN, JOHN DAVIDSON (1794-1830). Anatomy; Natural history. OCCUP: Physician; Professor. MSS: Ind NUCMC. WORKS ABOUT: Dict Am Biog; Dict Am Med Biog; Elliott; Natl Cycl Am Biog 7:284; Who Was Who Am h. See also: Biog Geneal Master Ind; Pelletier.

GOESSMANN, CHARLES ANTHONY (1827-1910). Chemistry. OCCUP: Professor. WORKS ABOUT: Am Men Sci 1 d3; Dict Am Biog; Elliott (s); Natl Cycl Am Biog 11; Who Was Who Am 1. See also: Barr; Biog Geneal Master Ind; Ireland Ind Sci; Pelletier.

GOETZ, GEORGE WASHINGTON (1856-1897). Metallurgy. OCCUP: Industrial manager; Consultant. WORKS ABOUT: Dict Am Biog; Natl Cycl Am Biog 24; Who Was Who Am h.

GOFF, EMMET STULL (1852-1902). Horticulture. OCCUP: Farmer; State government scientist; Professor. WORKS ABOUT: Dict Am Biog; Natl Cycl Am Biog 23; Who Was Who Am h. See also: Barr; Biog Geneal Master Ind.

GOLD, THEODORE SEDGWICK (1818-1906). Agriculture. OCCUP: Educator; Public official. WORKS ABOUT: Allibone Supp; Am Men Sci 1 d3; Wallace.

GOLDEN, MICHAEL JOSEPH (1862-1918). Mechanics. OCCUP: Professor. WORKS ABOUT: Sci 49(1919):42; Wallace; Who Was Who Am 4.

GOOD, ADOLPHUS CLEMENS (1856-1894). Natural history. OCCUP: Clergyman. WORKS ABOUT: Dict Am Biog; Elliott; Natl Cycl Am Biog 23; Who Was Who Am h. See also: Biog Geneal Master Ind.

GOODE, GEORGE BROWN (1851-1896). Natural history; Ichthyology; Science administration. OCCUP: Museum curator; U.S. government scientist. MSS: Ind NUCMC (61-2419, 72-1237). WORKS ABOUT: Dict Am Biog; Elliott; Natl Ac Scis Biog Mem 4; Natl Cycl Am Biog 3; Who Was Who Am h. See also: Barr; Biog Geneal Master Ind; Ireland Ind Sci.

GOODE, RICHARD URQUHART (1858-1903). Geography;
Topography. OCCUP: U.S. government scientist. WORKS ABOUT:
Am Men Sci 1 d3; Who Was Who Am 1.

GOODE, WILLIAM HENRY (1814-1897). Physical science. OCCUP:
Physician. Scientific work: see Roy Soc Cat. WORKS ABOUT:
Yale Obit Rec (1897):484.

GOODFELLOW, EDWARD (1828-1899). Geodesy. OCCUP: U.S.
government scientist. WORKS ABOUT [death dates may differ]:
Appleton's; Natl Cycl Am Biog 3; Twentieth Cent Biog Dict;
Who Was Who Am 1.

GOODRICH, CHAUNCEY ENOCH (1801-ca.1864). Agriculture;
Chemistry. OCCUP: Clergyman; Farmer. Scientific work: see
Roy Soc Cat. WORKS ABOUT: Country Gentleman 24(1864):19;
New York St Ag Soc Trans 23(1863):135-39.

GOODRICH, JOSEPH (1795-1852). Geology (volcanoes). OCCUP:
Clergyman; Farmer. Scientific work: see Roy Soc Cat. WORKS
ABOUT: Dexter Biog Notices.

GOODYEAR, CHARLES (1800-1860). Chemistry; Invention. OCCUP:
Inventor. MSS: Ind NUCMC. WORKS ABOUT: Dict Am Biog; Natl
Cycl Am Biog 3; Who Was Who Am h. See also: Barr; Biog
Geneal Master Ind; Ireland Ind Sci; Pelletier.

GOODYEAR, NELSON (1811-1857). Chemistry. OCCUP: Not
determined; Inventor. WORKS ABOUT: Sci 21(1905):876; Wolf.

GOODYEAR, WATSON ANDREWS (1838-1891). Geology. OCCUP: State
government scientist. MSS: Ind NUCMC. WORKS ABOUT: Wallace;
Yale Obit Rec (1891):64.

GORDON, ROBERT H. (1852?-1910). Geology. OCCUP: Not
determined; Collector. WORKS ABOUT: Am Journ Sci
180(1910):96; Sci 31(1910):854.

GORDY, JOHN PANCOAST (1851-1908). Psychology; History of
education. OCCUP: Professor (education and history of
education). WORKS ABOUT: Am Men Sci 1 d3; Dict Am Biog; Who
Was Who Am 1. See also: Barr; Biog Geneal Master Ind.

GORGAS, WILLIAM CRAWFORD (1854-1920). Medicine. OCCUP: Army
surgeon; Administrator. MSS: Ind NUCMC (62-4578). WORKS
ABOUT: Am Men Sci 2 d3; Dict Am Biog; Dict Am Med Biog;
Natl Cycl Am Biog 14, 32; Who Was Who Am 1. See also: Barr;
Biog Geneal Master Ind; Ireland Ind Sci; Pelletier.

GORHAM, JOHN (1783-1829). Chemistry. OCCUP: Physician;
Professor. MSS: Ind NUCMC (62-1245). WORKS ABOUT: Dict Am
Biog; Elliott; Natl Cycl Am Biog 12; Who Was Who Am h. See
also: Barr; Biog Geneal Master Ind; Ireland Ind Sci;
Pelletier.

GOSS, NATHANIEL S. (1826-1891). Ornithology. OCCUP: Public
official; Businessman. WORKS ABOUT: Auk 8(1891):245-47.

GOTTHEIL, WILLIAM SAMUEL (1859-1920). Dermatology. OCCUP: Dermatologist; Professor. WORKS ABOUT: Am Men Sci d3; Who Was Who Am 4.

GOULD, AUGUSTUS ADDISON (1805-1866). Conchology; Medicine. OCCUP: Physician. MSS: Ind NUCMC (81-520). WORKS ABOUT: Dict Am Biog; Dict Sci Biog; Elliott; Natl Ac Scis Biog Mem 5; Natl Cycl Am Biog 3; Who Was Who Am h. See also: Barr; Biog Geneal Master Ind.

GOULD, BENJAMIN APTHORP, JR. (1824-1896). Astronomy. OCCUP: Editor; U.S. government scientist; Observatory director. MSS: Ind NUCMC. WORKS ABOUT: Dict Am Biog; Dict Sci Biog; Elliott; Natl Ac Scis Biog Mem [Mem 17]; Natl Cycl Am Biog 5; Who Was Who Am h. See also: Barr; Biog Geneal Master Ind; Ireland Ind Sci.

GOW, JAMES ELLIS (1877-1914). Botany; Geology. OCCUP: U.S. government scientist; Professor. WORKS ABOUT: Am Men Sci 2 d3; Sci 40(1914):446.

GOWELL, GILBERT M. (1845?-1908). Zoology. OCCUP: Professor. WORKS ABOUT: Sci 27(1908):836-37.

GRADLE, HENRY (1855-1911). Medicine. OCCUP: Physician; Professor. WORKS ABOUT: Am Men Sci 2 d3; Dict Am Biog; Natl Cycl Am Biog 24; Who Was Who Am 1. See also: Biog Geneal Master Ind.

GRAF, ARNOLD (1870-1898). Biology; Morphology. OCCUP: Hospital scientist. WORKS ABOUT: Deutsche Akademie Nat Nova Acta 72(1899):217-18; Sci 8(1898):331.

GRAHAM, JAMES DUNCAN (1799-1865). Topographical engineering. OCCUP: Army officer; Engineer. MSS: Ind NUCMC (?). WORKS ABOUT: Dict Am Biog; Elliott; Natl Cycl Am Biog 23. See also: Biog Geneal Master Ind.

GRAHAM, ROBERT ORLANDO (1853-1911). Chemistry. OCCUP: Professor; Academic administrator. WORKS ABOUT: Am Men Sci 1-2 d3; Twentieth Cent Biog Dict; Who Was Who Am 1.

GRASSELLI, EUGENE RAMIRO (1810-1882). Chemistry; Instrumentation. OCCUP: Manufacturer. WORKS ABOUT: Am Chem and Chem Eng; Haynes p88-106; Natl Cycl Am Biog 21.

GRATACAP, LOUIS POPE (1851-1917). Conchology; Mineralogy. OCCUP: Museum curator; Writer. MSS: Ind NUCMC (84-1260). WORKS ABOUT: Adams O; Allibone Supp; Am Men Sci 1-2 d3; Appleton's; Nicholls; Twentieth Cent Biog Dict; Sci 46(1917):635; Wallace; Who Was Who Am 1.

GRAVES, HERBERT CORNELIUS (1869-1919). Hydrography. OCCUP: Civil engineer; U.S. government scientist. WORKS ABOUT: Sci 50(1919):208; Who Was Who Am 1.

GRAY, ASA (1810-1888). Botany. OCCUP: Professor. MSS: <u>Ind</u> <u>NUCMC</u> (60-2792, 69-1, 71-2019, 72-1887, 75-1978). WORKS ABOUT: <u>Dict Am Biog</u>; <u>Dict Sci Biog</u>; Elliott; Natl Ac Scis <u>Biog Mem</u> 3; <u>Natl Cycl Am Biog</u> 3; <u>Who Was Who Am</u> h. See also: Barr; <u>Biog Geneal Master Ind</u>; Ireland <u>Ind Sci</u>; Pelletier.

GRAY, ELISHA (1835-1901). Invention; Telegraphy. OCCUP: Inventor; Manufacturer. MSS: <u>Ind NUCMC</u> (80-201). WORKS ABOUT: <u>Dict Am Biog</u>; <u>Natl Cycl Am Biog</u> 4; <u>Who Was Who Am</u> 1. See also: Barr; <u>Biog Geneal Master Ind</u>; Ireland <u>Ind Sci</u>; Pelletier.

GRAY, LANDON CARTER (1850-1900). Medicine; Neurology. OCCUP: Physician. WORKS ABOUT: <u>Natl Cycl Am Biog</u> 5; <u>New York Med Journ</u> 72(1900):923-24; <u>Sci</u> 11(1900):795; Wallace.

GRAY, THOMAS (1850-1908). Physics. OCCUP: Electrical engineer; Professor. WORKS ABOUT: <u>Am Men Sci</u> 1 d3; <u>Sci</u> 29(1909):23; <u>Who Was Who Am</u> 1.

GREEN, BERNARD RICHARDSON (1843-1914). Engineering. OCCUP: U.S. government scientist. MSS: <u>Ind NUCMC</u>. WORKS ABOUT: <u>Am Men Sci</u> 1-2 d3; <u>Dict Am Lib Biog</u>; <u>Natl Cycl Am Biog</u> 20; <u>Sci</u> 40(1914):632; <u>Twentieth Cent Biog Dict</u>; <u>Who Was Who Am</u> 1.

GREEN, FRANCIS MATHEWS (1835-1902). Hydrography. OCCUP: Naval officer. WORKS ABOUT: <u>Dict Am Biog</u>; Elliott; <u>Natl Cycl Am Biog</u> 26; <u>Who Was Who Am</u> 1. See also: <u>Biog Geneal Master Ind</u>.

GREEN, G. W. (1857?-1902?). Mathematics. OCCUP: Professor. WORKS ABOUT: <u>Sci</u> 17(1903):37.

GREEN, GABRIEL MARCUS (1891-1919). Mathematics. OCCUP: Academic instructor. WORKS ABOUT: <u>Dict Am Biog</u>. See also: <u>Biog Geneal Master Ind</u>.

GREEN, JACOB (1790-1841). Natural history; Botany; Zoology; Chemistry; Dissemination of knowledge. OCCUP: Lawyer; Professor. MSS: <u>Ind NUCMC</u> (?). WORKS ABOUT: <u>Dict Am Biog</u>; <u>Dict Sci Biog</u>; Elliott; <u>Natl Cycl Am Biog</u> 13; <u>Who Was Who Am</u> h. See also: <u>Biog Geneal Master Ind</u>; Ireland <u>Ind Sci</u>; Pelletier.

GREEN, SAMUEL BOWDLEAR (1859-1910). Horticulture. OCCUP: Professor; Editor. WORKS ABOUT: <u>Am Men Sci</u> 1-2 d3; <u>Dict Am Biog</u>; <u>Who Was Who Am</u> 1. See also: <u>Biog Geneal Master Ind</u>.

GREEN, SETH (1817-1888). Pisciculture. OCCUP: Businessman; State government scientist. MSS: <u>Ind NUCMC</u> (84-959). WORKS ABOUT: <u>Dict Am Biog</u>; <u>Natl Cycl Am Biog</u> 6; <u>Who Was Who Am</u> h. See also: <u>Biog Geneal Master Ind</u>.

GREEN, TRAILL (1813-1897). Chemistry. OCCUP: Physician; Professor. WORKS ABOUT: <u>Am Chem and Chem Eng</u>; Appleton's; <u>Natl Cycl Am Biog</u> 11; <u>Sci</u> 5(1897):728, 34(1911):12; <u>Twentieth Cent Biog Dict</u>.

GREENE, BENJAMIN DANIEL (1793-1862). Botany. OCCUP: Physician (?); Engaged in independent study. WORKS ABOUT: Am Journ Sci 85(1863):449-50; Natl Cycl Am Biog 7:509.

GREENE, CHARLES EZRA (1842-1903). Civil engineering. OCCUP: Professor; Consultant. MSS: Ind NUCMC. WORKS ABOUT: Dict Am Biog; Natl Cycl Am Biog 26; Who Was Who Am 1. See also: Biog Geneal Master Ind.

GREENE, EDWARD LEE (1843-1915). Botany. OCCUP: Professor; Museum associate. MSS: Ind NUCMC (67-521, 76-1599). WORKS ABOUT: Am Men Sci 2 d3; Dict Am Biog; Elliott (s); Natl Cycl Am Biog 19. See also: Barr; Biog Geneal Master Ind; Ireland Ind Sci.

GREENE, SAMUEL STILLMAN (1810-1883). Mathematics; Natural philosophy. OCCUP: Educator; Professor. WORKS ABOUT: Dict Am Biog; Natl Cycl Am Biog 8; Who Was Who Am h. See also: Biog Geneal Master Ind.

GREENE, WILLIAM HOUSTON (1853-1918). Chemistry. OCCUP: Teacher; Businessman. WORKS ABOUT: Adams O; Allibone Supp; Am Chem and Chem Eng; Appleton's; Twentieth Cent Biog Dict; Wallace; Who Was Who Am 1.

GREENLEAF, RICHARD CRANCH (1809-1887). Botany (diatoms); Microscopy. OCCUP: Merchant. WORKS ABOUT: Boston Soc Nat Hist Proc 23(1888):529-31.

GREENWOOD, ISAAC (1702-1745). Mathematics; Natural philosophy. OCCUP: Professor; Teacher. MSS: Ind NUCMC (?). WORKS ABOUT: Dict Am Biog; Dict Sci Biog; Elliott; Who Was Who Am h. See also: Biog Geneal Master Ind; Pelletier.

GREGG, WILLIAM HENRY (1830-1915). Zoology. OCCUP: Physician (?); Public official. WORKS ABOUT: Am Men Sci 1-2 d3.

GREGORY, EMILY LOVIRA (1841-1897). Botany. OCCUP: Professor. WORKS ABOUT: Ogilvie (app); Sci 5(1897):728; Torrey Bot Club Bull 24(1897):221-24; Twentieth Cent Biog Dict.

GREW, THEOPHILUS (-1759). Astronomy; Mathematics. OCCUP: Teacher; Professor. MSS: Ind NUCMC (?). WORKS ABOUT: Dict Am Biog; Elliott; Who Was Who Am h.

GRIFFIN, BRADNEY BEVERLEY (1872?-1898). Zoology. OCCUP: Academic affiliate. WORKS ABOUT: Nature 57(1897-1898):612; New York Ac Scis Ann 11(1898):193-94; Sci 7(1898):523-24.

GRIFFITH, HERBERT EUGENE (1866-1920). Chemistry. OCCUP: Professor. WORKS ABOUT: Am Men Sci 1-2 d3; Who Was Who Am 1.

GRIFFITH, ROBERT EGLESFELD (1798-1850). Botany;
Conchology. OCCUP: Physician; Professor (materia medica and
medicine). MSS: Ind NUCMC. WORKS ABOUT: Elliott; Natl Cycl
Am Biog 12; Who Was Who Am h. See also: Biog Geneal Master
Ind.

GRIMES, JAMES STANLEY (1807-1903). Geology; Natural
philosophy; Phrenology. OCCUP: Lawyer; Professor
(medicine); Lecturer. WORKS ABOUT: Dict Am Biog; Elliott;
Who Was Who Am h. See also: Barr; Biog Geneal Master Ind.

GRISCOM, JOHN (1774-1852). Chemistry. OCCUP: Teacher;
Educator. MSS: Ind NUCMC (?). WORKS ABOUT: Dict Am Biog;
Elliott; Natl Cycl Am Biog 10; Who Was Who Am h. See also:
Biog Geneal Master Ind; Ireland Ind Sci; Pelletier.

GROSSBECK, JOHN ARTHUR (1883-1914). Entomology. OCCUP:
State government scientist. MSS: Ind NUCMC. WORKS ABOUT: Am
Men Sci 2 d3.

GROTE, AUGUSTUS RADCLIFFE (1841-1903). Entomology. OCCUP:
Not determined; Curator. MSS: Ind NUCMC (?). WORKS ABOUT:
Dict Am Biog; Dict Sci Biog; Elliott; Natl Cycl Am Biog 22.
See also: Barr; Biog Geneal Master Ind; Pelletier.

GRUNOW, WILLIAM (1830-1917). Instrumentation. OCCUP:
Instrument maker; Observatory affiliate. WORKS ABOUT: Sci
45(1917):42.

GUILBEAU, BRAXTON H. (-1909). Zoology. OCCUP:
Professor. WORKS ABOUT: Sci 29(1909):179.

GUITARAS, RAMON (1860?-1917). Surgery. OCCUP: Professor.
WORKS ABOUT: Sci 46(1917):614.

GULICK, LUTHER HALSEY (1865-1918). Hygiene; Physical
education. OCCUP: Educator; Editor. MSS: Ind NUCMC. WORKS
ABOUT: Am Men Sci 2 d3; Dict Am Biog; Natl Cycl Am Biog 26;
Who Was Who Am 1. See also: Biog Geneal Master Ind.

GULLEY, ALFRED GURDON (1848-1917). Horticulture. OCCUP:
Horticulturist; Professor. WORKS ABOUT: Am Men Sci 2 d3.

GULLIVER, FREDERIC PUTNAM (1865-1919). Geography; Geology.
OCCUP: U.S. and state government scientist; Teacher. WORKS
ABOUT: Am Men Sci 1-2 d3; Sci 49(1919):233; Who Was Who
Am 4.

GUMMERE, JOHN (1784-1845). Mathematics. OCCUP: Teacher;
Educator. WORKS ABOUT: Dict Am Biog; Elliott; Natl Cycl Am
Biog 23; Who Was Who Am h. See also: Biog Geneal Master
Ind.

GUNDER, HENRY (1837?-1916). Mathematics. OCCUP: Professor.
WORKS ABOUT: Sci 44(1916):782.

GUNNING, WILLIAM DICKEY (1830-1888). Biology. OCCUP:
Lecturer; Author. WORKS ABOUT [birth dates may differ]:
Ohio Authors; Pop Sci Mo 50(1896-1897):526-30; Wallace.

GUTHE, KARL EUGEN (1866-1915). Physics. OCCUP: Professor;
Academic administrator. MSS: Ind NUCMC. WORKS ABOUT: Am Men
Sci 1-2 d3; Dict Am Biog; Natl Cycl Am Biog 22; Who Was Who
Am 1. See also: Barr; Biog Geneal Master Ind.

GUTHRIE, SAMUEL (1782-1848). Chemistry. OCCUP: Physician;
Farmer; Manufacturer. WORKS ABOUT: Dict Am Biog; Dict Am
Med Biog; Elliott; Natl Cycl Am Biog 11; Who Was Who Am h.
See also: Barr; Biog Geneal Master Ind; Ireland Ind Sci;
Pelletier.

GUTTENBERG, GUSTAVE (-ca.1895?). Geology. OCCUP:
Professor (?); Author. WORKS ABOUT: Pop Sci Mo 59(1901):6-
7.

GUYOT, ARNOLD HENRY (1807-1884). Geography; Geology;
Glaciology. OCCUP: Professor. MSS: Ind NUCMC (76-910).
WORKS ABOUT: Dict Am Biog; Dict Sci Biog; Elliott; Natl Ac
Scis Biog Mem 2; Natl Cycl Am Biog 4; Who Was Who Am h. See
also: Barr; Biog Geneal Master Ind; Ireland Ind Sci.

H

HACKLEY, CHARLES WILLIAM (1809-1861). Astronomy; Mathematics. OCCUP: Clergyman; Professor; College president. WORKS ABOUT: Adams O; Allibone; Am Journ Sci 81(1861):303-04; Appleton's; Drake; Twentieth Cent Biog Dict; Wallace.

HADLEY, GEORGE (1814?-1877). Chemistry. OCCUP: Professor. WORKS ABOUT: Am Journ Sci 114(1877):499.

HAGEN, HERMANN AUGUST (1817-1893). Entomology. OCCUP: Physician; Museum curator; Professor. MSS: Ind NUCMC. WORKS ABOUT: Dict Am Biog; Elliott; Natl Cycl Am Biog 5; Who Was Who Am h. See also: Barr; Biog Geneal Master Ind; Ireland Ind Sci; Pelletier.

HAGER, ALBERT DAVID (1817-1888). Geology. OCCUP: Educator; State government scientist; Administrator. MSS: Ind NUCMC (?). WORKS ABOUT: Elliott; Natl Cycl Am Biog 3; Who Was Who Am h. See also: Barr; Biog Geneal Master Ind.

HAGUE, ARNOLD (1840-1917). Geology. OCCUP: U.S. government scientist. MSS: Ind NUCMC (?). WORKS ABOUT: Am Men Sci 1-2 d3; Dict Am Biog; Dict Sci Biog; Elliott (s); Natl Ac Scis Biog Mem 9; Natl Cycl Am Biog 3; Who Was Who Am 1. See also: Barr; Biog Geneal Master Ind.

HAGUE, JAMES DUNCAN (1836-1908). Geology; Mining engineering. OCCUP: U.S. government scientist; Consultant. MSS: Ind NUCMC (61-3003). WORKS ABOUT: Am Men Sci 1 d3; Dict Am Biog; Elliott (s); Natl Cycl Am Biog 2, 23; Who Was Who Am 1. See also: Barr; Biog Geneal Master Ind.

HALDEMAN, SAMUEL STEMAN (1812-1880). Zoology; Philology. OCCUP: Businessman; Engaged in independent study; Professor. MSS: Ind NUCMC (66-49). WORKS ABOUT: Dict Am Biog; Elliott; Natl Ac Scis Biog Mem 2; Natl Cycl Am Biog 9; Who Was Who Am h. See also: Barr; Biog Geneal Master Ind; Ireland Ind Sci; Pelletier.

HALE, ENOCH (1790-1848). Medicine. OCCUP: Physician. MSS: Ind NUCMC. WORKS ABOUT: Dict Am Biog; Natl Cycl Am Biog 3; Who Was Who Am h. See also: Biog Geneal Master Ind.

HALE, HORATIO EMMONS (1817-1896). Ethnology. OCCUP: Lawyer; U.S. government scientist. MSS: Ind NUCMC. WORKS ABOUT: Dict Am Biog; Natl Cycl Am Biog 3; Who Was Who Am h. See also: Barr; Biog Geneal Master Ind.

HALL, ASAPH (1829-1907). Astronomy. OCCUP: U.S. government scientist; Observatory affilaite; Academic instructor. MSS: Ind NUCMC (72-1723). WORKS ABOUT: Am Men Sci 1 d3; Dict Am Biog; Dict Sci Biog; Elliott (s); Natl Ac Scis Biog Mem 6; Natl Cycl Am Biog 11, 22; Who Was Who Am 1. See also: Barr; Biog Geneal Master Ind; Ireland Ind Sci; Pelletier.

HALL, CHARLES MARTIN (1863-1914). Chemistry; Commercial chemistry; Electrochemistry. OCCUP: Inventor; Industrialist. MSS: Ind NUCMC. WORKS ABOUT: Am Men Sci 1-2 d3; Dict Am Biog; Dict Sci Biog; Natl Cycl Am Biog 13; Who Was Who Am 1. See also: Barr; Biog Geneal Master Ind; Ireland Ind Sci; Pelletier.

HALL, CHRISTOPHER WEBBER (1845-1911). Geology. OCCUP: Professor; U.S. government scientist. WORKS ABOUT: Adams O; Am Men Sci 1-2 d3; Appleton's; Elliott (s); Natl Cycl Am Biog 9; Twentieth Cent Biog Dict; Wallace; Who Was Who Am 1.

HALL, ELIHU (1822-1882). Botany. OCCUP: Natural history explorer/ collector. MSS: Ind NUCMC (?). WORKS ABOUT: Am Journ Sci 127(1884):242; Bot Gaz 9&10(1884-1885):59-62.

HALL, FREDERICK (1780-1843). Chemistry; Mineralogy. OCCUP: Professor; College president. WORKS ABOUT: Adams O; Am Journ Sci 45(1843):404; Appleton's; Drake; Twentieth Cent Biog Dict; Wallace.

HALL, JAMES (1811-1898). Geology; Paleontology. OCCUP: Professor; State government scientist; Museum administrator. MSS: Ind NUCMC (79-970). WORKS ABOUT: Dict Am Biog; Dict Sci Biog; Elliott; Natl Cycl Am Biog 3; Who Was Who Am h. See also: Barr; Biog Geneal Master Ind; Ireland Ind Sci; Pelletier.

HALL, JAMES PIERRE (1849-1919). Astronomy. OCCUP: Journalist. WORKS ABOUT: Am Men Sci 1-2 d3; Who Was Who Am 4.

HALL, LYMAN (1859-1905). Mathematics. OCCUP: Professor; College president. WORKS ABOUT: Knight; Natl Cycl Am Biog 29; Sci 22(1905):253; Twentieth Cent Biog Dict; Wallace; Who Was Who Am 1.

HALLOCK, CHARLES (1834-1917). Natural history. OCCUP: Journalist; Research institution associate; Author. WORKS ABOUT: Dict Am Biog; Natl Cycl Am Biog 9; Who Was Who Am 1. See also: Biog Geneal Master Ind.

HALLOCK, WILLIAM (1857-1913). Physics. OCCUP: U.S. government scientist; Professor. WORKS ABOUT: Am Men Sci 1 d3; Sci 37(1914):830; Twentieth Cent Biog Dict; Wallace; Who Was Who Am 1.

HALLOWELL, BENJAMIN (1799-1877). Mathematics. OCCUP: Educator; Professor; College president. MSS: Ind NUCMC. WORKS ABOUT: Dict Am Biog; Natl Cycl Am Biog 22; Who Was Who Am h. See also: Barr; Biog Geneal Master Ind.

HALLOWELL, EDWARD (1808-1860). Herpetology. OCCUP: Physician. MSS: Ind NUCMC (66-43). WORKS ABOUT: Elliott.

HALLOWELL, SUSAN MARIA (1835-1911). Botany. OCCUP: Professor. WORKS ABOUT: Am Men Sci 1-2 d3; Sci 34(1911):911-12; Siegel and Finley.

HALSTED, BYRON DAVID (1852-1918). Botany. OCCUP: Editor; Professor. MSS: Ind NUCMC. WORKS ABOUT: Allibone Supp; Am Men Sci 1-2 d3; Appleton's; Natl Cycl Am Biog 10; Sci 48(1918):244; Twentieth Cent Biog Dict; Wallace; Who Was Who Am 1.

HAMILTON, ALLAN McLANE (1848-1919). Neurology. OCCUP: Physician; Professor. WORKS ABOUT: Dict Am Biog; Natl Cycl Am Biog 9; Who Was Who Am 1. See also: Barr; Biog Geneal Master Ind.

HAMILTON, HUGH (1847-   ). Medicine. OCCUP: Physician; State government scientist. WORKS ABOUT: Am Men Sci 3 d4.

HAMILTON, JOHN (1827-1897). Entomology. OCCUP: Physician. WORKS ABOUT: Ent News 8(1897):73-74; Ent Mo Mag 33(1897):108.

HAMILTON, JOHN BROWN (1847-1898). Medicine. OCCUP: Physician; Professor. WORKS ABOUT: Dict Am Med Biog; Natl Cycl Am Biog 23; Sci 9(1899):38; Twentieth Cent Biog Dict; Wallace.

HAMILTON, WILLIAM (-1871). Physical science. OCCUP: Research institution administrator. Scientific work: see Roy Soc Cat. WORKS ABOUT: Franklin Inst Journ 62(1871):360 (referred to), 165(1908):267 (listed); Sinclair.

HAMLIN, CHARLES EDWARD (1825-1886). Natural history. OCCUP: Professor; Museum affiliate. WORKS ABOUT: Am Ac Arts Scis Proc 21(1886):524-26; Sci 1st ser 7(1886):74; Twentieth Cent Biog Dict.

HAMMOND, WILLIAM ALEXANDER (1828-1900). Neurology; Physiology. OCCUP: Army surgeon; Physician; Professor. MSS: Ind NUCMC. WORKS ABOUT: Dict Am Biog; Dict Am Med Biog; Elliott; Natl Cycl Am Biog 9, 26; Who Was Who Am 1, 4 (add). See also: Barr; Biog Geneal Master Ind; Pelletier.

HANCOCK, EDWARD LEE (1873-1911). Mechanics. OCCUP: Professor. WORKS ABOUT: Am Men Sci 1-2 d3; Sci 34(1911):485; Wallace.

HANKS, HENRY GARBER (1826-1907). Chemistry; Geology. OCCUP: Businessman; State government scientist. WORKS ABOUT: Elliott; Natl Cycl Am Biog 13; Who Was Who Am 1. See also: Barr.

HARDIN, MARK BERNARD (1838-1910). Chemistry. OCCUP: Professor; State government scientist. WORKS ABOUT: Am Men Sci 1-2 d3.

HARDY, JOSEPH JOHNSTON (1844-1915). Mathematics. OCCUP: Professor. WORKS ABOUT: Am Men Sci 1-2 d3; Sci 41(1915):787; Who Was Who Am 1.

HARE, JOHN INNES CLARK (1816-1905). Chemistry. OCCUP: Judge; Professor (law). Scientific work: see Roy Soc Cat (as Clark Hare). WORKS ABOUT: Dict Am Biog; Who Was Who Am 4. See also: Biog Geneal Master Ind.

HARE, ROBERT (1781-1858). Chemistry. OCCUP: Businessman; Professor. MSS: Ind NUCMC (62-932, 80-299). WORKS ABOUT: Dict Am Biog; Dict Sci Biog; Elliott; Natl Cycl Am Biog 5; Who Was Who Am h. See also: Barr; Biog Geneal Master Ind; Ireland Ind Sci; Pelletier.

HARFORD, W. G. W. (1825-1911). Conchology. OCCUP: Academic affiliate; Curator; Collector. WORKS ABOUT: Essig p650-51; Sci 33(1911):452.

HARGER, OSCAR (1843-1887). Geology; Paleontology. OCCUP: Academic affiliate. WORKS ABOUT: Am Journ Sci 134(1887):496, 135(1888):425-26; Am Nat 21(1887):1133-34; Appleton's Supp 1901; Sci 1st ser 10(1887):238-39; Twentieth Cent Biog Dict; Wallace.

HARKNESS, WILLIAM (1837-1903). Astronomy. OCCUP: U.S. government scientist; Observatory affiliate. WORKS ABOUT: Dict Am Biog; Dict Sci Biog; Elliott; Natl Cycl Am Biog 8; Who Was Who Am 1. See also: Barr; Biog Geneal Master Ind.

HARKNESS, WILLIAM H. (1821?-1901). Entomology. OCCUP: Physician; Researcher. WORKS ABOUT: Sci 14(1901):117, 158.

HARLAN, RICHARD (1796-1843). Natural history; Zoology; Comparative anatomy. OCCUP: Physician; Teacher. MSS: Ind NUCMC. WORKS ABOUT: Dict Am Biog; Dict Sci Biog; Elliott; Who Was Who Am h. See also: Barr; Biog Geneal Master Ind; Pelletier.

HARRINGTON, CHARLES (1856-1908). Hygiene. OCCUP: Physician; Professor; City and state government scientist. WORKS ABOUT: Am Men Sci 1 d3; Dict Am Biog; Who Was Who Am 1. See also: Barr; Biog Geneal Master Ind.

HARRINGTON, NATHAN RUSSELL (1870-1899). Zoology. OCCUP: Academic instructor. WORKS ABOUT: Sci 10(1899):157, 529-31.

HARRIS, EDWARD (1799-1863). Natural history; Geology; Ornithology. OCCUP: Inherited wealth. MSS: Ind NUCMC. WORKS ABOUT: Cassinia 6(1902):1-5.

HARRIS, ELIJAH PADDOCK (1832-1920). Chemistry. OCCUP: Professor. WORKS ABOUT: Am Men Sci 1-2 d3; Wallace; Who Was Who Am 1.

HARRIS, GEORGE H. (-1905). Entomology. OCCUP: U.S. government scientist. WORKS ABOUT: Sci 21(1905):158.

HARRIS, JOSEPH SMITH (1836-1910). Astronomy. OCCUP: U.S. government scientist; Industrial engineer and officer. WORKS ABOUT: Sci 31(1910):947; Who Was Who Am 1.

HARRIS, ROLLIN ARTHUR (1863-1918). Mathematics; Oceanography; Physics. OCCUP: U.S. government scientist. WORKS ABOUT: Am Men Sci 1-2 d3; Dict Am Biog; Natl Cycl Am Biog 22; Who Was Who Am 1. See also: Barr; Biog Geneal Master Ind.

HARRIS, THADDEUS WILLIAM (1795-1856). Entomology. OCCUP: Physician; Librarian; Academic instructor. MSS: Ind NUCMC. WORKS ABOUT: Dict Am Biog; Elliott; Who Was Who Am h. See also: Barr; Biog Geneal Master Ind; Ireland Ind Sci; Pelletier.

HARRISON, JOHN (1773-1833). Chemistry; Manufacturing chemistry. OCCUP: Manufacturer. MSS: Ind NUCMC. WORKS ABOUT: Dict Am Biog; Elliott; Natl Cycl Am Biog 13; Who Was Who Am h. See also: Biog Geneal Master Ind; Pelletier.

HARRISON, JOSEPH, JR. (1810-1874). Engineering. OCCUP: Industrial engineer; Manufacturer. MSS: Ind NUCMC (62-4729). WORKS ABOUT: Dict Am Biog; Natl Cycl Am Biog 12; Who Was Who Am h. See also: Barr; Biog Geneal Master Ind.

HART, CHARLES ARTHUR (1859-1918). Entomology. OCCUP: State government scientist. WORKS ABOUT: Am Men Sci 1-2 d3; Osborn Herb p237, 368; Sci 47(1918):236; Who Was Who Am 3.

HARTER, NOBLE (-1907). Psychology. OCCUP: Not determined. WORKS ABOUT: Sci 25(1907):517.

HARTLEY, EDWARD (1847?-1870). Geology. OCCUP: Mining engineer; Government scientist (foreign). WORKS ABOUT: Am Journ Sci 101(1871):74.

HARTLEY, FRANK (1856-1913). Surgery. OCCUP: Surgeon; Professor. WORKS ABOUT: Dict Am Biog; Natl Cycl Am Biog 15; Who Was Who Am 1. See also: Barr; Ireland Ind Sci.

HARTMAN, ROBERT NELSON (-1903). Chemistry. OCCUP: Professor. WORKS ABOUT: Sci 17(1903):799.

HARTMAN, WILLIAM DELL (1817-1899). Natural history; Conchology. OCCUP: Physician. WORKS ABOUT: Appleton's; Nautilus 13(1899-1900):61-63; Twentieth Cent Biog Dict; Wallace.

HARTSHORNE, HENRY (1823-1897). Biology; Medicine. OCCUP: Physician; Professor. MSS: Ind NUCMC (62-4365). WORKS ABOUT: Dict Am Biog; Natl Cycl Am Biog 8; Who Was Who Am h. See also: Biog Geneal Master Ind.

HARTT, CHARLES FREDERICK (1840-1878). Geology. OCCUP: Professor; Government scientist (foreign). MSS: Ind NUCMC (61-2509, 70-1088). WORKS ABOUT: Elliott; Natl Cycl Am Biog 11; Who Was Who Am h. See also: Barr; Biog Geneal Master Ind; Ireland Ind Sci.

HARTWELL, CHARLES (1825-   ). General science. OCCUP: Clergyman. Scientific work: see Roy Soc Cat. WORKS ABOUT: Amherst Biog Rec 1871 p233 (listed).

HARTWELL, GEORGE WILBER (1881-1917). Mathematics. OCCUP: Professor; Academic administrator. WORKS ABOUT: Am Men Sci 2 d3; Sci 46(1917):234.

HARVEY, FRANCIS LEROY (1850-1900). Biology; Entomology. OCCUP: Professor. WORKS ABOUT: Ent News 11(1900):451-52; Sci 11(1900):436-37.

HARWOOD, WILLIAM SUMNER (1857-1908). Biology. OCCUP: Journalist; Author. WORKS ABOUT: Sci 28(1908):752; Upham and Dunlap; Wallace; Who Was Who Am 1.

HASKINS, CARYL DAVIS (1867-1911). Electrical and military engineering. OCCUP: Industrial engineer and manager. WORKS ABOUT: Adams O; Am Men Sci 1-2 d3; Natl Cycl Am Biog 28; Wallace; Who Was Who Am 1.

HASSLER, FERDINAND RUDOLPH (1770-1843). Geodesy; Mathematics. OCCUP: U.S. government scientist. MSS: Ind NUCMC (71-1221). WORKS ABOUT: Dict Am Biog; Dict Sci Biog; Elliott; Natl Cycl Am Biog 3; Who Was Who Am h. See also: Barr; Biog Geneal Master Ind; Ireland Ind Sci.

HASTINGS, JOHN WALTER (-1908). Ethnology. OCCUP: Museum affiliate. WORKS ABOUT: Sci 27(1908):798.

HASWELL, CHARLES HAYNES (1809-1907). Engineering. OCCUP: Civil engineer; Naval engineer. WORKS ABOUT: Dict Am Biog; Natl Cycl Am Biog 9; Who Was Who Am 1. See also: Biog Geneal Master Ind.

HATCH, DANIEL G. (-1895). Zoology. OCCUP: U.S. government scientist. WORKS ABOUT: Sci 2(1895):163.

HATCHER, JOHN BELL (1861-1904). Paleontology. OCCUP: Natural history collector; Museum curator. WORKS ABOUT: Elliott; Natl Cycl Am Biog 21; Who Was Who Am 1. See also: Barr; Biog Geneal Master Ind.

HAUPT, HERMAN (1817-1905). Engineering. OCCUP: Industrial engineer and manager. MSS: Ind NUCMC (62-647). WORKS ABOUT: Dict Am Biog; Natl Cycl Am Biog 10; Who Was Who Am 1. See also: Biog Geneal Master Ind.

HAVARD, FRANCIS T. (1878?-1913). Metallurgy. OCCUP: Mining expert; Professor. WORKS ABOUT: Sci 37(1913):904.

HAWES, GEORGE WESSON (1848-1882). Mineralogy. OCCUP: Academic instructor; Museum curator. WORKS ABOUT: Am Journ Sci 124(1882):80, 159-60.

HAY, ROBERT (1835-1895). Geology. OCCUP: Teacher; U.S. and state government scientist. MSS: Ind NUCMC (?). WORKS ABOUT: Geol Soc Am Bull 8(1897):370-74; Kansas Ac Sci Trans 15(1898):131-34.

HAYDEN, FERDINAND VANDIVEER (1829-1887). Geology. OCCUP: U.S. government scientist; Professor. MSS: Ind NUCMC (75-1534). WORKS ABOUT: Dict Am Biog; Dict Sci Biog; Elliott; Natl Ac Scis Biog Mem 3; Natl Cycl Am Biog 11; Who Was Who Am h. See also: Barr; Biog Geneal Master Ind; Ireland Ind Sci; Pelletier.

HAYDEN, HORACE H. (1769-1844). Geology; Dentistry. OCCUP: Dentist; Professor (dentistry). WORKS ABOUT: Dict Am Biog; Dict Am Med Biog; Elliott; Natl Cycl Am Biog 13; Who Was Who Am h. See also: Biog Geneal Master Ind; Ireland Ind Sci.

HAYES, AUGUSTUS ALLEN (1806-1882). Chemistry. OCCUP: Industrial scientist and manager; Consultant; State government scientist. WORKS ABOUT: Dict Am Biog; Elliott; Natl Cycl Am Biog 11; Who Was Who Am h. See also: Barr; Biog Geneal Master Ind; Pelletier.

HAYES, CHARLES WILLARD (1858-1916). Geology. OCCUP: U.S. government scientist; Industrial officer. MSS: Ind NUCMC. WORKS ABOUT: Am Men Sci 1-2 d3; Dict Am Biog; Who Was Who Am 1. See also: Barr; Biog Geneal Master Ind.

HAYES, ISAAC ISRAEL (1832-1881). Arctic exploration. OCCUP: Physician; Explorer. MSS: Ind NUCMC. WORKS ABOUT: Dict Am Biog; Natl Cycl Am Biog 3; Who Was Who Am h. See also: Biog Geneal Master Ind.

HAYES, JOHN LORD (1812-1887). Natural history; Geology. OCCUP: Lawyer. MSS: Ind NUCMC. WORKS ABOUT: Dict Am Biog; Natl Cycl Am Biog 14; Who Was Who Am h. See also: Biog Geneal Master Ind.

HAYMOND, RUFUS (1805?-1886). Zoology. OCCUP: Physician (?). WORKS ABOUT: Sci 1st ser 8(1886):123.

HAYNES, ARTHUR EDWIN (1849-1915). Mathematics. OCCUP: Professor. WORKS ABOUT: Am Men Sci 1-2 d3; Sci 41(1915):606; Who Was Who Am 1.

HAYNES, HENRY WILLIAMSON (1831-1912). Archeology. OCCUP: Professor (languages); Researcher. WORKS ABOUT: Am Men Sci 1; Appleton's; Natl Cycl Am Biog 8; Twentieth Cent Biog Dict; Who Was Who Am 1.

HAYS, ISAAC (1796-1879). Fossil studies; Medicine; Ornithology. OCCUP: Ophthalmologist; Editor. MSS: Ind NUCMC (69-1373). WORKS ABOUT: Dict Am Biog; Dict Am Med Biog; Natl Cycl Am Biog 11; Who Was Who Am h. See also: Biog Geneal Master Ind; Ireland Ind Sci.

HAZARD, ROWLAND (1829-1898). Chemical manufacturing; Textiles. OCCUP: Manufacturer. MSS: Ind NUCMC. WORKS ABOUT: Am Chem and Chem Eng; Appleton's Supp 1918-31; Natl Cycl Am Biog 12; Twentieth Cent Biog Dict.

HAZEN, HENRY ALLEN (1849-1900). Meteorology. OCCUP: Academic instructor; U.S. government scientist. WORKS ABOUT: Dict Am Biog; Elliott; Natl Cycl Am Biog 8. See also: Barr; Biog Geneal Master Ind.

HAZEN, JOHN VOSE (1850-1919). Engineering. OCCUP: Professor. WORKS ABOUT: Am Men Sci 1 d3; Natl Cycl Am Biog 31; Who Was Who Am 1.

HAZEN, WILLIAM BABCOCK (1830-1887). Meteorology. OCCUP: Army officer; Administrator. MSS: Ind NUCMC. WORKS ABOUT: Dict Am Biog; Natl Cycl Am Biog 3; Who Was Who Am h. See also: Barr; Biog Geneal Master Ind.

HEAD, W. R. (1829?-1910). Biology. OCCUP: Not determined. WORKS ABOUT: Sci 31(1910):854.

HEIDEMANN, OTTO (1842-1916). Entomology. OCCUP: Artist; U.S. government scientist; Museum curator. MSS: Ind NUCMC. WORKS ABOUT: Am Men Sci 2 d3; Essig p653-54; Mallis; Osborn Herb p203, 349.

HEILPRIN, ANGELO (1853-1907). Geography; Natural history; Geology; Paleontology. OCCUP: Teacher; Explorer; Curator. MSS: Ind NUCMC (66-46). WORKS ABOUT: Am Men Sci 1 d3; Dict Am Biog; Elliott (s); Natl Cycl Am Biog 12; Who Was Who Am 1. See also: Barr; Biog Geneal Master Ind.

HEISKELL, HENRY LEE (1850-1914). Meteorology. OCCUP: U.S. government scientist. WORKS ABOUT: Who Was Who Am 1.

HEITZMAN, CHARLES (1836-1896). Histology. OCCUP: Physician; Academic instructor; Researcher. WORKS ABOUT: Adams O; Allibone Supp; Appleton's; Sci 5(1897):103; Wallace.

HEMPHILL, HENRY (1830-1914). Biology; Conchology. OCCUP: Craftsman. WORKS ABOUT: Sci 40(1914):265-66.

HENCK, JOHN BENJAMIN (1815-1903). Engineering. OCCUP: Engineer; Professor. WORKS ABOUT: <u>Dict Am Biog</u>. See also: Barr; <u>Biog Geneal Master Ind</u>.

HENDRICKSON, W. F. (1876?-1902). Pathology. OCCUP: Academic instructor. WORKS ABOUT: <u>Sci</u> 16(1902):357.

HENDRICKSON, WILLIAM WOODBURY (1844-1920). Mathematics. OCCUP: Naval officer; Professor. WORKS ABOUT: <u>Am Men Sci</u> 1-2 d3; <u>Who Was Who Am</u> 1.

HENRY, FREDERICK PORTEOUS (1844-   ). Medicine. OCCUP: Physician; Professor. WORKS ABOUT: <u>Am Men Sci</u> 1-2 d3; <u>Who Was Who Am</u> 4.

HENRY, JOSEPH (1797-1878). Physics. OCCUP: Professor; Research institution administrator. MSS: <u>Ind NUCMC</u> (76-911). WORKS ABOUT: <u>Dict Am Biog</u>; <u>Dict Sci Biog</u>; Elliott; Natl Ac Scis <u>Biog Mem</u> 5; <u>Natl Cycl Am Biog</u> 3; <u>Who Was Who Am</u> h. See also: Barr; <u>Biog Geneal Master Ind</u>; Ireland <u>Ind Sci</u>; Pelletier.

HENRY, THOMAS CHARLTON (1825-1877). Ornithology. OCCUP: Physician; Army surgeon. Scientific work: see Roy Soc <u>Cat</u>. WORKS ABOUT: Barnhart; Durfee.

HENRY, WILLIAM (1729-1786). Study of heat. OCCUP: Manufacturer; Inventor; Public official. MSS: <u>Ind NUCMC</u> (60-2223). WORKS ABOUT: <u>Dict Am Biog</u>; <u>Natl Cycl Am Biog</u> 11; <u>Who Was Who Am</u> h. See also: <u>Biog Geneal Master Ind</u>.

HENSHAW, MARSHALL (1820-1900). Astronomy; Physics. OCCUP: Educator; Professor. WORKS ABOUT: <u>Sci</u> 12(1900):971; <u>Twentieth Cent Biog Dict</u>; Wallace.

HENSHAW, SAMUEL (-1907). Horticulture. OCCUP: Horticulturist. WORKS ABOUT: <u>Sci</u> 26(1907):326.

HENTZ, NICHOLAS MARCELLUS (1797-1856). Arachnology; Entomology. OCCUP: Teacher; Educator; Professor (languages and belles lettres). MSS: <u>Ind NUCMC</u>. WORKS ABOUT: Elliott; <u>Natl Cycl Am Biog</u> 9; <u>Who Was Who Am</u> h. See also: Barr; <u>Biog Geneal Master Ind</u>; Pelletier.

HERDMAN, WILLIAM JAMES (1848-1906). Neurology. OCCUP: Professor (medicine). MSS: <u>Ind NUCMC</u>. WORKS ABOUT: <u>Am Men Sci</u> 1 d3; <u>Who Was Who Am</u> 1.

HERRICK, CLARENCE LUTHER (1858-1904). Neurology; Comparative neurology; Psychobiology. OCCUP: Professor; College president. MSS: <u>Ind NUCMC</u>. WORKS ABOUT: <u>Dict Sci Biog</u>; <u>Natl Cycl Am Biog</u> 26; <u>Who Was Who Am</u> 1. See also: Barr; <u>Biog Geneal Master Ind</u>.

HERRICK, EDWARD CLAUDIUS (1811-1862). Astronomy; Entomology. OCCUP: Librarian; Administrator. MSS: Ind NUCMC. WORKS ABOUT: Dict Am Biog; Elliott; Natl Cycl Am Biog 11; Who Was Who Am h. See also: Barr; Biog Geneal Master Ind; Pelletier.

HERRMAN, AUGUSTINE (1605-1686). Cartography. OCCUP: Businessman; Map-maker. MSS: Ind NUCMC (67-1498). WORKS ABOUT: Dict Am Biog; Who Was Who Am h. See also: Biog Geneal Master Ind.

HERRMAN, ESTHER (1822?-1911). Philanthropy. OCCUP: Not determined; Benefactor (of scientific societies). WORKS ABOUT: Sci 34(1911):49; Who Was Who Am 1.

HERTER, CHRISTIAN ARCHIBALD (1865-1910). Biochemistry; Pathology. OCCUP: Physician; Professor. MSS: Ind NUCMC (65-1948). WORKS ABOUT: Am Men Sci 1 d3; Dict Am Biog; Dict Am Med Biog; Who Was Who Am 1. See also: Barr; Biog Geneal Master Ind; Pelletier.

HESS, HOWARD DRYSDALE (1872?-1916). Engineering. OCCUP: Professor. WORKS ABOUT: Sci 43(1916):639.

HESSE, CONRAD E. (1866-1910). Meteorology. OCCUP: Mining engineer; U.S. government scientist. WORKS ABOUT: Am Men Sci 1-2 d3.

HESSEL, RUDOLPH (1825-1900). Ichthyology. OCCUP: U.S. government scientist. WORKS ABOUT: Leopoldina 36(1900):153; Sci 12(1900):317.

HICKS, JOHN F. (-1903). Botany. OCCUP: State government scientist. WORKS ABOUT: Sci 17(1903):956.

HIDDEN, WILLIAM EARL (1853-1918). Mineralogy. OCCUP: Natural history explorer/ collector; Consultant; Mining manager. MSS: Ind NUCMC (?). WORKS ABOUT: Am Journ Sci 196(1918):480; Am Men Sci 1-2 d3; Sci 48(1918):188; Who Was Who Am 1.

HIGGINS, MILTON PRINCE (1842-1912). Mechanical engineering. OCCUP: Academic affiliate; Manufacturer. MSS: Ind NUCMC (83-1560). WORKS ABOUT: Am Men Sci 2 d3; Appleton's Supp 1918-31; Natl Cycl Am Biog 31; Who Was Who Am 1.

HIGGINSON, THOMAS WENTWORTH (1823-1911). Natural science. OCCUP: Clergyman; Writer. MSS: Ind NUCMC (61-1092, 61-1405, 61-2155, 61-2531, 81-547). WORKS ABOUT: Am Men Sci 1-2 d3; Dict Am Biog; Natl Cycl Am Biog 1; Who Was Who Am 1. See also: Biog Geneal Master Ind.

HILDER, F. F. (1827?-1901). Ethnology. OCCUP: U.S. government employee. WORKS ABOUT: Sci 13(1901):198.

HILDRETH, SAMUEL PRESCOTT (1783-1863). Natural history. OCCUP: Physician. MSS: Ind NUCMC (75-1031). WORKS ABOUT: Dict Am Biog; Elliott; Who Was Who Am h. See also: Barr; Biog Geneal Master Ind; Pelletier.

HILGARD, EUGENE WOLDEMAR (1833-1916). Agriculture; Agricultural chemistry; Geology; Soil studies. OCCUP: Professor; Laboratory administrator. MSS: Ind NUCMC (80-2266). WORKS ABOUT: Am Men Sci 1 d3; Dict Am Biog; Elliott (s); Natl Ac Scis Biog Mem 9; Natl Cycl Am Biog 10; Who Was Who Am 1. See also: Barr; Biog Geneal Master Ind; Ireland Ind Sci.

HILGARD, JULIUS ERASMUS (1825-1891). Geodesy. OCCUP: U.S. government scientist. MSS: Ind NUCMC. WORKS ABOUT: Dict Am Biog; Elliott; Natl Ac Scis Biog Mem 3; Natl Cycl Am Biog 10; Who Was Who Am h. See also: Barr; Biog Geneal Master Ind.

HILGARD, THEODORE CHARLES (1828-1875). Botany; Geodesy; Zoology. OCCUP: Physician. MSS: Ind NUCMC (?). WORKS ABOUT: Appleton's; Natl Cycl Am Biog 22; Pop Sci Mo 74(1909):130-33.

HILL, CHARLES BARTON (1863?-1910). Geodesy. OCCUP: Observatory affiliate; U.S. government scientist. WORKS ABOUT: Sci 32(1910):303.

HILL, ELLSWORTH JEROME (1833-1917). Botany. OCCUP: Clergyman; Teacher. WORKS ABOUT: Am Men Sci 1-2 d3.

HILL, GEORGE ANTHONY (1842-1916). Physics. OCCUP: Professor; Teacher, Author. WORKS ABOUT: Sci 44(1916):270; Wallace; Who Was Who Am 1.

HILL, GEORGE WILLIAM (1838-1914). Mathematics; Mathematical astronomy. OCCUP: U.S. government scientist. MSS: Ind NUCMC. WORKS ABOUT: Am Men Sci 1-2 d3; Dict Am Biog; Dict Sci Biog; Elliott (s); Natl Ac Scis Biog Mem 8; Natl Cycl Am Biog 13; Who Was Who Am 1. See also: Barr; Biog Geneal Master Ind; Ireland Ind Sci.

HILL, HAMPDEN (1886-1918). Chemistry. OCCUP: Industrial and state government scientist; Academic instructor. WORKS ABOUT: Am Men Sci 2 d3; Sci 48(1918):392.

HILL, HENRY BARKER (1849-1903). Chemistry. OCCUP: Professor; Consultant. WORKS ABOUT: Dict Am Biog; Elliott; Natl Ac Scis Biog Mem 5; Who Was Who Am 1. See also: Barr; Biog Geneal Master Ind; Ireland Ind Sci.

HILL, NATHANIEL PETER (1832-1900). Metallurgy. OCCUP: Professor; Industrialist; Legislator. MSS: Ind NUCMC (68-1192). WORKS ABOUT: Dict Am Biog; Natl Cycl Am Biog 6; Who Was Who Am h, 1. See also: Biog Geneal Master Ind.

HILL, THOMAS (1818-1891). Astronomy; Mathematics; Science and religion. OCCUP: Clergyman; College president; Professor (theology). MSS: Ind NUCMC (65-1252). WORKS ABOUT: Dict Am Biog; Elliott; Natl Cycl Am Biog 6; Who Was Who Am h. See also: Biog Geneal Master Ind.

HILL, WALTER NICKERSON (1846-1884). Chemistry. OCCUP: Professor; Industrial and U.S. government scientist. MSS: Ind NUCMC (78-83). WORKS ABOUT: Natl Cycl Am Biog 12.

HIMES, CHARLES FRANCIS (1838-1918). Physics. OCCUP: Professor. MSS: Ind NUCMC (66-469). WORKS ABOUT: Am Men Sci 1-2 d3; Dict Am Biog; Natl Cycl Am Biog 4; Who Was Who Am h, 3. See also: Biog Geneal Master Ind.

HINTON, CHARLES H. (1844-1907). Mathematics. OCCUP: U.S. government scientist. WORKS ABOUT: Sci 25(1907):758.

HIRSCHFELDER, JOSEPH OAKLAND (1854-1920). Medicine. OCCUP: Physician; Professor. WORKS ABOUT: Am Men Sci 2 d3; Who Was Who Am 1.

HISS, PHILIP HANSON, JR. (1868-1913). Bacteriology; Pathology. OCCUP: Professor. WORKS ABOUT: Am Men Sci 1-2 d3; Bulloch p373; Natl Cycl Am Biog 16; Sci 37(1913):368.

HITCHCOCK, CHARLES HENRY (1836-1919). Geology. OCCUP: Professor; State government scientist. WORKS ABOUT: Am Men Sci 1-2 d3; Dict Am Biog; Elliott (s); Natl Cycl Am Biog 12; Who Was Who Am 1. See also: Barr; Biog Geneal Master Ind; Ireland Ind Sci.

HITCHCOCK, EDWARD (1793-1864). Geology. OCCUP: Professor; College president; State government scientist. MSS: Ind NUCMC (75-14). WORKS ABOUT: Dict Am Biog; Dict Sci Biog; Elliott; Natl Ac Scis Biog Mem 1; Natl Cycl Am Biog 5; Who Was Who Am h. See also: Barr; Biog Geneal Master Ind; Ireland Ind Sci; Pelletier.

HITCHCOCK, EDWARD (1828-1911). Physical education; Anthropometry. OCCUP: Professor; Academic administrator. MSS: Ind NUCMC (75-15). WORKS ABOUT: Am Men Sci 1-2 d3; Dict Am Biog; Dict Am Med Biog; Natl Cycl Am Biog 13; Who Was Who Am 1. See also: Barr; Biog Geneal Master Ind.

HITCHCOCK, ETHAN ALLEN (1798-1870). Alchemy. OCCUP: Army officer. MSS: Ind NUCMC (64-1569, 67-573). WORKS ABOUT: Dict Am Biog; Natl Cycl Am Biog 11; Who Was Who Am h. See also: Biog Geneal Master Ind; Pelletier.

HOBBS, PERRY LYNES (1861-1912). Chemistry. OCCUP: Professor; Consultant. WORKS ABOUT: Natl Cycl Am Biog 16; Sci 35(1912):616; Who Was Who Am 4.

HOCH, AUGUST (1868-1919). Psychiatry. OCCUP: Physician; Professor; Administrator. WORKS ABOUT: Am Men Sci 2 d3; Dict Am Biog; Natl Cycl Am Biog 23; Who Was Who Am 1. See also: Barr; Biog Geneal Master Ind.

HOCKLEY, THOMAS (-1892). Archeology. OCCUP: Not determined. MSS: Ind NUCMC (?). WORKS ABOUT: Sci 1st ser 19(1892):158.

HODENPYL, EUGENE (1863-1910). Pathology. OCCUP: Physician; Professor. WORKS ABOUT: Am Men Sci 1 d3; Sci 31(1910):736-37.

HODGE, JAMES THATCHER (1816-1871). Geology. OCCUP: State government scientist; Mining engineer. WORKS ABOUT: Appleton's; Natl Cycl Am Biog 4.

HODGEN, JOHN THOMPSON (1826-1882). Surgery. OCCUP: Surgeon; Professor; Acadmic administrator. WORKS ABOUT: Dict Am Biog; Natl Cycl Am Biog 8; Who Was Who Am h. See also: Barr.

HODGSON, RICHARD (1855-1905). Psychical research. OCCUP: Administrator; Researcher. WORKS ABOUT: Biog Dict Parapsychology; Ency Occultism; Sci 22(1905):886; Who Was Who Am 1.

HOEN, AUGUST (1817-1886). Map making. OCCUP: Printer; Map-maker. WORKS ABOUT: Dict Am Biog; Natl Cycl Am Biog 24; Who Was Who Am h.

HOEVE, HEIKOBUS JOHANNES HUBERTUS (1882-1918). Anatomy. OCCUP: Professor (anatomy and surgery). WORKS ABOUT: Am Men Sci 2 d3.

HOFFMAN, WALTER JAMES (1846-1899). Ethnology. OCCUP: Physician; Army surgeon; U.S. government scientist. MSS: Ind NUCMC (?). WORKS ABOUT: Sci 10(1899):782; Smithsonian Bur Eth Rep (1899-1900):xxxviii-xxxix; Twentieth Cent Biog Dict.

HOFFMANN, FRIEDRICH (1832-1904). Pharmacy. OCCUP: State government scientist; Editor. WORKS ABOUT: Allibone Supp; Sci 20(1904):933; Wallace.

HOLBROOK, JOHN EDWARDS (1794-1871). Zoology; Herpetology; Ichthyology. OCCUP: Physician; Professor. MSS: Ind NUCMC. WORKS ABOUT: Dict Am Biog; Dict Sci Biog; Elliott; Natl Ac Scis Biog Mem 5; Natl Cycl Am Biog 13; Who Was Who Am h. See also: Barr; Biog Geneal Master Ind.

HOLBROOK, MARTIN LUTHER (1831-1902). Zoology; Microscopy. OCCUP: Editor; Physician. WORKS ABOUT: Adams O; Allibone Supp; Appleton's Supp 1901; Natl Cycl Am Biog 12; Ohio Authors; Wallace; Who Was Who Am 1.

HOLCOMB, AMASA (1787-1875). Instrumentation; Manufacture of telescopes. OCCUP: Instrument maker; Public official. MSS: Ind NUCMC (80-300). WORKS ABOUT: Dict Am Biog; Natl Cycl Am Biog 3; Who Was Who Am h. See also: Barr; Biog Geneal Master Ind.

HOLDEN, EDWARD SINGLETON (1846-1914). Astronomy. OCCUP: College president; Observatory director; Librarian. MSS: Ind NUCMC. WORKS ABOUT: Am Men Sci 1-2 d3; Dict Am Biog; Dict Sci Biog; Elliott (s); Natl Ac Scis Biog Mem 8; Natl Cycl Am Biog 7:229; Who Was Who Am 1. See also: Barr; Biog Geneal Master Ind.

HOLDER, CHARLES FREDERICK (1851-1915). Natural history. OCCUP: Museum associate; Author. WORKS ABOUT: Am Men Sci 1; Dict Am Biog; Natl Cycl Am Biog 7:402; Who Was Who Am 1. See also: Barr; Biog Geneal Master Ind.

HOLDER, JOSEPH BASSETT (1824-1888). Natural history; Zoology. OCCUP: Physician; Army surgeon; Museum curator. WORKS ABOUT: Dict Am Biog; Elliott; Natl Cycl Am Biog 7:402; Who Was Who Am h. See also: Biog Geneal Master Ind.

HOLLADAY, WALLER (1840-   ). Mathematics. OCCUP: Professor; Educator; Industrial scientist. WORKS ABOUT: Twentieth Cent Biog Dict; Who Was Who Am 4.

HOLLAND, JOHN PHILIP (1840-1914). Invention. OCCUP: Teacher; Inventor; Businessman. WORKS ABOUT: Dict Am Biog; Natl Cycl Am Biog 15; Who Was Who Am h (add), 4. See also: Biog Geneal Master Ind; Ireland Ind Sci; Pelletier.

HOLLEY, ALEXANDER LYMAN (1832-1882). Metallurgy. OCCUP: Editor; Industrial engineer. MSS: Ind NUCMC (64-197, 80-238). WORKS ABOUT: Dict Am Biog; Natl Cycl Am Biog 11; Who Was Who Am h. See also: Biog Geneal Master Ind; Ireland Ind Sci.

HOLMAN, DAVID SHEPARD (-1901). Microscopy. OCCUP: Administrator; Industrial scientist. WORKS ABOUT: Sci 13(1901):838.

HOLMAN, SILAS WHITCOMB (1856-1900). Physics. OCCUP: Professor. MSS: Ind NUCMC (?). WORKS ABOUT: Sci 11(1900):556, 13(1901):857-59; Wallace.

HOLME, THOMAS (1624-1695). Map making; Surveying. OCCUP: Surveyor; Public official. MSS: Ind NUCMC. WORKS ABOUT: Dict Am Biog; Natl Cycl Am Biog 24; Who Was Who Am h. See also: Biog Geneal Master Ind.

HOLMES, EZEKIEL (1801-1865). Agriculture. OCCUP: Educator; Editor; Public official. MSS: Ind NUCMC (65-1971). WORKS ABOUT: Dict Am Biog; Natl Cycl Am Biog 24; Who Was Who Am h.

HOLMES, FRANCIS SIMMONS (1815-1882). Natural history; Geology. OCCUP: Planter; Professor; Manufacturer. MSS: Ind NUCMC (?). WORKS ABOUT: Elliott. See also: Biog Geneal Master Ind.

HOLMES, JOSEPH AUSTIN (1859-1915). Geology; Mining engineering. OCCUP: Professor; U.S. and state government scientist. MSS: Ind NUCMC (70-355). WORKS ABOUT: Am Men Sci 1-2 d3; Dict Am Biog; Natl Cycl Am Biog 23; Who Was Who Am 1. See also: Barr.

HOLMES, MARY EMILEE (1849-1906). Geology. OCCUP: Professor; Administrator. WORKS ABOUT [birth dates may differ]: Am Men Sci 1-2; Am Wom; Siegel and Finley; Who Was Who Am 1.

HOLT, JACOB FARNUM (-1908). Anatomy. OCCUP: Teacher. WORKS ABOUT: Sci 28(1908):338.

HOLTON, ISAAC FARWELL (1812-1874). Botany. OCCUP: Professor; Clergyman; Journalist. WORKS ABOUT: Am Journ Sci 107(1874):240; Wallace.

HOLTON, NINA (-1908). Botany. OCCUP: U.S. government scientist. WORKS ABOUT: Sci 27(1908):798.

HOLYOKE, EDWARD AUGUSTUS (1728-1829). Meteorology. OCCUP: Physician. MSS: Ind NUCMC. WORKS ABOUT: Dict Am Biog; Dict Am Med Biog; Natl Cycl Am Biog 7:488; Who Was Who Am h. See also: Biog Geneal Master Ind.

HOOKER, CHARLES W. (1883?-1913). Entomology. OCCUP: U.S. government scientist. WORKS ABOUT: Sci 37(1913):368.

HOOKER, JOHN DAGGETT (1838-1911). Astronomy. OCCUP: Manufacturer; Benefactor. WORKS ABOUT: Natl Cycl Am Biog 22; Sci 34(1911):74; Who Was Who Am 1.

HOOKER, WORTHINGTON (1806-1867). Science writing. OCCUP: Physician; Professor (medicine). WORKS ABOUT: Dict Am Biog; Dict Am Med Biog; Natl Cycl Am Biog 13; Who Was Who Am h. See also: Biog Geneal Master Ind; Pelletier.

HOOPER, FRANKLIN WILLIAM (1851-1914). Geology; Zoology. OCCUP: Professor; Administrator. WORKS ABOUT: Am Journ Sci 188(1914):370; Am Men Sci 1-2 d3; Natl Cycl Am Biog 13; Sci 40(1914):205; Twentieth Cent Biog Dict; Who Was Who Am 1.

HOOPER, WILLIAM LESLIE (1855-1918). Electrical engineering. OCCUP: Professor. WORKS ABOUT: Am Men Sci 1-2 d3; Sci 48(1918):442-43; Wallace; Who Was Who Am 1.

HOOPES, JOSHUA (1788?-1874). Botany. OCCUP: Not determined. WORKS ABOUT: Am Journ Sci 107(1874):600-01.

HOOPES, JOSIAH (1832-1904). Botany; Ornithology. OCCUP: Businessman; Horticulturist. MSS: Ind NUCMC (?). WORKS ABOUT: Adams O; Allibone Supp; Am Men Sci 1 d3; Appleton's; Twentieth Cent Biog Dict; Wallace; Who Was Who Am 1.

HOPKINS, ALBERT (1807-1872). Astronomy. OCCUP: Professor. WORKS ABOUT: Appleton's; Natl Cycl Am Biog 6; Twentieth Cent Biog Dict.

HOPKINS, CYRIL GEORGE (1866-1919). Agronomy; Agricultural chemistry. OCCUP: Professor. MSS: Ind NUCMC (?). WORKS ABOUT: Am Men Sci 1-2 d3; Dict Am Biog; Who Was Who Am 1. See also: Barr; Biog Geneal Master Ind; Ireland Ind Sci; Pelletier.

HOPKINS, GEORGE MILTON (1842-1902). Science writing. OCCUP: Author; Editor. WORKS ABOUT: Sci 16(1902):357; Wallace.

HORN, GEORGE HENRY (1840-1897). Entomology; Coleopterology. OCCUP: Physician. MSS: Ind NUCMC. WORKS ABOUT: Dict Am Biog; Dict Sci Biog; Elliott; Natl Cycl Am Biog 7:502; Who Was Who Am h. See also: Barr; Ireland Ind Sci; Pelletier.

HORNER, WILLIAM EDMONDS (1793-1853). Anatomy. OCCUP: Physician; Professor; Academic administrator (medicine). MSS: Ind NUCMC (?). WORKS ABOUT: Dict Am Biog; Dict Am Med Biog; Elliott; Natl Cycl Am Biog 6; Who Was Who Am h. See also: Biog Geneal Master Ind; Pelletier.

HORNUNG, CHRISTIAN (1845-1918). Mathematics. OCCUP: Professor. WORKS ABOUT: Am Men Sci 2 d3; Sci 47(1918):236; Who Was Who Am 4.

HORSFIELD, THOMAS (1773-1859). Natural history; Botany. OCCUP: Physician; Natural history explorer/ collector; Museum curator. WORKS ABOUT: Dict Am Biog; Elliott; Who Was Who Am h. See also: Barr; Biog Geneal Master Ind.

HORSFORD, EBEN NORTON (1818-1893). Chemistry. OCCUP: Professor; Industrial scientist; Manufacturer. MSS: Ind NUCMC (70-1513, 76-1591). WORKS ABOUT: Dict Am Biog; Dict Sci Biog; Elliott; Natl Cycl Am Biog 6; Who Was Who Am h. See also: Barr; Biog Geneal Master Ind; Ireland Ind Sci; Pelletier.

HORTON, WILLIAM (-1845). Mineralogy. OCCUP: Physician (?); State government scientist. WORKS ABOUT: Am Journ Sci 51(1846):152.

HORWITZ, ORVILLE (1858?-1913). Surgery. OCCUP: Physician; Academic affiliate. WORKS ABOUT: Sci 37(1913):250.

HOSACK, DAVID (1769-1835). Botany; Medicine. OCCUP: Physician; Professor. MSS: Ind NUCMC (60-2672, 61-773). WORKS ABOUT: Dict Am Biog; Dict Am Med Biog; Dict Sci Biog; Natl Cycl Am Biog 3, 9; Who Was Who Am h. See also: Barr; Biog Geneal Master Ind; Ireland Ind Sci; Pelletier.

HOSKINS, THOMAS H. (1828?-1902). Agriculture; Anatomy. OCCUP: Physician; Farmer; Author. WORKS ABOUT: Sci 16(1902):158.

HOTCHKISS, JEDEDIAH (1828-1899). Geology. OCCUP: Engineer; Mining expert; Educator. MSS: Ind NUCMC (74-1136, 80-2056). WORKS ABOUT [birth dates may differ]: Allibone Supp; Knight; Sci 9(1899):158; Twentieth Cent Biog Dict; Wakelyn; Wallace.

HOUGH, FRANKLIN BENJAMIN (1822-1885). Botany; Forestry. OCCUP: Physician; Public official; U.S. government scientist. MSS: Ind NUCMC (68-1246). WORKS ABOUT: Dict Am Biog; Elliott; Natl Cycl Am Biog 13; Who Was Who Am h. See also: Biog Geneal Master Ind.

HOUGH, GEORGE WASHINGTON (1836-1909). Astronomy; Meteorology. OCCUP: Professor; Observatory director. MSS: Ind NUCMC. WORKS ABOUT: Am Men Sci 1 d3; Dict Am Biog; Dict Sci Biog; Elliott (s); Natl Cycl Am Biog 8; Who Was Who Am 1. See also: Barr; Biog Geneal Master Ind.

HOUGH, WILLISTON SAMUEL (1860-1912). Philosophy; Psychology. OCCUP: Professor; Academic administrator. WORKS ABOUT: Am Men Sci 2 d3; Sci 36(1912):399; Who Was Who Am 1.

HOUGHTON, DOUGLASS (1809-1845). Geology; Medicine. OCCUP: Physician; State government scientist; Professor. MSS: Ind NUCMC (65-327). WORKS ABOUT: Dict Am Biog; Dict Sci Biog; Elliott; Natl Cycl Am Biog 5; Who Was Who Am h. See also: Barr; Biog Geneal Master Ind; Ireland Ind Sci; Pelletier.

HOUSE, ROYAL EARL (1814-1895). Invention; Telegraphy. OCCUP: Inventor. WORKS ABOUT: Dict Am Biog; Natl Cycl Am Biog 12; Who Was Who Am h. See also: Biog Geneal Master Ind; Pelletier.

HOUSTON, EDWIN JAMES (1847-1914). Electrical engineering. OCCUP: Teacher; Engineer. MSS: Ind NUCMC. WORKS ABOUT: Am Men Sci 1-2 d3; Dict Am Biog; Natl Cycl Am Biog 13; Who Was Who Am 1. See also: Barr; Biog Geneal Master Ind.

HOVEY, HORACE CARTER (1833-1914). Geology. OCCUP: Clergyman. MSS: Ind NUCMC. WORKS ABOUT: Adams O; Allibone Supp; Am Journ Sci 188(1914):370; Am Men Sci 1-2 d3; Indiana Authors 1816; Natl Cycl Am Biog 12; Sci 40(1914):205; Twentieth Cent Biog Dict; Wallace; Who Was Who Am 1.

HOWARD, WILLIAM (1793-1834). Anatomy; Engineering; Physical science. OCCUP: Physician; Professor; U.S. government engineer. Scientific work: see Roy Soc Cat. WORKS ABOUT: Kelly and Burrage.

HOWE, ELLIOT C. (1828-1899). Botany. OCCUP: Physician; Teacher. WORKS ABOUT: Torrey Bot Club Bull 26(1899):251-53.

HOWELL, EDWIN EUGENE (1845-1911). Geology. OCCUP: U.S. government scientist; Artist. MSS: Ind NUCMC (76-1350). WORKS ABOUT: Am Journ Sci 181(1911):468; Am Men Sci 1-2 d3; Sci 33(1911):720-21; Who Was Who Am 1.

HOWELL, THOMAS JEFFERSON (1842-1912). Botany. OCCUP: Farmer; Engaged in independent study. MSS: Ind NUCMC. WORKS ABOUT: Am Men Sci 1-2 d3; Dict Am Biog. See also: Barr; Biog Geneal Master Ind.

HOWISON, GEORGE HOLMES (1834-1916). Philosophy. OCCUP: Professor. MSS: Ind NUCMC (71-703, 75-305. WORKS ABOUT: Am Men Sci 1-2 d3; Dict Am Biog; Natl Cycl Am Biog 23; Who Was Who Am 1. See also: Barr; Biog Geneal Master Ind.

HOY, PHILO ROMAYNE (1816-1892). Natural history; Entomology. OCCUP: Physician. MSS: Ind NUCMC (?). WORKS ABOUT: Appleton's Supp 1901; Auk 10(1893):95-96; Insect Life 5(1893):286; Kelly and Burrage; Natl Cycl Am Biog 15; Osborn Herb p180.

HUBBARD, G. C. (-1902). Chemistry. OCCUP: Academic affiliate. WORKS ABOUT: Sci 15(1902):917.

HUBBARD, GARDINER GREENE (1822-1897). Geography. OCCUP: Lawyer; Businessman; Administrator. MSS: Ind NUCMC. WORKS ABOUT: Dict Am Biog; Natl Cycl Am Biog 5; Who Was Who Am h. See also: Barr; Biog Geneal Master Ind.

HUBBARD, HENRY GUERNSEY (1850-1899). Entomology. OCCUP: Engaged in independent study; U.S. government scientist. MSS: Ind NUCMC (74-977). WORKS ABOUT: Dict Am Biog; Elliott; Who Was Who Am h. See also: Ireland Ind Sci; Pelletier.

HUBBARD, JOSEPH STILLMAN (1823-1863). Astronomy. OCCUP: U.S. government scientist; Observatory affiliate. WORKS ABOUT: Dict Am Biog; Elliott; Natl Ac Scis Biog Mem 1; Natl Cycl Am Biog 9; Who Was Who Am h. See also: Barr; Biog Geneal Master Ind.

HUBBARD, OLIVER PAYSON (1809-1900). Geology; Mineralogy. OCCUP: Professor; Educator. MSS: Ind NUCMC. WORKS ABOUT: Elliott; Natl Cycl Am Biog 9; Who Was Who Am 1. See also: Barr; Biog Geneal Master Ind; Ireland Ind Sci.

HUDDE, ANDRIES (1608-1663). Surveying. OCCUP: Surveyor; Public official. WORKS ABOUT: Dict Am Biog; Who Was Who Am h. See also: Biog Geneal Master Ind.

HUDDLESTON, JOHN HENRY (1864-1915). Medicine. OCCUP: Physician; City government scientist. WORKS ABOUT: Am Men Sci 2 d3; Who Was Who Am 1.

HUDSON, THOMSON JAY (1834-1903). Psychology. OCCUP: Editor; U.S. government scientist; Author. WORKS ABOUT: Dict Am Biog; Natl Cycl Am Biog 14; Who Was Who Am 1. See also: Barr; Biog Geneal Master Ind.

HUEY, EDMUND BURKE (1870-1913). Psychology. OCCUP: Professor; Psychologist. MSS: Ind NUCMC (?). WORKS ABOUT: Am Men Sci 1-2 d3; Sci 39(1914):136-37; Wallace.

HUFF, HELEN SCHAEFFER (1883-1913). Mathematics; Physics. OCCUP: Academic affiliate. WORKS ABOUT: Sci 37(1913):172; Wom Whos Who Am.

HULBERT, EDWIN JAMES (1829-1910). Mining engineering; Surveying. OCCUP: Mining engineer; Businessman. WORKS ABOUT: Dict Am Biog. See also: Biog Geneal Master Ind.

HULST, GEORGE DURYEA (1846-1900). Entomology. OCCUP: Clergyman. WORKS ABOUT: Canadian Ent 32(1900):369; Ent News 11(1900):613-15; Mallis; New York Ent Soc Journ 8(1900):248-54; Pop Sci Mo 76(1910):472.

HUME, WILLIAM (1801-1870). General science; Medicine. OCCUP: Physician; Professor; Planter. Scientific work: see Roy Soc Cat. WORKS ABOUT: Thomas p239-41.

HUMPHREY, JAMES ELLIS (1861-1897). Biology; Botany. OCCUP: State government scientist; Professor. WORKS ABOUT: Am Nat 31(1897):920-22; Johns Hopkins Circ [17](1897-1898):17-19; Nature 57(1897-1898):60; Sci 6(1897):363, 7(1898):277; Wood's Hole Biol Lectures (1898):341-43.

HUMPHREYS, ANDREW ATKINSON (1810-1883). Hydraulic engineering; Promotion of science. OCCUP: Army officer; Engineer. MSS: Ind NUCMC (60-2611). WORKS ABOUT: Dict Am Biog; Elliott; Natl Ac Scis Biog Mem 2; Natl Cycl Am Biog 7:34; Who Was Who Am h. See also: Barr; Biog Geneal Master Ind.

HUNICKE, HENRY AUGUST (1861-1909). Chemistry. OCCUP: Professor; Industrial scientist; Consultant. WORKS ABOUT: Sci 29(1909):696.

HUNT, ALFRED EPHRAIM (1855-1899). Metallurgy. OCCUP: Industrial scientist and manager; Consultant; Businessman. MSS: Ind NUCMC. WORKS ABOUT: Dict Am Biog; Natl Cycl Am Biog 25; Who Was Who Am h. See also: Biog Geneal Master Ind.

HUNT, EDWARD BISSELL (1822-1863). Physics. OCCUP: Army officer; Engineer. WORKS ABOUT: Elliott; Natl Ac Scis Biog Mem 3; Natl Cycl Am Biog 11. See also: Barr; Biog Geneal Master Ind.

HUNT, JAMES GIBBONS (ca.1826-1893). Botany; Microscopy. OCCUP: Physician. WORKS ABOUT: Am Micr Soc Proc 14(1892):166-67.

HUNT, MARY HANNAH HANCHETT (1830-1906). Chemistry. OCCUP: Teacher; Activist (temperance). MSS: Ind NUCMC (71-1235). WORKS ABOUT: Dict Am Biog; Natl Cycl Am Biog 9; Notable Am Wom; Who Was Who Am 1. See also: Biog Geneal Master Ind; Pelletier.

HUNT, THOMAS STERRY (1826-1892). Chemistry; Geology. OCCUP: Government scientist (foreign); Professor; Consultant. MSS: Ind NUCMC. WORKS ABOUT: Dict Am Biog; Dict Sci Biog; Elliott; Natl Ac Scis Biog Mem 15; Natl Cycl Am Biog 3; Who Was Who Am h. See also: Barr; Biog Geneal Master Ind; Ireland Ind Sci; Pelletier.

HUSMANN, GEORGE (1827-1902). Viticulture. OCCUP: Horticulturist; Author; Professor. WORKS ABOUT: Dict Am Biog. See also: Barr; Biog Geneal Master Ind.

HUSTED, ALBERT N. (1833-1912). Mathematics. OCCUP: Professor. WORKS ABOUT: Sci 36(1912):627-28.

HUTCHINS, THOMAS (1730-1789). Geography; Military engineering. OCCUP: Army officer; Engineer; U.S. government scientist. MSS: Ind NUCMC (60-2226). WORKS ABOUT: Dict Am Biog; Elliott; Natl Cycl Am Biog 9; Who Was Who Am h. See also: Biog Geneal Master Ind; Ireland Ind Sci.

HUTTON, FREDERICK REMSEN (1853-1918). Mechanical engineering. OCCUP: Professor; Academic administrator; Consultant. WORKS ABOUT: Am Men Sci 2 d3; Dict Am Biog; Natl Cycl Am Biog 16; Who Was Who Am 1. See also: Barr; Biog Geneal Master Ind.

HYATT, ALPHEUS (1838-1902). Zoology; Invertebrate paleontology. OCCUP: Curator; Professor; Educator. MSS: Ind NUCMC (61-1277, 67-1523). WORKS ABOUT: Dict Am Biog; Dict Sci Biog; Elliott; Natl Ac Scis Biog Mem 6; Natl Cycl Am Biog 3, 23; Who Was Who Am 1. See also: Barr; Biog Geneal Master Ind; Ireland Ind Sci; Pelletier.

HYATT, JOHN WESLEY (1837-1920). Invention. OCCUP: Inventor; Manufacturer. WORKS ABOUT: Dict Am Biog; Natl Cycl Am Biog 12; Who Was Who Am 1. See also: Barr; Biog Geneal Master Ind; Ireland Ind Sci; Pelletier.

HYATT, JONATHAN DEUEL (1825-1912). Microscopy. OCCUP: Educator. WORKS ABOUT: Natl Cycl Am Biog 13; Sci 37(1913):172.

HYDE, JAMES NEVINS (1840-1910). Dermatology. OCCUP: Physician; Professor. WORKS ABOUT: Am Men Sci 1-2 d3; Dict Am Biog; Who Was Who Am 1. See also: Barr; Biog Geneal Master Ind.

HYSLOP, JAMES HERVEY (1854-1920). Philosophy; Psychology; Psychical research. OCCUP: Professor; Administrator. MSS: Ind NUCMC (77-100). WORKS ABOUT: Am Men Sci 1-2 d3; Dict Am Biog; Natl Cycl Am Biog 10, 14, 26; Who Was Who Am 1. See also: Biog Geneal Master Ind.

I

IDDINGS, JOSEPH PAXSON (1857-1920). Geology; Petrology. OCCUP: Professor; U.S. government scientist. WORKS ABOUT: Am Men Sci 1-2 d3; Dict Am Biog; Dict Sci Biog; Elliott (s); Natl Cycl Am Biog 15; Who Was Who Am 1. See also: Barr; Biog Geneal Master Ind.

INGALS, EPHRAIM FLETCHER (1848-1918). Medicine. OCCUP: Physician; Professor. MSS: Ind NUCMC. WORKS ABOUT: Am Men Sci 1-2 d3; Dict Am Biog; Who Was Who Am 1. See also: Barr; Biog Geneal Master Ind.

IORNS, MARTIN JOSHUA (-1909). Horticulture. OCCUP: Horticulturist; U.S. government scientist. WORKS ABOUT: Sci 29(1909):893.

IRONS, DAVID (1870-1907). Philosophy; Psychology. OCCUP: Professor (philosophy). WORKS ABOUT: Am Men Sci 1 d3; Sci 25(1907):198.

IRVING, JOHN DUER (1874-1918). Geology; Mining geology. OCCUP: U.S. government scientist; Professor. WORKS ABOUT: Am Men Sci 1-2 d3; Dict Am Biog; Natl Cycl Am Biog 23; Who Was Who Am 1. See also: Barr.

IRVING, ROLAND DUER (1847-1888). Geology; Mining engineering. OCCUP: Professor; U.S. and state government scientist. MSS: Ind NUCMC (?). WORKS ABOUT: Dict Am Biog; Elliott; Who Was Who Am h. See also: Barr; Biog Geneal Master Ind.

ISAACS, CHARLES EDWARD (1811-1860). Medicine. OCCUP: Physician; Professor. WORKS ABOUT: Dict Sci Biog.

ISHERWOOD, BENJAMIN FRANKLIN (1822-1915). Mechanical engineering. OCCUP: Naval officer; Engineer. MSS: Ind NUCMC. WORKS ABOUT: Dict Am Biog; Natl Cycl Am Biog 12; Who Was Who Am 1. See also: Biog Geneal Master Ind.

IVES, ELI (1778-1861). Botany. OCCUP: Physician; Professor. MSS: <u>Ind NUCMC</u>. WORKS ABOUT: <u>Dict Am Biog</u>; <u>Natl Cycl Am Biog</u> 12; <u>Who Was Who Am</u> h. See also: Barr; <u>Biog Geneal Master Ind</u>.

J

JACKMAN, ALONZO (1809-1879). Mathematics. OCCUP: Professor. WORKS ABOUT: <u>Natl Cycl Am Biog</u> 16; Wallace.

JACKMAN, WILBUR SAMUEL (1855-1907). Nature study; Science education. OCCUP: Educator; Professor. WORKS ABOUT: <u>Am Men Sci</u> 1 d3; <u>Dict Am Biog</u>; <u>Natl Cycl Am Biog</u> 27; <u>Who Was Who Am</u> 1. See also: Barr; <u>Biog Geneal Master Ind</u>.

JACKSON, CHARLES THOMAS (1805-1880). Chemistry; Geology; Mineralogy; Medicine. OCCUP: Physician; U.S. and state government scientist; Consultant. MSS: <u>Ind NUCMC</u> (66-1424, 71-300). WORKS ABOUT: <u>Dict Am Biog</u>; <u>Dict Am Med Biog</u>; <u>Dict Sci Biog</u>; Elliott; <u>Natl Cycl Am Biog</u> 3; <u>Who Was Who Am</u> h. See also: Barr; <u>Biog Geneal Master Ind</u>; Ireland <u>Ind Sci</u>; Pelletier.

JACKSON, GEORGE THOMAS (1852-1916). Medicine; Dermatology. OCCUP: Physician; Professor. WORKS ABOUT: <u>Dict Am Biog</u>; <u>Natl Cycl Am Biog</u> 11; <u>Who Was Who Am</u> 1. See also: Barr; <u>Biog Geneal Master Ind</u>.

JACKSON, JOHN BARNARD SWETT (1806-1879). Pathological anatomy. OCCUP: Physician; Professor; Museum curator. MSS: <u>Ind NUCMC</u> (62-3639). WORKS ABOUT: <u>Dict Am Med Biog</u>; Elliott. See also: <u>Biog Geneal Master Ind</u>.

JACKSON, JOHN HENRY (1838-1908). Medicine. OCCUP: Physician; Professor. WORKS ABOUT: <u>Am Men Sci</u> 1 d3; <u>Who Was Who Am</u> 1.

JACKSON, SAMUEL (1787-1872). Medicine; Pharmacy; Physiology. OCCUP: Physician; Professor. MSS: <u>Ind NUCMC</u>. WORKS ABOUT: <u>Dict Am Biog</u>; <u>Dict Am Med Biog</u>; <u>Natl Cycl Am Biog</u> 11; <u>Who Was Who Am</u> h. See also: <u>Biog Geneal Master Ind</u>; Ireland <u>Ind Sci</u>.

JACOBI, ABRAHAM (1830-1919). Medicine. OCCUP: Physician; Professor. MSS: Ind NUCMC. WORKS ABOUT: Am Men Sci 1-2 d3; Dict Am Biog; Dict Am Med Biog; Natl Cycl Am Biog 9; Who Was Who Am 1. See also: Barr; Biog Geneal Master Ind; Ireland Ind Sci; Pelletier.

JACOBI, MARY CORINNA PUTNAM (1842-1906). Medicine. OCCUP: Physician; Professor. MSS: Ind NUCMC (62-2487). WORKS ABOUT: Dict Am Biog; Dict Am Med Biog; Natl Cycl Am Biog 8; Notable Am Wom; Who Was Who Am 1. See also: Barr; Biog Geneal Master Ind; Ireland Ind Sci; Pelletier.

JACOBS, MICHAEL (1808-1871). Meteorology. OCCUP: Professor. MSS: Ind NUCMC (61-2514). WORKS ABOUT: Dict Am Biog; Elliott; Natl Cycl Am Biog 11; Who Was Who Am h. See also: Biog Geneal Master Ind.

JACOBS, WILLIAM STEPHEN (1772-1843). Chemistry. OCCUP: Physician. WORKS ABOUT: Am Chem and Chem Eng.

JAMES, BUSHROD WASHINGTON (1836-1903). Ophthalmology. OCCUP: Physician. WORKS ABOUT: Adams O; Natl Cycl Am Biog 3; Sci 17(1903):159; Twentieth Cent Biog Dict; Wallace; Who Was Who Am 1.

JAMES, EDWIN (1797-1861). Natural history; Botany. OCCUP: Army surgeon; Farmer. MSS: Ind NUCMC. WORKS ABOUT: Dict Am Biog; Elliott; Who Was Who Am h. See also: Barr; Biog Geneal Master Ind; Ireland Ind Sci.

JAMES, JOSEPH FRANCIS (1857-1897). Botany; Geology. OCCUP: Curator; Professor; U.S. government scientist. WORKS ABOUT: Am Geol 21(1898):1-11; Am Journ Sci 153(1897):428; Appleton's; Geol Soc Am Bull 9(1898):408-12; Sci 5(1897):580; Twentieth Cent Biog Dict.

JAMES, THOMAS POTTS (1803-1882). Botany. OCCUP: Pharmacist; Businessman; Researcher. MSS: Ind NUCMC (60-2802). WORKS ABOUT: Dict Am Biog; Elliott; Who Was Who Am h. See also: Barr; Biog Geneal Master Ind.

JAMES, URIAH PIERSON (1811-1889). Paleontology. OCCUP: Businessman; Publisher. MSS: Ind NUCMC (75-1048). WORKS ABOUT: Am Geol 3(1889):281-87; Am Journ Sci 137(1889):322; Cincinnati Soc Nat Hist Journ 12([1890]):5-7.

JAMES, WILLIAM (1842-1910). Philosophy; Psychology. OCCUP: Professor. MSS: Ind NUCMC (83-944). WORKS ABOUT: Am Men Sci 1-2 d3; Dict Am Biog; Dict Sci Biog; Elliott (s); Natl Cycl Am Biog 18; Who Was Who Am 1. See also: Barr; Biog Geneal Master Ind; Ireland Ind Sci; Pelletier.

JANEWAY, EDWARD GAMALIEL (1841-1911). Pathology. OCCUP: Physician; Professor; Academic administrator. MSS: Ind NUCMC (?). WORKS ABOUT: Am Men Sci 1-2 d3; Dict Am Biog; Dict Am Med Biog; Natl Cycl Am Biog 13; Who Was Who Am 1. See also: Barr; Biog Geneal Master Ind.

JANEWAY, THEODORE CALDWELL (1872-1917). Medicine. OCCUP:
Physician; Professor. WORKS ABOUT: Am Men Sci 2 d3; Dict Am
Biog; Natl Cycl Am Biog 17; Who Was Who Am 1. See also:
Barr; Biog Geneal Master Ind; Ireland Ind Sci.

JANSEN, A. M. (-1914). Medicine. OCCUP: Academic
instructor. WORKS ABOUT: Sci 39(1914):137.

JAQUES, WILLIAM HENRY (1848-1916). Engineering. OCCUP:
Engineer; Naval officer; Industrial engineer. WORKS ABOUT:
Adams O; Allibone Supp; Am Men Sci 1-2 d3; Wallace; Who Was
Who Am 1.

JARVES, DEMING (1790-1869). Chemistry. OCCUP: Manufacturer;
Inventor. WORKS ABOUT: Dict Am Biog; Natl Cycl Am Biog 27.
See also: Biog Geneal Master Ind.

JARVIS, EDWARD (1803-1884). Statistics. OCCUP: Physician.
MSS: Ind NUCMC. WORKS ABOUT: Dict Am Biog; Dict Am Med
Biog; Natl Cycl Am Biog 12; Who Was Who Am h. See also:
Biog Geneal Master Ind.

JARVIS, WILLIAM CHAPMAN (1855-1895). Laryngology. OCCUP:
Physician; Professor. WORKS ABOUT: Dict Am Biog; Dict Am
Med Biog; Who Was Who Am h. See also: Barr.

JAYNE, HARRY WALKER (1857-1910). Chemistry. OCCUP:
Industrialist. WORKS ABOUT: Natl Cycl Am Biog 19; Sci
31(1910):576.

JAYNE, HORACE FORT (1859-1913). Zoology; Anatomy. OCCUP:
Professor; Academic administrator. MSS: Ind NUCMC. WORKS
ABOUT: Am Men Sci 2 d3; Dict Am Biog; Elliott (s); Natl
Cycl Am Biog 13; Who Was Who Am 1. See also: Barr; Biog
Geneal Master Ind.

JEFFERS, WILLIAM NICHOLSON (1824-1883). Naval ordnance.
OCCUP: Naval officer. WORKS ABOUT: Dict Am Biog; Natl Cycl
Am Biog 4; Who Was Who Am h. See also: Biog Geneal Master
Ind.

JEFFERSON, THOMAS (1743-1826). Natural history;
Agriculture; Botany; Zoology; Paleontology; Cartography;
Diplomacy; Ethnology; Meteorology; Surveying;
Technology. OCCUP: Lawyer; Public official; President. MSS:
Ind NUCMC (60-262, 60-2752, 61-276, 61-696, 62-523, 66-310,
68-1655, 69-1213, 71-1725, 72-690, 77-1537, 78-2074). WORKS
ABOUT: Dict Am Biog; Dict Sci Biog; Elliott; Natl Cycl Am
Biog 3; Who Was Who Am h. See also: Barr; Biog Geneal
Master Ind; Ireland Ind Sci; Pelletier.

JEFFRIES, BENJAMIN JOY (1833-1915). Medicine;
Ophthalmological surgery. OCCUP: Physician. MSS: Ind
NUCMC. WORKS ABOUT: Am Men Sci 1-2 d3; Dict Am Biog; Natl
Cycl Am Biog 24; Who Was Who Am 1. See also: Biog Geneal
Master Ind.

JEFFRIES, JOHN (1744/45-1819). Aeronautics; Meteorology. OCCUP: Physician. MSS: Ind NUCMC (82-631). WORKS ABOUT: Dict Am Biog; Elliott; Natl Cycl Am Biog 24; Who Was Who Am h. See also: Biog Geneal Master Ind; Ireland Ind Sci; Pelletier.

JEFFRIES, JOHN AMORY (1859-1892). Bacteriology; Ornithology. OCCUP: Physician. WORKS ABOUT: Auk 9(1892):311-12.

JEGI, JOHN I. (1866-1904). Psychology. OCCUP: Professor. WORKS ABOUT: Sci 19(1904):517.

JENISON, WILLIAM (-1853). Entomology. OCCUP: Not determined. WORKS ABOUT: Am Journ Sci 67(1854):444.

JENKS, JOHN WHIPPLE POTTER (1819-1895). Zoology. OCCUP: Educator; Museum curator; Professor. WORKS ABOUT: Elliott; Natl Cycl Am Biog 10; Who Was Who Am h. See also: Biog Geneal Master Ind.

JENNE, ELDRED LLEWELLYN (1885-1912). Entomology. OCCUP: U.S. government scientist. WORKS ABOUT: Am Men Sci 2 d3.

JENNINGS, EDWARD PAYSON (1853-1915). Geology. OCCUP: Mining manager; Consultant. MSS: Ind NUCMC (82-2055). WORKS ABOUT: Am Men Sci 2 d3.

JERVIS, JOHN BLOOMFIELD (1795-1885). Engineering. OCCUP: Engineer. MSS: Ind NUCMC (66-1176, 72-218). WORKS ABOUT: Dict Am Biog; Natl Cycl Am Biog 9; Who Was Who Am h. See also: Barr; Biog Geneal Master Ind.

JESUP, HENRY GRISWOLD (1826-1903). Botany. OCCUP: Clergyman; Professor. MSS: Ind NUCMC. WORKS ABOUT: Allibone Supp; Sci 17(1903):1020, 18(1903):158; Twentieth Cent Biog Dict; Wallace; Who Was Who Am 1.

JESUP, MORRIS KETCHUM (1830-1908). Biology; Philanthropy. OCCUP: Banker. MSS: Ind NUCMC. WORKS ABOUT: Dict Am Biog; Natl Cycl Am Biog 11; Who Was Who Am 1. See also: Barr; Biog Geneal Master Ind.

JEWELL, LEWIS ELLSWORTH (fl. ca.1910). Physics; Astrophysics. OCCUP: Academic instructor; Industrial employee (?). WORKS ABOUT: Am Men Sci 2; Johns Hopkins Half Cent Dir.

JEWETT, EZEKIEL (1791-1877). Conchology. OCCUP: Soldier; Collector. WORKS ABOUT: Am Journ Sci 114(1877):80; Appleton's Supp 1901; Twentieth Cent Biog Dict.

JOHN, JOHN PRICE DURBIN (1843-1916). Astronomy; Mathematics. OCCUP: Professor; College president. MSS: Ind NUCMC (?). WORKS ABOUT: Indiana Authors 1816; Natl Cycl Am Biog 7:384; Sci 44(1916):270; Wallace; Who Was Who Am 1.

JOHNSON, ALEXANDER BRYAN (1860-1917). Surgery. OCCUP: Surgeon; Professor. WORKS ABOUT: Am Men Sci 2 d3; Wallace.

JOHNSON, ALEXANDER SMITH (1817-1878). Microscopy. OCCUP: Lawyer; Judge. MSS: Ind NUCMC (65-1155). Scientific work: Likely author of works in Roy Soc Cat v3. WORKS ABOUT: Dict Am Biog; Natl Cycl Am Biog 5; Who Was Who Am h. See also: Biog Geneal Master Ind.

JOHNSON, ARNOLD BURGES (1834-1915). Physics. OCCUP: Lawyer; U.S. government employee. WORKS ABOUT: Natl Cycl Am Biog 16.

JOHNSON, EDWIN FERRY (1803-1872). Civil engineering. OCCUP: Civil engineer. MSS: Ind NUCMC. WORKS ABOUT: Dict Am Biog; Natl Cycl Am Biog 17; Who Was Who Am h. See also: Biog Geneal Master Ind.

JOHNSON, JOHN BUTLER (1850-1902). Civil engineering. OCCUP: Professor; Academic administrator; Consultant. WORKS ABOUT: Dict Am Biog; Natl Cycl Am Biog 11; Who Was Who Am 1. See also: Barr; Biog Geneal Master Ind.

JOHNSON, JOHN E., JR. (1860?-1919). Metallurgy. OCCUP: Engineer (?); Administrator. WORKS ABOUT: Sci 49(1919):377.

JOHNSON, LAWRENCE CLEMENT (1822-   ). Biology; Geology. OCCUP: Lawyer; U.S. government scientist. MSS: Ind NUCMC (?). WORKS ABOUT: Am Men Sci 1.

JOHNSON, LORENZO N. (-1897). Botany. OCCUP: Academic instructor. WORKS ABOUT: Sci 5(1897):442.

JOHNSON, ORSON BENNETT (1848?-1917). Exploration. OCCUP: Professor; Natural history explorer/ collector. WORKS ABOUT: Sci 45(1917):360.

JOHNSON, OTIS COE (1839-1912). Chemistry. OCCUP: Professor. WORKS ABOUT: Am Men Sci 1-2 d3; Natl Cycl Am Biog 19; Wallace; Who Was Who Am 1.

JOHNSON, SAMUEL WILLIAM (1830-1909). Chemistry; Agricultural chemistry. OCCUP: Professor; State government scientist. WORKS ABOUT: Am Men Sci 1 d3; Dict Am Biog; Elliott (s); Natl Ac Scis Biog Mem 7; Natl Cycl Am Biog 6; Who Was Who Am 1. See also: Barr; Biog Geneal Master Ind; Ireland Ind Sci; Pelletier.

JOHNSON, THOMAS HUMRICKHOUSE (1841-1914). Engineering. OCCUP: Civil engineer. WORKS ABOUT: Am Men Sci 1-2 d3; Who Was Who Am 1.

JOHNSON, WALTER ROGERS (1794-1852). Chemistry; Geology; Physics. OCCUP: Educator; Professor; Consultant. MSS: Ind NUCMC. WORKS ABOUT: Elliott; Natl Cycl Am Biog 12. See also: Biog Geneal Master Ind; Pelletier.

JOHNSON, WILLARD DRAKE (1859-1917). Geomorphology. OCCUP: U.S. government scientist. MSS: Ind NUCMC (?). WORKS ABOUT: Dict Sci Biog.

JOHNSON, WILLIS GRANT (1866-1908). Agriculture. OCCUP: State government scientist; Professor; Editor. WORKS ABOUT: Am Men Sci 1 d3; Who Was Who Am 1.

JOHNSTON, CHARLES HUGHES (1877-1917). Psychology. OCCUP: Professor (psychology, education); Academic administrator. WORKS ABOUT: Am Men Sci 2 d3; Sci 46(1917):358; Wallace; Who Was Who Am 1.

JOHNSTON, CHRISTOPHER (1822-1891). Anatomy; Microscopy. OCCUP: Physician; Professor (anatomy and surgery). WORKS ABOUT: Appleton's; Johns Hopkins Circ [11](1891-1892):31; Kelly and Burrage.

JOHNSTON, JOHN (1806-1879). Chemistry; Physics. OCCUP: Professor. WORKS ABOUT: Elliott. See also: Barr; Biog Geneal Master Ind.

JOLLIFFE, WILLIAM (-1902). Engineering. OCCUP: Civil engineer. WORKS ABOUT: Sci 15(1902):838.

JONES, DAVID PHILLIPS (1841-1903). Engineering. OCCUP: Naval officer; Engineer. WORKS ABOUT: Sci 17(1903):238; Twentieth Cent Biog Dict.

JONES, GEORGE (1800-1870). Astronomy (studies of Zodiacal light). OCCUP: Clergyman; Professor (English). MSS: Ind NUCMC. WORKS ABOUT: Dict Am Biog; Who Was Who Am h. See also: Barr; Biog Geneal Master Ind.

JONES, GEORGE WILLIAM (1837-1911). Mathematics. OCCUP: Professor. WORKS ABOUT: Am Men Sci 1-2 d3; Sci 34(1911):635; Wallace; Who Was Who Am 1.

JONES, HARRY CLARY (1865-1916). Chemistry; Physical chemistry. OCCUP: Professor. WORKS ABOUT: Am Men Sci 1-2 d3; Dict Am Biog; Dict Sci Biog; Who Was Who Am 1. See also: Barr; Biog Geneal Master Ind.

JONES, HERBERT LYON (1866?-1898). Botany. OCCUP: Professor. WORKS ABOUT: Sci 8(1898):441.

JONES, HUGH (ca.1692-1760). Mathematics. OCCUP: Professor; Clergyman. MSS: Ind NUCMC (?). WORKS ABOUT: Dict Am Biog; Elliott; Natl Cycl Am Biog 24; Who Was Who Am h. See also: Biog Geneal Master Ind.

JONES, JOSEPH (1833-1896). Chemistry; Medicine. OCCUP: Physician; Professor. MSS: Ind NUCMC. WORKS ABOUT: Dict Am Biog; Dict Am Med Biog; Natl Cycl Am Biog 10; Who Was Who Am h. See also: Biog Geneal Master Ind; Pelletier.

JONES, RICHARD WATSON (1837-1914). Chemistry. OCCUP: Professor; College president. WORKS ABOUT: <u>Am Men Sci</u> 2 d3; Knight; <u>Natl Cycl Am Biog</u> 19; <u>Twentieth Cent Biog Dict</u>; <u>Who Was Who Am</u> 1.

JONES, SAMUEL J. (1836-1901). Otology. OCCUP: Physician; Professor. WORKS ABOUT: <u>Natl Cycl Am Biog</u> 10; <u>Sci</u> 14(1901):621-22; <u>Twentieth Cent Biog Dict</u>; <u>Who Was Who Am</u> 1.

JONES, THOMAS P. (1774-1848). Chemistry. OCCUP: Editor. MSS: <u>Ind NUCMC</u> (82-1820). WORKS ABOUT: <u>Dict Am Biog</u>; <u>Who Was Who Am</u> h. See also: Barr; <u>Biog Geneal Master Ind</u>; Pelletier.

JONES, WILLIAM (1871-1909). Anthropology. OCCUP: Museum affiliate; Explorer. WORKS ABOUT: <u>Dict Am Biog</u>; <u>Natl Cycl Am Biog</u> 24. See also: Barr; <u>Biog Geneal Master Ind</u>.

JONES, WILLIAM JAMES, JR. (1870-1917). Chemistry. OCCUP: State government scientist; Professor. WORKS ABOUT: <u>Am Men Sci</u> 2 d3; <u>Who Was Who Am</u> 1.

JONES, WILLIAM LOUIS (1827-1914). Agriculture; Chemistry; Geology. OCCUP: Professor; Editor. WORKS ABOUT: Elliott; <u>Natl Cycl Am Biog</u> 9. See also: <u>Biog Geneal Master Ind</u>.

JONES, WILLIAM RALPH (1883-1915). Botany. OCCUP: U.S. government scientist. WORKS ABOUT: <u>Sci</u> 41(1915):859.

JOOR, JOSEPH F. (1848-1892). Botany. OCCUP: Physician; Museum curator; Professor. WORKS ABOUT: <u>Bot Gaz</u> 26(1898):270-74.

JOSLIN, BENJAMIN FRANKLIN (1796-1861). Medicine; Meteorology; Physics. OCCUP: Physician; Professor. WORKS ABOUT: Elliott. See also: <u>Biog Geneal Master Ind</u>.

JOSSELYN, JOHN (ca.1608-1675). Natural history. OCCUP: Physician (?); Natural history explorer. WORKS ABOUT: <u>Dict Am Biog</u>; Elliott; <u>Natl Cycl Am Biog</u> 7:214; <u>Who Was Who Am</u> h. See also: <u>Biog Geneal Master Ind</u>.

JOUY, PIERRE LOUIS (1856-1894). Ornithology. OCCUP: Museum associate; Collector. WORKS ABOUT: <u>Auk</u> 11(1894):262-63; <u>Ibis</u> 6(1894):581-82.

JOY, CHARLES ARAD (1823-1891). Chemistry; Mineralogy. OCCUP: Professor. MSS: <u>Ind NUCMC</u> (73-85). WORKS ABOUT: <u>Am Chem and Chem Eng</u>; <u>Am Journ Sci</u> 142(1891):78; Appleton's; <u>Natl Cycl Am Biog</u> 5; <u>Pop Sci Mo</u> 43(1893):289, 405-09; <u>Twentieth Cent Biog Dict</u>.

JUDD, ORANGE (1822-1892). Agriculture. OCCUP: Editor and publisher. WORKS ABOUT: <u>Dict Am Biog</u>; Elliott; <u>Natl Cycl Am Biog</u> 8; <u>Who Was Who Am</u> h. See also: <u>Biog Geneal Master Ind</u>.

JUDD, SYLVESTER DWIGHT (1871-1905). Biology. OCCUP: U.S. government scientist; Professor. WORKS ABOUT: <u>Sci</u> 22(1905):575; Wallace; <u>Whos Who Am</u> (1903-1905).

JULICH, WILHELM (ca.1839-1893). Entomology. OCCUP: Not determined. WORKS ABOUT: <u>Ent News</u> 5(1894):32; <u>Insect Life</u> 6(1894):280-81.

JULIEN, ALEXIS ANASTAY (1840-1919). Geology; Petrography. OCCUP: Curator; Academic instructor. WORKS ABOUT: <u>Am Journ Sci</u> 197(1919):454; <u>Am Men Sci</u> 1-2 d3; Appleton's; Elliott (s); <u>Natl Cycl Am Biog</u> 18; <u>Sci</u> 49(1919):493; <u>Twentieth Cent Biog Dict</u>; <u>Who Was Who Am</u> 1.

K

KALBFLEISCH, MARTIN (1804-1873). Chemistry. OCCUP: Manufacturer; Public official. WORKS ABOUT: <u>Am Chem and Chem Eng</u>; Appleton's; <u>Biog Dir Am Congress</u>; Haynes p42-56; Lanman; Morris and Morris; <u>Natl Cycl Am Biog</u> 15; <u>Twentieth Cent Biog Dict</u>; <u>Who Was Who Am</u> h.

KANE, ELISHA KENT (1820-1857). Exploration. OCCUP: Naval officer; Surgeon. MSS: <u>Ind NUCMC</u> (69-1377, 81-16). WORKS ABOUT: <u>Dict Am Biog</u>; <u>Natl Cycl Am Biog</u> 3; <u>Who Was Who Am</u> h. See also: Barr; <u>Biog Geneal Master Ind</u>; Ireland <u>Ind Sci</u>.

KASTLE, JOSEPH HOEING (1864-1916). Chemistry. OCCUP: Professor; U.S. and state government scientist. WORKS ABOUT: <u>Am Chem and Chem Eng</u>; <u>Am Men Sci</u> 1-2 d3; <u>Natl Cycl Am Biog</u> 15; <u>Sci</u> 44(1916):461; Wallace; <u>Who Was Who Am</u> 1.

KATO, FREDERICK (1866-1909). Mineralogy. OCCUP: Manufacturer. WORKS ABOUT: <u>Sci</u> 29(1909):696; <u>Who Was Who Am</u> 1.

KEANY, FRANCIS J. (1866?-1916). Dermatology. OCCUP: Professor. WORKS ABOUT: <u>Sci</u> 44(1916):782.

KEARFOTT, WILLIAM DUNHAM (1864-1917). Entomology. OCCUP: Mechanical engineer. WORKS ABOUT: <u>Am Men Sci</u> 2 d3.

KEATING, WILLIAM HYPOLITUS (1799-1840). Chemistry; Mineralogy. OCCUP: Professor; Industrial manager; Lawyer. MSS: <u>Ind NUCMC</u>. WORKS ABOUT: <u>Dict Am Biog</u>; Elliott; <u>Natl Cycl Am Biog</u> 24; <u>Who Was Who Am</u> h. See also: <u>Biog Geneal Master Ind</u>; Pelletier.

KEDZIE, ROBERT CLARK (1823-1902). Chemistry. OCCUP: Physician; Professor; State government scientist. WORKS ABOUT: <u>Dict Am Biog</u>; Elliott; <u>Natl Cycl Am Biog</u> 8, 10; <u>Who Was Who Am</u> 1. See also: Barr; <u>Biog Geneal Master Ind</u>; Pelletier.

KEELER, JAMES EDWARD (1857-1900). Astronomy. OCCUP: Observatory director. MSS: <u>Ind NUCMC</u>. WORKS ABOUT: <u>Dict Am Biog</u>; <u>Dict Sci Biog</u>; Elliott; Natl Ac Scis <u>Biog Mem</u> 5; <u>Natl Cycl Am Biog</u> 10; <u>Who Was Who Am</u> 1. See also: Barr; <u>Biog Geneal Master Ind</u>; Ireland <u>Ind Sci</u>; Pelletier.

KEELY, JOHN ERNST WORRELL (1827-1898). Invention; Scientific fraud. OCCUP: Inventor; Businessman. MSS: <u>Ind NUCMC</u>. WORKS ABOUT: <u>Dict Am Biog</u>; <u>Natl Cycl Am Biog</u> 9; <u>Who Was Who Am</u> h. See also: Barr; <u>Biog Geneal Master Ind</u>.

KEEP, JOSIAH (1849-1911). Zoology. OCCUP: Professor. WORKS ABOUT: Adams O; Allibone <u>Supp</u>; <u>Am Men Sci</u> 2 d3; <u>Sci</u> 34(1911):371; Wallace.

KEEP, WILLIAM JOHN (1842-1918). Engineering. OCCUP: Manufacturer. WORKS ABOUT: <u>Am Men Sci</u> 1-2 d3; <u>Natl Cycl Am Biog</u> 18; <u>Sci</u> 48(1918):437-38; <u>Who Was Who Am</u> 1.

KELLERMAN, WILLIAM ASHBROOK (1850-1908). Botany. OCCUP: Professor. MSS: <u>Ind NUCMC</u> (?). WORKS ABOUT: Adams O; <u>Am Men Sci</u> 1 d3; Elliott (s); <u>Natl Cycl Am Biog</u> 9, 26; <u>Ohio Authors</u>; <u>Sci</u> 27(1908):'118, 479, 858; <u>Twentieth Cent Biog Dict</u>; Wallace; <u>Who Was Who Am</u> 1.

KELLICOTT, DAVID SIMONS (1842-1898). Zoology. OCCUP: Professor. WORKS ABOUT: Am Micr Soc <u>Trans</u> 20(1899):21-24; <u>Canadian Ent</u> 30(1898):166-67; <u>Ent News</u> 9(1898):160; <u>Natl Cycl Am Biog</u> 13; Osborn Herb p232, 350; <u>Sci</u> 7(1898):596; <u>Twentieth Cent Biog Dict</u>.

KELLICOTT, WILLIAM ERSKINE (1878-1919). Zoology. OCCUP: Professor. WORKS ABOUT: <u>Am Men Sci</u> 1-2 d3; <u>Sci</u> 49(1919):146, 322-23; Wallace; <u>Who Was Who Am</u> 1.

KELLOGG, ALBERT (1813-1887). Botany. OCCUP: Physician. MSS: <u>Ind NUCMC</u>. WORKS ABOUT: <u>Dict Am Biog</u>; <u>Dict Sci Biog</u>; Elliott; <u>Natl Cycl Am Biog</u> 25; <u>Who Was Who Am</u> h. See also: Barr; <u>Biog Geneal Master Ind</u>.

KELLY, ALOYSIUS OLIVER JOSEPH (1870-1911). Medicine. OCCUP: Physician; Professor. WORKS ABOUT: <u>Am Men Sci</u> 1-2 d3; <u>Dict Am Biog</u>; <u>Who Was Who Am</u> 1. See also: Barr; <u>Biog Geneal Master Ind</u>.

KEMP, JOHN (1763-1812). Mathematics; Natural philosophy. OCCUP: Professor. WORKS ABOUT: <u>Dict Am Biog</u>; <u>Natl Cycl Am Biog</u> 6; <u>Who Was Who Am</u> h. See also: <u>Biog Geneal Master Ind</u>.

KENDALL, EZRA OTIS (1818-1899). Astronomy; Mathematics. OCCUP: Professor; Academic administrator. WORKS ABOUT: Elliott; <u>Natl Cycl Am Biog</u> 2. See also: Barr; <u>Biog Geneal Master Ind</u>.

KENDALL, THOMAS, JR. (ca.1786-1831). General science; Atmospherical phenomena. OCCUP: Instrument maker. Scientific work: Likely author of works in Roy Soc <u>Cat</u> v3. WORKS ABOUT: Ellis p316.

KENNEDY, ALFRED L. (1818-1896). Chemistry. OCCUP: Physician; Professor; Educator. WORKS ABOUT: <u>Am Chem and Chem Eng</u>; Appleton's; <u>Sci</u> 3(1896):200; <u>Twentietch Cent Biog Dict</u>.

KENNEDY, JOSEPH CAMP GRIFFITH (1813-1887). Statistics. OCCUP: Publisher; Editor; U.S. government employee. MSS: <u>Ind NUCMC</u>. WORKS ABOUT: <u>Dict Am Biog</u>; <u>Natl Cycl Am Biog</u> 11, 16; <u>Who Was Who Am</u> h. See also: <u>Biog Geneal Master Ind</u>.

KENNICOTT, ROBERT (1835-1866). Natural history; Zoology. OCCUP: Natural history explorer/ collector; Curator. MSS: <u>Ind NUCMC</u> (60-655). WORKS ABOUT: <u>Dict Am Biog</u>; Elliott; <u>Natl Cycl Am Biog</u> 24; <u>Who Was Who Am</u> h. See also: Barr; <u>Biog Geneal Master Ind</u>.

KENT, WALTER HENRY (1851-1907). Chemistry. OCCUP: Professor; City government scientist. WORKS ABOUT: <u>Natl Cycl Am Biog</u> 7:536; <u>Who Was Who Am</u> 1.

KENT, WILLIAM (1851-1918). Mechanical engineering. OCCUP: Industrial engineer; Consultant; Academic administrator. WORKS ABOUT: <u>Am Men Sci</u> 1-2; <u>Dict Am Biog</u>; <u>Who Was Who Am</u> 1. See also: <u>Biog Geneal Master Ind</u>.

KEPPEL, HERBERT GOVERT (1866-1918). Mathematics. OCCUP: Professor. WORKS ABOUT: <u>Am Men Sci</u> 2 d3; <u>Sci</u> 48(1918):415.

KERR, MARK BRICKELL (1860-1917). Mining. OCCUP: Mining engineer; U.S. government scientist. WORKS ABOUT: <u>Twentieth Cent Biog Dict</u>; <u>Who Was Who Am</u> 1.

KERR, WALTER CRAIG (1858-1910). Mechanical engineering; Natural history. OCCUP: Professor; Industrial engineer and officer. MSS: <u>Ind NUCMC</u>. WORKS ABOUT: <u>Dict Am Biog</u>; <u>Who Was Who Am</u> 1.

KERR, WASHINGTON CARUTHERS (1827-1885). Geology. OCCUP: Professor; U.S. and state government scientist. WORKS ABOUT: <u>Dict Am Biog</u>; Elliott; <u>Natl Cycl Am Biog</u> 7:450; <u>Who Was Who Am</u> h. See also: Barr; <u>Biog Geneal Master Ind</u>; Ireland <u>Ind Sci</u>.

KEYSER, PETER DIRCK (1835-1897). Ophthalmology. OCCUP: Surgeon; Professor. WORKS ABOUT: Adams O; Allibone <u>Supp</u>; Appleton's; <u>Natl Cycl Am Biog</u> 4; <u>Sci</u> 5(1897):473; <u>Twentieth Cent Biog Dict</u>; Wallace.

KEYT, ALONZO THRASHER (1827-1885). Physiology. OCCUP: Physician. WORKS ABOUT: <u>Dict Am Biog</u>; Elliott; <u>Natl Cycl Am Biog</u> 15; <u>Who Was Who Am</u> h. See also: <u>Biog Geneal Master Ind</u>.

KIDDER, JEROME HENRY (1842-1889). Natural history; Surgery. OCCUP: Naval surgeon; U.S. government scientist. WORKS ABOUT: Appleton's <u>Supp</u> 1901; <u>Auk</u> 6(1889):282; <u>Natl Cycl Am Biog</u> 15; <u>New York Med Journ</u> 49(1889):437; Phil Soc Washington <u>Bull</u> 11(1892):480-88; <u>Sci</u> 41(1915):749; Smithsonian <u>Rep</u> (1889):66-67.

KIMBALL, ALONZO SMITH (1843-1897). Physics. OCCUP: Professor. WORKS ABOUT: Am Ac Arts Scis <u>Proc</u> 33(1898):524-26; <u>Sci</u> 6(1897):876, 7(1898):54-56.

KIMBALL, EDWIN A. (1834?-1898). Mechanics. OCCUP: Inventor; Administrator. WORKS ABOUT: <u>Sci</u> 8(1898):744.

KIMBALL, JAMES PUTNAM (1836-1913). Geology. OCCUP: Mining engineer; U.S. and state government scientist; Professor. WORKS ABOUT: Appleton's; <u>Natl Cycl Am Biog</u> 11; <u>Who Was Who Am</u> 1.

KIMBALL, RODNEY G. (-1900). Mathematics. OCCUP: Professor. WORKS ABOUT: <u>Sci</u> 11(1900):717.

KING, ALBERT FREEMAN AFRICANUS (1841-1914). Medicine. OCCUP: Physician; Professor; Academic administrator. WORKS ABOUT: <u>Am Men Sci</u> 1-2 d3; <u>Dict Am Biog</u>; <u>Natl Cycl Am Biog</u> 24; <u>Who Was Who Am</u> 1. See also: Barr; <u>Biog Geneal Master Ind</u>; Ireland <u>Ind Sci</u>.

KING, ALFRED T. (1813-1858?). General science; Geology. OCCUP: Physician; Professor (medicine). Scientific work: see Roy Soc <u>Cat</u>. WORKS ABOUT: Albert p353-57; Am Assoc Advancement Sci <u>Proc</u> 1(1848) (membership list).

KING, CLARENCE RIVERS (1842-1901). Geology. OCCUP: U.S. government scientist; Mining engineer. MSS: <u>Ind NUCMC</u> (61-2321). WORKS ABOUT: <u>Dict Am Biog</u>; <u>Dict Sci Biog</u>; Elliott; Natl Ac Scis <u>Biog Mem</u> 6; <u>Natl Cycl Am Biog</u> 13; <u>Who Was Who Am</u> 1. See also: Barr; <u>Biog Geneal Master Ind</u>; Ireland <u>Ind Sci</u>; Pelletier.

KING, FRANKLIN HIRAM (1848-1911). Agricultural chemistry; Agricultural physics. OCCUP: Professor; U.S. government scientist. MSS: <u>Ind NUCMC</u> (62-2512). WORKS ABOUT: <u>Am Men Sci</u> 1-2 d3; <u>Dict Am Biog</u>; <u>Natl Cycl Am Biog</u> 19; <u>Who Was Who Am</u> 1. See also: <u>Biog Geneal Master Ind</u>; Ireland <u>Ind Sci</u>.

KING, GEORGE B. (1848-1911). Entomology. OCCUP: Laborer. WORKS ABOUT: <u>Am Men Sci</u> 1-2 d3; <u>Who Was Who Am</u> 4.

KING, HERBERT MAXON (1864-1917). Medicine. OCCUP: Physician. MSS: <u>Ind NUCMC</u> (?). WORKS ABOUT: <u>Am Men Sci</u> 2 d3.

KING, VERNON (-1918). Entomology. OCCUP: U.S. government scientist. MSS: <u>Ind NUCMC</u> (?). WORKS ABOUT: <u>Sci</u> 48(1918):67.

KINNERSLEY, EBENEZER (1711-1778). Electricity. OCCUP: Lecturer; Professor (English and oratory). MSS: Ind NUCMC (?). WORKS ABOUT: Dict Am Biog; Dict Sci Biog; Elliott; Natl Cycl Am Biog 1; Who Was Who Am h. See also: Biog Geneal Master Ind; Ireland Ind Sci.

KINNEY, ABBOT (1850-1920). Forestry. OCCUP: Businessman; Publisher; Horticulturist. WORKS ABOUT: Adams O; Am Lit Yearbook; Am Men Sci 1-3 d4; Twentieth Cent Biog Dict; Wallace; Who Was Who Am 1.

KINNICUTT, FRANCIS PARKER (1846-1913). Medicine. OCCUP: Physician; Professor. WORKS ABOUT: Am Men Sci 1-2 d3; Natl Cycl Am Biog 15; Who Was Who Am 1. See also: Biog Geneal Master Ind.

KINNICUTT, LEONARD PARKER (1854-1911). Chemistry. OCCUP: Professor. WORKS ABOUT: Am Men Sci 1-2 d3; Dict Am Biog; Elliott (s); Natl Cycl Am Biog 25; Who Was Who Am 1. See also: Barr; Biog Geneal Master Ind; Pelletier.

KINO, EUSEBIO FRANCISCO (ca.1645-1711). Cartography. OCCUP: Clergyman; Explorer. MSS: Ind NUCMC (62-939). WORKS ABOUT: Dict Am Biog; Who Was Who Am h. See also: Biog Geneal Master Ind.

KINYOUN, JOSEPH JAMES (1860-1919). Pathology. OCCUP: Physician; Professor. WORKS ABOUT: Am Men Sci 1-2 d3; Dict Am Med Biog; Natl Cycl Am Biog 23; Who Was Who Am 1.

KIRCHHOFF, CHARLES WILLIAM HENRY (1853-1916). Metallurgy; Mining engineering. OCCUP: Editor. WORKS ABOUT: Am Men Sci 1-2 d3; Dict Am Biog; Natl Cycl Am Biog 10; Who Was Who Am 1. See also: Barr; Biog Geneal Master Ind.

KIRKALDY, GEORGE WILLIS (1873-1910). Entomology. OCCUP: Industrial and state government scientist. WORKS ABOUT: Osborn Herb p233, 376; Sci 31(1910):454.

KIRKPATRICK, JOHN (1819-1869). Natural history; Entomology. OCCUP: Skilled laborer; Editor; Administrator. WORKS ABOUT: Cleveland Ac Nat Scis Proc p174-78.

KIRKWOOD, DANIEL (1814-1895). Astronomy. OCCUP: Educator; Professor. WORKS ABOUT: Dict Am Biog; Dict Sci Biog; Elliott; Natl Cycl Am Biog 4; Who Was Who Am h. See also: Barr; Biog Geneal Master Ind; Ireland Ind Sci; Pelletier.

KIRSCH, PHILIP H. (-1900?). Ichthyology. OCCUP: Educator. Scientific work: see Roy Soc Cat. WORKS ABOUT: Sci 13(1901):38.

KIRTLAND, JARED POTTER (1793-1877). Natural history; Zoology. OCCUP: Physician; Legislator; Professor (medicine). MSS: Ind NUCMC. WORKS ABOUT: Dict Am Biog; Elliott; Natl Ac Scis Biog Mem 2; Natl Cycl Am Biog 11; Who Was Who Am h. See also: Barr; Biog Geneal Master Ind; Pelletier.

KLEIN, JOSEPH FREDERIC (1849-1918). Mechanical engineering. OCCUP: Professor; Academic administrator. WORKS ABOUT: <u>Am Men Sci</u> 1-2 d3; <u>Dict Am Biog</u>; <u>Natl Cycl Am Biog</u> 18; <u>Who Was Who Am</u> 1. See also: Barr; <u>Biog Geneal Master Ind</u>.

KLEINHAUS, FRANK B. (1869?-1908). Engineering. OCCUP: Mechanical engineer. WORKS ABOUT: <u>Sci</u> 28(1908):338.

KLIPPART, JOHN HANCOCK (1823-1878). Agriculture. OCCUP: Public official; State government scientist. MSS: <u>Ind NUCMC</u> (75-1067). WORKS ABOUT: <u>Dict Am Biog</u>; <u>Natl Cycl Am Biog</u> 17; <u>Who Was Who Am</u> h. See also: <u>Biog Geneal Master Ind</u>.

KNAB, FREDERICK (1865-1918). Entomology. OCCUP: Artist; U.S. government scientist. WORKS ABOUT: <u>Am Men Sci</u> 1-2 d3; <u>Dict Am Biog</u>; <u>Natl Cycl Am Biog</u> 24. See also: <u>Biog Geneal Master Ind</u>; Ireland <u>Ind Sci</u>; Pelletier.

KNAPP, (JACOB) HERMANN (1832-1911). Ophthalmology. OCCUP: Physician; Professor. WORKS ABOUT: <u>Am Men Sci</u> 1-2 d3; <u>Dict Am Biog</u>; <u>Dict Am Med Biog</u>; <u>Who Was Who Am</u> 1. See also: Barr; <u>Biog Geneal Master Ind</u>.

KNAPP, PHILIP COOMBS (1858-1920). Neurology. OCCUP: Physician; Academic instructor. WORKS ABOUT: <u>Dict Am Biog</u>; <u>Natl Cycl Am Biog</u> 24; <u>Who Was Who Am</u> 1. See also: <u>Biog Geneal Master Ind</u>.

KNEELAND, SAMUEL (1821-1888). Zoology. OCCUP: Physician; Professor. MSS: <u>Ind NUCMC</u>. WORKS ABOUT: <u>Dict Am Biog</u>; Elliott; <u>Natl Cycl Am Biog</u> 26; <u>Who Was Who Am</u> h. See also: <u>Biog Geneal Master Ind</u>.

KNIESKERN, PETER D. (1798-1871). Botany. OCCUP: Physician. WORKS ABOUT: <u>Am Journ Sci</u> 102(1871):388-89.

KNIGHT, ORA WILLIS (1874-1913). Chemistry; Geology. OCCUP: State government scientist. WORKS ABOUT: <u>Am Men Sci</u> 2 d3; Wallace; <u>Who Was Who Am</u> 1.

KNIGHT, WILBUR CLINTON (1858-1903). Geology. OCCUP: Professor; State government scientist; Consultant. WORKS ABOUT: Elliott; <u>Who Was Who Am</u> 1. See also: Barr; <u>Biog Geneal Master Ind</u>.

KOCH, CHARLES RUDOLPH EDWARD (1844-1916). Dentistry. OCCUP: Dentist; Academic administrator and instructor. WORKS ABOUT: <u>Natl Cycl Am Biog</u> 19; <u>Sci</u> 44(1916):165-66; <u>Who Was Who Am</u> 1.

KOCH, WALDEMAR (1875-1912). Pharmacology. OCCUP: Professor. WORKS ABOUT: <u>Am Men Sci</u> 1-2 d3; <u>Sci</u> 35(1912):214.

KOENIG, GEORGE AUGUSTUS (1844-1913). Chemistry; Mineralogy. OCCUP: Professor. WORKS ABOUT: Am Men Sci 1-2 d3; Dict Am Biog; Elliott (s); Natl Cycl Am Biog 17; Who Was Who Am 1. See also: Barr; Biog Geneal Master Ind.

KOONS, BENJAMIN FRANKLIN (1848-1903). Natural history. OCCUP: Professor; Museum curator. WORKS ABOUT: Sci 18(1903):838.

KORTRIGHT, FREDERIC LAWRENCE (1867-1914). Chemistry. OCCUP: Professor. WORKS ABOUT: Am Men Sci 1-2 d3; Sci 40(1914):205.

KRAUSS, WILLIAM CHRISTOPHER (1863-1909). Medicine; Microscopy. OCCUP: Physician. WORKS ABOUT: Natl Cycl Am Biog 12; Who Was Who Am 1.

KRAUTER, LOUIS (-1910). Botany. OCCUP: Professor. WORKS ABOUT: Sci 31(1910):105.

KUHN, ADAM (1741-1817). Botany; Medicine. OCCUP: Professor; Physician. MSS: Ind NUCMC. WORKS ABOUT: Dict Am Biog; Dict Am Med Biog; Elliott; Natl Cycl Am Biog 21; Who Was Who Am h. See also: Biog Geneal Master Ind; Ireland Ind Sci.

KUICHLING, EMIL (1848-1914). Engineering. OCCUP: City government engineer; Consultant. WORKS ABOUT: Am Men Sci 1-2 d3; Natl Cycl Am Biog 16; Who Was Who Am 1.

KUMLIEN, LUDWIG (1853-1902). Biology. OCCUP: Natural history explorer; U.S. government science associate. MSS: Ind NUCMC (?). WORKS ABOUT: Sci 16(1902):997.

KUMLIEN, THURE LUDWIG THEODOR (1819-1888). Botany; Ornithology. OCCUP: Natural history collector; Teacher; Museum curator. MSS: Ind NUCMC (62-2443). WORKS ABOUT: Auk 6(1889):205.

KUMMELL, CHARLES HUGO (1836-1897). Mathematics. OCCUP: U.S. government scientist. WORKS ABOUT: Phil Soc Washington Bull 13(1900):404-05.

KUNZE, RICHARD ERNEST (1838-1919). Natural history. OCCUP: Physician; Planter. WORKS ABOUT: Dict Am Biog; Natl Cycl Am Biog 3; Who Was Who Am 1. See also: Biog Geneal Master Ind.

L

LACOE, R. D. (-1901). Paleontology. OCCUP: Not determined. WORKS ABOUT: <u>Sci</u> 13(1901):279.

LANE, JONATHAN HOMER (1819-1880). Physics. OCCUP: U.S. government scientist. WORKS ABOUT: <u>Dict Sci Biog</u>; Elliott; Natl Ac Scis <u>Biog Mem</u> 3; <u>Natl Cycl Am Biog</u> 3. See also: Barr; <u>Biog Geneal Master Ind</u>.

LANGFORD, NATHANIEL PITT (1832-1911). Conservation. OCCUP: Explorer; Public official. MSS: <u>Ind NUCMC</u> (79-446). WORKS ABOUT: <u>Dict Am Biog</u>; <u>Natl Cycl Am Biog</u> 25; <u>Who Was Who Am</u> 1. See also: <u>Biog Geneal Master Ind</u>; Pelletier.

LANGLEY, JOHN WILLIAMS (1841-1918). Chemistry. OCCUP: Professor. MSS: <u>Ind NUCMC</u> (65-358). WORKS ABOUT: <u>Am Men Sci</u> 1-2; <u>Dict Am Biog</u>; <u>Natl Cycl Am Biog</u> 10, 15; <u>Who Was Who Am</u> 4. See also: <u>Biog Geneal Master Ind</u>.

LANGLEY, SAMUEL PIERPONT (1834-1906). Astronomy; Physics; Astrophysics. OCCUP: Professor; Observatory director; Research institution administrator. MSS: <u>Ind NUCMC</u> (72-1242). WORKS ABOUT: <u>Am Men Sci</u> 1 d3; <u>Dict Am Biog</u>; <u>Dict Sci Biog</u>; Elliott (s); Natl Ac Scis <u>Biog Mem</u> 7; <u>Natl Cycl Am Biog</u> 3, 15; <u>Who Was Who Am</u> 1. See also: Barr; <u>Biog Geneal Master Ind</u>; Ireland <u>Ind Sci</u>; Pelletier.

LANGTON, DANIEL WEBSTER (1864-1909). Geology. OCCUP: State government scientist; Consultant; Architect (landscape). WORKS ABOUT: <u>Am Men Sci</u> 1 d3; <u>Who Was Who Am</u> 1.

LANSING, ODELL E. (1867?-1918). Botany. OCCUP: Museum affiliate. WORKS ABOUT: <u>Sci</u> 48(1918):292.

LANTZ, DAVID ERNEST (1855-1918). Biology; Zoology. OCCUP: Professor (mathematics); U.S. and state government scientist. WORKS ABOUT: <u>Am Men Sci</u> 1-2 d3; <u>Sci</u> 48(1918):391, 49(1919):84-85; <u>Who Was Who Am</u> 1.

LAPHAM, INCREASE ALLEN (1811-1875). Natural history; Geology; Meteorology. OCCUP: Businessman; Public official; U.S. and state government scientist. MSS: Ind NUCMC (62-2911, 75-1071). WORKS ABOUT: Dict Am Biog; Elliott; Natl Cycl Am Biog 8; Who Was Who Am h. See also: Barr; Biog Geneal Master Ind; Ireland Ind Sci.

LATHROP, JOHN (1740-1816). General science (lightning effects, study of wells and springs). OCCUP: Clergyman. Scientific work: see Roy Soc Cat. WORKS ABOUT: Appleton's; Drake.

LATROBE, BENJAMIN HENRY (1764-1820). Engineering. OCCUP: Architect; Engineer. MSS: Ind NUCMC (63-367, 78-808). WORKS ABOUT: Dict Am Biog; Natl Cycl Am Biog 9; Who Was Who Am h. See also: Biog Geneal Master Ind; Ireland Ind Sci.

LATTIMORE, SAMUEL ALLEN (1828-1913). Chemistry. OCCUP: Professor. MSS: Ind NUCMC (?). WORKS ABOUT: Am Chem and Chem Eng; Am Men Sci 1-2 d3; Appleton's; Elliott (s); Natl Cycl Am Biog 12; Sci 37(1913):330; Twentieth Cent Biog Dict; Who Was Who Am 1.

LAUBACH, CHARLES (1836-1904). Geology. OCCUP: Physician; Farmer; Researcher. WORKS ABOUT: Who Was Who Am 1.

LAUDY, LOUIS HYACINTH (ca.1842-1905). Chemistry. OCCUP: Academic instructor; Photographer. WORKS ABOUT: Am Chem and Chem Eng; Sci 22(1905):253.

LAW, ANNIE E. (1840?-1889). Conchology. OCCUP: Not determined; Collector. MSS: Ind NUCMC (?). WORKS ABOUT: Am Journ Sci 137(1889):422; Ogilvie (app).

LAWRENCE, GEORGE NEWBOLD (1806-1895). Ornithology. OCCUP: Businessman. MSS: Ind NUCMC (76-914). WORKS ABOUT: Dict Am Biog; Elliott; Natl Cycl Am Biog 2; Who Was Who Am h. See also: Barr; Biog Geneal Master Ind; Pelletier.

LAWRENCE, ROBERT HOE (1861-1897). Ornithology. OCCUP: Not determined. WORKS ABOUT: Auk 14(1897):342.

LAWSON, JOHN (-1711). Natural history; Travels. OCCUP: Surveyor. MSS: Ind NUCMC. WORKS ABOUT: Dict Am Biog; Elliott; Natl Cycl Am Biog 7:115; Who Was Who Am h. See also: Biog Geneal Master Ind; Pelletier.

LAWTON, GEORGE K. (1873-1901). Astronomy. OCCUP: Observatory affiliate. WORKS ABOUT: Sci 14(1901):191, 215-16.

LAZEAR, JESSE WILLIAM (1866-1900). Medicine; Pathology. OCCUP: Physician; Army surgeon. MSS: Ind NUCMC. WORKS ABOUT: Dict Am Biog; Dict Am Med Biog; Natl Cycl Am Biog 15; Who Was Who Am h. See also: Barr; Biog Geneal Master Ind; Ireland Ind Sci; Pelletier.

LAZENBY, WILLIAM RANE (1850-1916). Forestry; Horticulture. OCCUP: Professor; State government scientist. MSS: Ind NUCMC (?). WORKS ABOUT: Am Men Sci 1-2 d3; Appleton's; Natl Cycl Am Biog 10; Sci 44(1916):912-13; Twentieth Cent Biog Dict; Who Was Who Am 1.

LEA, HENRY CHARLES (1825-1909). Chemistry; Conchology. OCCUP: Businessman; Publisher. MSS: Ind NUCMC (?). WORKS ABOUT: Dict Am Biog; Natl Cycl Am Biog 5; Who Was Who Am 1. See also: Biog Geneal Master Ind.

LEA, ISAAC (1792-1886). Natural history; Malacology. OCCUP: Businessman; Publisher. MSS: Ind NUCMC (66-55, 72-1243). WORKS ABOUT: Dict Am Biog; Dict Sci Biog; Elliott; Natl Cycl Am Biog 6; Who Was Who Am h. See also: Barr; Biog Geneal Master Ind; Ireland Ind Sci; Pelletier.

LEA, MATHEW CAREY (1823-1897). Chemistry. OCCUP: Lawyer; Independent wealth. MSS: Ind NUCMC (?). WORKS ABOUT: Dict Am Biog; Elliott; Natl Ac Scis Biog Mem 5; Natl Cycl Am Biog 10; Who Was Who Am h, 4. See also: Barr; Biog Geneal Master Ind; Ireland Ind Sci; Pelletier.

LEACH, ALBERT ERNEST (1864-1910). Chemistry. OCCUP: U.S. and state government scientist. WORKS ABOUT: Am Men Sci 1-2 d3; Natl Cycl Am Biog 19; Wallace; Who Was Who Am 1.

LEAMING, EDWARD (1861-1916). Photography; Photomicrography. OCCUP: Academic instructor. WORKS ABOUT: Am Men Sci 1-2 d3; Sci 43(1916):710.

LEAVENWORTH, MELINES CONKLIN (1796-1862). Botany. OCCUP: Physician; Army surgeon; Pharmacist. WORKS ABOUT: Am Journ Sci 85(1863):306, 451; Kelly and Burrage.

LEAVITT, DUDLEY (1772-1851). Almanac making; Mathematics. OCCUP: Editor and publisher; Teacher. MSS: Ind NUCMC. WORKS ABOUT: Dict Am Biog; Natl Cycl Am Biog 25; Who Was Who Am h. See also: Biog Geneal Master Ind.

LEAVITT, ERASMUS DARWIN (1836-1916). Engineering. OCCUP: Mechanical engineer. MSS: Ind NUCMC (80-242). WORKS ABOUT: Dict Am Biog; Natl Cycl Am Biog 12, 24; Who Was Who Am 1. See also: Barr.

LeBARON, WILLIAM (1814-1876). Entomology. OCCUP: Physician; State government scientist. WORKS ABOUT: Mallis; Osborn Herb p221, 354; Pop Sci Mo 76(1910):474-76.

LeCONTE, JOHN (1818-1891). Natural history; Physics. OCCUP: Physician; Professor; College president. MSS: Ind NUCMC. WORKS ABOUT: Dict Am Biog; Dict Sci Biog; Elliott; Natl Ac Scis Biog Mem 3; Natl Cycl Am Biog 7:228; Who Was Who Am h. See also: Barr; Biog Geneal Master Ind; Pelletier.

LeCONTE, JOHN EATTON, JR. (1784-1860). Natural history; Entomology. OCCUP: Army engineer. MSS: Ind NUCMC. WORKS ABOUT: Elliott. See also: Barr; Biog Geneal Master Ind; Pelletier.

LeCONTE, JOHN LAWRENCE (1825-1883). Entomology. OCCUP: Inherited wealth; Collector; Administrator. MSS: Ind NUCMC. WORKS ABOUT: Dict Am Biog; Elliott; Natl Ac Scis Biog Mem 2; Natl Cycl Am Biog 11; Who Was Who Am h. See also: Barr; Biog Geneal Master Ind; Ireland Ind Sci; Pelletier.

LeCONTE, JOSEPH (1823-1901) Natural history; Geology; Physiology. OCCUP: Professor. MSS: Ind NUCMC (64-543, 66-567). WORKS ABOUT: Dict Am Biog; Dict Sci Biog; Elliott; Natl Ac Scis Biog Mem 6; Natl Cycl Am Biog 7:231; Who Was Who Am 1. See also: Barr; Biog Geneal Master Ind; Ireland Ind Sci.

LeCONTE, LOUIS (1782-1838). Natural history. OCCUP: Planter. WORKS ABOUT [some as Lewis]: Appleton's; Knight; Natl Cycl Am Biog 11; Stephens p2-8; Twentieth Cent Biog Dict.

LeCOUNT, LLEWELLYN (1878?-1900). Engineering. OCCUP: Academic affiliate. WORKS ABOUT: Sci 12(1900):814.

LEE, CHARLES ALFRED (1801-1872). Medicine; Medical botany. OCCUP: Physician; Professor. WORKS ABOUT: Dict Am Biog; Natl Cycl Am Biog 15; Who Was Who Am h. See also: Biog Geneal Master Ind.

LEE, LESLIE ALEXANDER (1852-1908). Natural history; Zoology. OCCUP: Professor; State government scientist. WORKS ABOUT: Am Men Sci 1 d3; Natl Cycl Am Biog 20; Sci 41(1915):750-51; Who Was Who Am 1.

LEEDOM, EDWIN CONOVER (1805-   ). Astronomy; Optics; Medicine. OCCUP: Physician. Scientific work: see Roy Soc Cat. WORKS ABOUT: Atkinson.

LEEDS, ALBERT RIPLEY (1843-1902). Chemistry. OCCUP: Professor; Consultant. WORKS ABOUT: Am Chem and Chem Eng; Am Journ Sci 163(1902):330; Browne and Weeks p498; Sci 15(1902):557.

LEEDS, DANIEL (1652-1720). Almanac making; Surveying. OCCUP: Surveyor; Public official. WORKS ABOUT: Dict Am Biog; Natl Cycl Am Biog 18; Who Was Who Am h. See also: Biog Geneal Master Ind.

LEEDS, JOHN (1705-1790). Astronomy; Mathematics. OCCUP: Public official; Surveyor. WORKS ABOUT: Dict Am Biog; Elliott; Natl Cycl Am Biog 25; Who Was Who Am h. See also: Biog Geneal Master Ind.

LEFAVOUR, EDWARD B. (1854-1889). Physics. OCCUP: U.S. government scientist; Academic affiliate. WORKS ABOUT: Phil Soc Washington <u>Bull</u> 11(1892):488-90.

LeFEVRE, EGBERT (1858-1914). Medicine. OCCUP: Physician; Professor. WORKS ABOUT: <u>Am Men Sci</u> 2 d3; <u>Sci</u> 39(1914):530; Wallace; <u>Who Was Who Am</u> 1.

LEGGETT, WILLIAM HENRY (1816-1882). Botany. OCCUP: Teacher; Editor. WORKS ABOUT: Appleton's; <u>Natl Cycl Am Biog</u> 6; Torrey Bot Club <u>Bull</u> 9(1882):85-86.

LEHMAN, AMBROSE EDWIN (1851-1917). Geology. OCCUP: Civil engineer; State government engineer. WORKS ABOUT: <u>Am Men Sci</u> 1-2 d3; <u>Who Was Who Am</u> 1.

LEHMANN, GUSTAV WILLIAM (1844-1906). Chemistry. OCCUP: City and U.S. government scientist; Industrial scientist and manager. WORKS ABOUT: <u>Am Men Sci</u> 1 d3; <u>Sci</u> 24(1906):221.

LEIDY, JOSEPH (1823-1891). Natural history; Zoology; Paleontology. OCCUP: Professor. MSS: <u>Ind NUCMC</u> (66-33, 66-1024). WORKS ABOUT: <u>Dict Am Biog</u>; <u>Dict Am Med Biog</u>; <u>Dict Sci Biog</u>; Elliott; Natl Ac Scis <u>Biog Mem</u> 7; <u>Natl Cycl Am Biog</u> 5; <u>Who Was Who Am</u> h. See also: Barr; <u>Biog Geneal Master Ind</u>; Ireland <u>Ind Sci</u>; Pelletier.

LEMMON, JOHN GILL (1832-1908). Botany. OCCUP: Teacher; Collector; State government scientist. MSS: <u>Ind NUCMC</u>. WORKS ABOUT: <u>Dict Am Biog</u>; Elliott; <u>Who Was Who Am</u> 1. See also: <u>Biog Geneal Master Ind</u>.

LENNON, WILLIAM H. (1838-1913). General science. OCCUP: Professor; Educator. WORKS ABOUT: <u>Am Men Sci</u> 1-2 d3; <u>Sci</u> 37(1913):443.

LEONARD CHARLES LESTER (1861-1913). Roentgenology. OCCUP: Physician. WORKS ABOUT: <u>Dict Am Biog</u>; <u>Dict Am Med Biog</u>. See also: Barr.

LESLEY, J. PETER (1819-1903). Geology. OCCUP: State government scientist; Professor; Consultant. MSS: <u>Ind NUCMC</u> (60-2793). WORKS ABOUT [Some as John or Peter]: <u>Dict Am Biog</u>; <u>Dict Sci Biog</u>; Elliott; Natl Ac Scis <u>Biog Mem</u> 8; <u>Natl Cycl Am Biog</u> 8; <u>Who Was Who Am</u> 1. See also: Barr; <u>Biog Geneal Master Ind</u>; Ireland <u>Ind Sci</u>.

LESQUEREUX, LEO (1806-1889). Botany; Paleobotany. OCCUP: Craftsman; Government scientist (foreign); Museum affiliate. MSS: <u>Ind NUCMC</u> (75-320). WORKS ABOUT: <u>Dict Am Biog</u>; <u>Dict Sci Biog</u>; Elliott; Natl Ac Scis <u>Biog Mem</u> 3; <u>Natl Cycl Am Biog</u> 9; <u>Who Was Who Am</u> h. See also: Barr; <u>Biog Geneal Master Ind</u>; Ireland <u>Ind Sci</u>.

LESUEUR, CHARLES ALEXANDRE (1778-1846). Natural history; Zoology. OCCUP: Artist; Natural history explorer/ collector; Museum administrator. MSS: Ind NUCMC (61-1743, 66-56). WORKS ABOUT: Dict Am Biog; Dict Sci Biog; Elliott; Natl Cycl Am Biog 8; Who Was Who Am h. See also: Barr; Biog Geneal Master Ind; Ireland Ind Sci; Pelletier.

LEVY, LOUIS EDWARD (1846-1919). Chemistry; Photo-chemistry. OCCUP: Editor and publisher; Businessman; Inventor. WORKS ABOUT: Am Men Sci 2 d3; Dict Am Biog; Natl Cycl Am Biog 13; Who Was Who Am 1. See also: Biog Geneal Master Ind.

LEWIS, ENOCH (1776-1856). Mathematics. OCCUP: Teacher; Surveyor; Editor. MSS: Ind NUCMC. WORKS ABOUT: Dict Am Biog; Elliott; Natl Cycl Am Biog 10; Who Was Who Am h. See also: Biog Geneal Master Ind.

LEWIS, GRACEANNA (1821-1912). Ornithology. OCCUP: Teacher; Author; Illustrator. MSS: Ind NUCMC (82-1749). WORKS ABOUT: Adams O; Am Wom; Natl Cycl Am Biog 9; Ogilvie (app); Twentieth Cent Biog Dict.

LEWIS, HENRY CARVILL (1853-1888). Geology; Mineralogy. OCCUP: State government scientist; Professor. MSS: Ind NUCMC (61-493, 66-57). WORKS ABOUT: Am Geol 2(1888):371-79; Am Journ Sci 136(1888):226; Am Nat 22(1888):667; Appleton's; Franklin Inst Journ 126(1888):154-55; Geol Mag 5(1888):428-30, 6(1889):155-60; Geol Soc London Q Journ 45(1889)(Proc):41-42; Johns Hopkins Circ 8(1888-1889):13; Natl Cycl Am Biog 5; Nature 38(1888):346-47; Pop Sci Mo 35(1889):289, 401-08; Preston; Rail Eng Journ 62(1888):429; Sci 1st ser 12(1888):37; Twentieth Cent Biog Dict.

LEWIS, MERIWETHER (1774-1809). Exploration. OCCUP: Explorer; Public official. MSS: Ind NUCMC (62-2513, 62-4849, 68-1301, 69-1282). WORKS ABOUT: Dict Am Biog; Natl Cycl Am Biog 5; Who Was Who Am h. See also: Barr; Biog Geneal Master Ind.

LICHTENTHALER, G. W. (-1893). Conchology. OCCUP: Not determined; Collector. WORKS ABOUT: Sci 1st ser 21(1893):187.

LIDDLE, LEONARD MERRITT (1885-1920). Chemistry. OCCUP: Academic instructor; Research institution scientist. WORKS ABOUT: Natl Cycl Am Biog 18.

LIEBER, OSCAR MONTGOMERY (1830-1862). Geology. OCCUP: State government scientist. MSS: Ind NUCMC (?). WORKS ABOUT: Adams O; Allibone; Appleton's; Drake; Elisha Mitchell Sci Soc Journ 7(1890):103-13, 116-17; Merrill p323-25; Natl Cycl Am Biog 13; Twentieth Cent Biog Dict; Wallace.

LILLEY, GEORGE (ca.1851-1904). Engineering; Mathematics. OCCUP: Professor; Educator. WORKS ABOUT: Elliott; Natl Cycl Am Biog 25; Who Was Who Am 1. See also: Biog Geneal Master Ind.

LINCECUM, GIDEON (1793-1874). Natural history; Entomology. OCCUP: Businessman; Public official; Physician. MSS: Ind NUCMC (70-866). WORKS ABOUT: Dict Am Biog; Elliott; Natl Cycl Am Biog 25; Who Was Who Am h. See also: Biog Geneal Master Ind; Ireland Ind Sci.

LINDAHL, (JOHAN HARALD) JOSUA (1844-1912). Zoology. OCCUP: Natural history explorer; Professor; Museum curator. MSS: Ind NUCMC. WORKS ABOUT: Am Men Sci 1-2 d3; Natl Cycl Am Biog 13; Who Was Who Am 1.

LINDEN, CHARLES (ca.1832-1888). Natural history; Ornithology. OCCUP: Teacher; Curator. WORKS ABOUT: Auk 5(1888):220-21.

LINDENKOHL, ADOLPH (1833-1904). Cartography; Oceanography; Topography. OCCUP: U.S. government scientist. WORKS ABOUT: Dict Am Biog; Elliott; Natl Cycl Am Biog 26; Who Was Who Am 1.

LINDHEIMER, FERDINAND JACOB (1801-1879). Botany. OCCUP: Teacher; Natural history collector; Editor. MSS: Ind NUCMC (?). WORKS ABOUT: Dict Am Biog; Elliott; Natl Cycl Am Biog 24; Who Was Who Am h. See also: Barr; Ireland Ind Sci.

LINELL, MARTIN LARSSON (1849-1897). Entomology. OCCUP: Museum affiliate. WORKS ABOUT: Ent News 8(1897):159-60; Osborn Herb p222-23; Sci 5(1897):766.

LINING, JOHN (1708-1760). Electricity; Meteorology; Physiology. OCCUP: Physician. WORKS ABOUT: Dict Am Biog; Dict Am Med Biog; Elliott; Natl Cycl Am Biog 25; Who Was Who Am h. See also: Biog Geneal Master Ind.

LINSLEY, JAMES HARVEY (1787-1843). Natural history. OCCUP: Teacher; Clergyman. WORKS ABOUT: Am Journ Sci 46(1844):216; Appleton's; Drake; Natl Cycl Am Biog 4; Twentieth Cent Biog Dict.

LINSLEY, JOSEPH HATCH (1859-1901). Bacteriology; Hygiene. OCCUP: Physician; Professor; State government scientist. WORKS ABOUT [some as John]: Am Med Assoc Journ 36(1901):585; Appleton's Supp 1901; Kelly and Burrage; Natl Cycl Am Biog 15; Sci 13(1901):316.

LINTNER, JOSEPH ALBERT (1822-1898). Entomology. OCCUP: Businessman; Museum curator; State government scientist. WORKS ABOUT: Dict Am Biog; Elliott; Natl Cycl Am Biog 5; Who Was Who Am h. See also: Barr; Biog Geneal Master Ind; Pelletier.

LINVILLE, JACOB HAYS (1825-1906). Engineering. OCCUP: Industrial engineer and officer; Consultant. WORKS ABOUT: Am Men Sci 1 d3.

LIPPINCOTT, JAMES STARR (1819-1885). Horticulture; Meteorology. OCCUP: Farmer; Author. WORKS ABOUT: Dict Am Biog; Elliott; Who Was Who Am h. See also: Biog Geneal Master Ind.

LITTON, ABRAM (1814-1901). Chemistry. OCCUP: Professor. WORKS ABOUT: Natl Cycl Am Biog 10; Sci 14(1901):542.

LLOYD, RACHEL (1839-1900). Chemistry. OCCUP: Professor; State government scientist. WORKS ABOUT: Siegel and Finley.

LOCKE, JOHN (1792-1856). Botany; Geology; Physics. OCCUP: Educator; Physician; U.S. and state government scientist. MSS: Ind NUCMC. WORKS ABOUT: Dict Am Biog; Elliott; Natl Cycl Am Biog 15; Who Was Who Am h. See also: Barr; Biog Geneal Master Ind; Pelletier.

LOCKINGTON, WILLIAM NEALE (1842?-1902). Zoology. OCCUP: Not determined; Curator. WORKS ABOUT: Sci 16(1902):559.

LOCKWOOD, SAMUEL (1819-1894). Biology; Geology. OCCUP: Clergyman; Educator. WORKS ABOUT: Adams O; Allibone Supp; Am Mo Micr Journ 16(1895):127-28; Am Nat 28(1894):289; Appleton's; Auk 11(1894):189-90; Drake; Pop Sci Mo 51(1897):577, 692-99; Wallace.

LODEMAN, ERNEST GUSTAVUS (1867-1896). Botany; Horticulture. OCCUP: Academic instructor and scientist. WORKS ABOUT: Gard and Forest 9(1896):500.

LOEB, MORRIS (1863-1912). Chemistry. OCCUP: Professor; Laboratory administrator. MSS: Ind NUCMC (?). WORKS ABOUT: Am Men Sci 1-2 d3; Dict Am Biog; Natl Cycl Am Biog 26; Who Was Who Am 1. See also: Barr; Pelletier.

LOGAN, JAMES (1674-1751), Botany; Optics. OCCUP: Administrator; Public official. MSS: Ind NUCMC (61-1266). WORKS ABOUT: Dict Am Biog; Dict Sci Biog; Elliott; Natl Cycl Am Biog 2; Who Was Who Am h. See also: Biog Geneal Master Ind; Ireland Ind Sci; Pelletier.

LOGAN, MARTHA DANIELL (1704-1779). Horticulture. OCCUP: Teacher; Horticulturist. WORKS ABOUT: Notable Am Wom. See also: Biog Geneal Master Ind.

LOGAN, THOMAS MULDRUP (1808-1876). Climatology; Public health. OCCUP: Physician. WORKS ABOUT: Dict Am Biog; Natl Cycl Am Biog 12; Who Was Who Am h. See also: Biog Geneal Master Ind; Ireland Ind Sci.

LOMB, HENRY (1828-1908). Optics. OCCUP: Manufacturer; Benefactor. WORKS ABOUT: Natl Cycl Am Biog 23; Sci 28(1908):46.

LOMBARD, GUY DAVENPORT (1872?-1907). Histology. OCCUP: Academic instructor. WORKS ABOUT: Sci 25(1907):879.

LONG, CRAWFORD WILLIAMSON (1815-1878). Surgery. OCCUP: Physician. MSS: Ind NUCMC (59-202). WORKS ABOUT: Dict Am Biog; Dict Am Med Biog; Natl Cycl Am Biog 13; Who Was Who Am h. See also: Barr; Biog Geneal Master Ind; Ireland Ind Sci; Pelletier.

LONG, JOHN HARPER (1856-1918). Chemistry. OCCUP: Professor. WORKS ABOUT: Am Men Sci 1-2 d3; Dict Am Biog; Natl Cycl Am Biog 13, 19; Who Was Who Am 1. See also: Barr; Biog Geneal Master Ind; Ireland Ind Sci.

LONG, STEPHEN HARRIMAN (1784-1864). Exploration; Topographical engineering. OCCUP: Army officer; Engineer. MSS: Ind NUCMC. WORKS ABOUT: Dict Am Biog; Natl Cycl Am Biog 11; Who Was Who Am h. See also: Biog Geneal Master Ind.

LONGSTRETH, MIERS FISHER (1819-1891). Astronomy. OCCUP: Merchant; Physician. WORKS ABOUT: Appleton's; Natl Ac Scis Biog Mem 8.

LONGSTRETH, MORRIS (1846-1914). Anatomy. OCCUP: Physician; Professor. WORKS ABOUT: Sci 40(1914):478; Who Was Who Am 4.

LONSDALE, ELSTON HOLMES (1868-1898). Geology. OCCUP: State government scientist. WORKS ABOUT: Am Geol 21(1898):264-65.

LOOMIS, EBEN JENKS (1828-1912). Astronomy. OCCUP: U.S. government scientist. MSS: Ind NUCMC. WORKS ABOUT: Adams O; Am Journ Sci 185(1913):120; Am Men Sci 1-2 d3; Natl Cycl Am Biog 40; Sci 36(1912):822; Wallace; Who Was Who Am 1.

LOOMIS, ELIAS (1811-1889). Astronomy; Mathematics; Meteorology. OCCUP: Professor. MSS: Ind NUCMC (80-760). WORKS ABOUT: Dict Am Biog; Dict Sci Biog; Elliott; Natl Ac Scis Biog Mem 3; Natl Cycl Am Biog 7:233. See also: Barr; Biog Geneal Master Ind; Pelletier.

LOOMIS, MAHLON (1826-1886). Experimentation in electricity. OCCUP: Dentist; Industrial scientist; Researcher. MSS: Ind NUCMC (62-4526). WORKS ABOUT: Dict Am Biog; Elliott; Natl Cycl Am Biog 25; Who Was Who Am h. See also: Biog Geneal Master Ind; Ireland Ind Sci.

LOOS, HERMANN A. (1876?-1900). Chemistry. OCCUP: Academic affiliate (?). WORKS ABOUT: Sci 12(1900):277.

LOPER, SAMUEL WARD (1835-1910). Paleontology. OCCUP: Author; U.S. government scientist; Museum curator. WORKS ABOUT: Am Journ Sci 179(1910):464; Am Men Sci 1 d3; Sci 31(1910):535; Shaw; Wallace.

LORD, NATHANIEL WRIGHT (1854-1911). Chemistry; Metallurgy; Mineralogy. OCCUP: Professor; State government scientist. MSS: Ind NUCMC (61-1322). WORKS ABOUT: Am Men Sci 1-2 d3; Sci 33(1911):849; Wallace; Who Was Who Am 1.

LORD, WILLIAM ROGERS (1847-1916). Ornithology. OCCUP: Clergyman. WORKS ABOUT: Sci 46(1917):450; Wallace.

LOTHROP, HENRY W. (1844?-1904). Entomology. OCCUP: Not determined. WORKS ABOUT: Sci 19(1904):117.

LOUGHRIDGE, ROBERT HILLS (1843-1917). Chemistry. OCCUP: Professor; State government scientist. MSS: Ind NUCMC (75-325). WORKS ABOUT: Am Men Sci 1-2 d3; Who Was Who Am 4.

LOVERING, JOSEPH (1813-1892). Mathematics; Physics. OCCUP: Professor; U.S. government science associate. MSS: Ind NUCMC. WORKS ABOUT: Dict Am Biog; Elliott; Natl Ac Scis Biog Mem 6; Natl Cycl Am Biog 6; Who Was Who Am h. See also: Barr; Biog Geneal Master Ind.

LOVEWELL, JOSEPH TAPLIN (1833-1918). Chemistry; Physics. OCCUP: Professor. WORKS ABOUT: Am Men Sci 2 d3; Who Was Who Am 1.

LOWE, THADDEUS SOBIESKI COULINCOURT (1832-1913). Aeronautics (ballooning); Atmospheric studies; Engineering; Meteorology. OCCUP: Inventor; Businessman. MSS: Ind NUCMC. WORKS ABOUT: Am Men Sci 2 d3; Dict Am Biog; Natl Cycl Am Biog 9; Who Was Who Am 1. See also: Barr; Biog Geneal Master Ind; Ireland Ind Sci; Pelletier.

LOWELL, FRANCIS CABOT (1775-1817). Invention. OCCUP: Merchant; Manufacturer. MSS: Ind NUCMC (84-2026). WORKS ABOUT: Dict Am Biog; Natl Cycl Am Biog 7:151; Who Was Who Am h. See also: Biog Geneal Master Ind; Ireland Ind Sci; Pelletier.

LOWELL, PERCIVAL (1855-1916). Astronomy. OCCUP: Businessman; Author; Observatory director. MSS: Ind NUCMC. WORKS ABOUT: Am Men Sci 1-2 d3; Dict Am Biog; Dict Sci Biog; Natl Cycl Am Biog 8; Who Was Who Am 1. See also: Barr; Biog Geneal Master Ind; Ireland Ind Sci; Pelletier.

LUGGER, OTTO (1844-1901). Entomology. OCCUP: U.S. and state government scientist; Curator. WORKS ABOUT: Mallis; Sci 13(1901):877, 980-81.

LUKENS, ISAIAH (1779-1846). Mechanics. OCCUP: Craftsman. WORKS ABOUT: Am Journ Sci 53(1847):144; Franklin Inst Journ 42(1846):423-25; Savage.

LUNDIN, CARL AXEL ROBERT (1851-1915). Astronomy; Telescope manufacturing. OCCUP: Instrument maker. WORKS ABOUT: Am Men Sci 2 d3; Dict Am Biog; Who Was Who Am 1. See also: Barr.

LUSK, WILLIAM THOMPSON (1838-1897). Gynecology. OCCUP: Physician; Professor. MSS: Ind NUCMC (62-3477). WORKS ABOUT: Dict Am Biog; Natl Cycl Am Biog 9, 27; Who Was Who Am h. See also: Barr; Biog Geneal Master Ind.

LYMAN, BENJAMIN SMITH (1835-1920). Geology; Mining engineering. OCCUP: Government scientist (foreign); State government scientist; Mining engineer. MSS: Ind NUCMC (61-129, 61-1124, 62-3746). WORKS ABOUT: Am Men Sci 1-2 d3; Dict Am Biog; Dict Sci Biog; Elliott (s); Natl Cycl Am Biog 9; Who Was Who Am 1. See also: Biog Geneal Master Ind.

LYMAN, CHARLES PARKER (1848?-1918). Veterinary medicine. OCCUP: Academic administrator. WORKS ABOUT: Sci 47(1917):266.

LYMAN, CHESTER SMITH (1814-1890). Astronomy; Geology; Physics. OCCUP: Clergyman; Surveyor; Professor. MSS: Ind NUCMC. WORKS ABOUT: Dict Am Biog; Dict Sci Biog; Elliott; Natl Cycl Am Biog 25; Who Was Who Am h. See also: Barr; Biog Geneal Master Ind.

LYMAN, THEODORE (1833-1897). Zoology. OCCUP: Museum affiliate; Public official. MSS: Ind NUCMC (84-2031). WORKS ABOUT: Dict Am Biog; Elliott; Natl Ac Scis Biog Mem 5; Natl Cycl Am Biog 24; Who Was Who Am h. See also: Barr; Biog Geneal Master Ind.

LYON, SIDNEY SMITH (1807-1872). Geology; Fossil studies. OCCUP: Engineer; State government scientist. WORKS ABOUT: Am Nat 6(1892):505-06.

LYON, WILLIAM SCRUGHAM (1852-1916). Botany. OCCUP: Government scientist; Collector. WORKS ABOUT: Sci 44(1916):306; Wallace.

M

MABBOTT, DOUGLAS C. (1893?-1918). Biology. OCCUP: U.S. government scientist. WORKS ABOUT: Sci 48(1918):544.

McARTHUR, WILLIAM POPE (1814-1850). Hydrography. OCCUP: Naval officer. WORKS ABOUT: Dict Am Biog; Elliott; Natl Cycl Am Biog 25; Who Was Who Am h. See also: Biog Geneal Master Ind.

McBRIDE, JAMES (1784-1817). Botany. OCCUP: Physician. MSS: Ind NUCMC. WORKS ABOUT: Knight; Natl Cycl Am Biog 11.

McBURNEY, CHARLES (1845-1913). Anatomy; Surgery. OCCUP: Surgeon; Professor. WORKS ABOUT: Dict Am Biog; Dict Am Med Biog; Natl Cycl Am Biog 13, 14, 26; Who Was Who Am 1. See also: Barr; Biog Geneal Master Ind; Ireland Ind Sci; Pelletier.

McCALL, GEORGE ARCHIBALD (1802-1868). Zoology. OCCUP: Army officer. MSS: Ind NUCMC (61-705). WORKS ABOUT: Adams O; Appleton's; Cassinia 16(1912):1-6; Drake; Twentieth Cent Biog Dict; Wallace.

McCALLA, ALBERT (1846-1918). Microscopy. OCCUP: Professor (natural sciences and mathematics); Manufacturer. WORKS ABOUT: Natl Cycl Am Biog 13; Sci 48(1918):9; Who Was Who Am 1.

McCALLEY, HENRY (1852-1904). Geology. OCCUP: Academic affiliate; State government scientist. WORKS ABOUT: Elliott; Who Was Who Am 1. See also: Barr.

MacCALLUM, JOHN BRUCE (1876-1906). Physiology. OCCUP: Physician; Professor. WORKS ABOUT: Am Men Sci 1 d3; Natl Cycl Am Biog 15; Who Was Who Am 5.

McCARTNEY, WASHINGTON (1812-1856). Mathematics. OCCUP: Professor; Lawyer; Public official. WORKS ABOUT: Dict Am Biog; Natl Cycl Am Biog 11; Who Was Who Am h. See also: Biog Geneal Master Ind.

McCAY, CHARLES FRANCIS (1810-1889). Actuarial science; Mathematics. OCCUP: Professor; Actuary; Businessman, MSS: Ind NUCMC (?). WORKS ABOUT: Dict Am Biog; Elliott; Natl Cycl Am Biog 11; Who Was Who Am h. See also: Biog Geneal Master Ind.

McCHEENEY,   (-1876). Geology. OCCUP: Professor. WORKS ABOUT: Am Journ Sci 112(1876):244.

McCLELLAN, GEORGE (1796-1847). Anatomy; Surgery. OCCUP: Surgeon; Professor. WORKS ABOUT: Dict Am Biog; Dict Am Med Biog; Natl Cycl Am Biog 4; Who Was Who Am h. See also: Biog Geneal Master Ind; Pelletier.

McCLELLAN, GEORGE (1849-1913). Anatomy. OCCUP: Physician; Professor. MSS: Ind NUCMC. WORKS ABOUT: Dict Am Biog; Natl Cycl Am Biog 15. See also: Barr; Biog Geneal Master Ind.

McCLINTIC, T. B. (1873?-1912). Sanitation. OCCUP: Physician (?); U.S. government scientist. WORKS ABOUT: Sci 36(1912):238.

McCLINTOCK, EMORY (1840-1916). Actuarial science; Mathematics. OCCUP: Actuary. MSS: Ind NUCMC (?). WORKS ABOUT: Am Men Sci 1-2 d3; Dict Am Biog; Elliott (s); Natl Cycl Am Biog 12; Who Was Who Am 1. See also: Biog Geneal Master Ind; Ireland Ind Sci.

McCLURE, EDGAR (-1897). Geology. OCCUP: Professor. WORKS ABOUT: Sci 6(1897):211.

McCOLLOM, JOHN HILDRETH (1843-1915). Medicine. OCCUP: Physician; Professor. WORKS ABOUT: Am Men Sci 1-2 d3; Natl Cycl Am Biog 16; Sci 41(1915):937; Who Was Who Am 1.

McCONNELL, J. C. (-1904). Anatomy. OCCUP: Physician (?); Museum affiliate; Artist. WORKS ABOUT: Sci 20(1904):188.

McCONNELL, WILBUR ROSS (1881-1920). Zoology. OCCUP: Professor; U.S. government scientist. WORKS ABOUT: Am Men Sci 2 d3; Osborn Herb p215.

McCOOK, HENRY CHRISTOPHER (1837-1911). Natural history; Entomology. OCCUP: Clergyman. MSS: Ind NUCMC (66-59). WORKS ABOUT: Am Men Sci 1-2 d3; Dict Am Biog; Natl Cycl Am Biog 4; Who Was Who Am 1. See also: Barr; Biog Geneal Master Ind; Ireland Ind Sci; Pelletier.

McCORMICK, LEANDER JAMES (1819-1900). Astronomy. OCCUP: Manufacturer; Benefactor. MSS: Ind NUCMC (?). WORKS ABOUT: Dict Am Biog; Natl Cycl Am Biog 1; Who Was Who Am 1. See also: Barr; Biog Geneal Master Ind.

McCRADY, JOHN (1831-1881). Zoology. OCCUP: Professor. WORKS ABOUT: Adams O; Natl Cycl Am Biog 11; Twentieth Cent Biog Dict.

McCRAY, ARTHUR HOWARD (1880-1919). Entomology. OCCUP: U.S. government scientist. WORKS ABOUT: Am Men Sci 2 d3.

McCULLOH, RICHARD SEARS (1818-1894). Chemistry; Physics. OCCUP: Professor. WORKS ABOUT: Elliott. See also: Pelletier.

McDONALD, MARSHALL (1835-1895). Ichthyology. OCCUP: Professor; U.S. and state government scientist. MSS: Ind NUCMC (61-3673). WORKS ABOUT: Am Nat 29(1895):1042-43; Appleton's Supp 1901; Natl Cycl Am Biog 13; Sci 2(1895):341.

MacDONALD, WILLIS GOSS (1863-1910). Surgery. OCCUP: Surgeon; Professor. WORKS ABOUT: Sci 33(1911):22; Who Was Who Am 1.

McELROY, GEORGE BEAMISH (1824-1907). Mathematics. OCCUP: Professor; College president. WORKS ABOUT: Sci 25(1907):238; Twentieth Cent Biog Dict.

McFARLAND, ROBERT WHITE (1825-1910). Mathematics. OCCUP: Professor. MSS: Ind NUCMC (75-1090). WORKS ABOUT: Ohio Authors; Sci 32(1910):626; Twentieth Cent Biog Dict; Who Was Who Am 1.

MacFARLANE, ALEXANDER (1851-1913). Mathematics; Physics. OCCUP: Professor. WORKS ABOUT: Allibone Supp; Am Journ Sci 186(1913):576; Am Men Sci 1-2 d3; Nature 92(1913):103-04; Roy Soc Edinburgh Proc 34(1913-1914):8; Sci 38(1913):440; Twentieth Cent Biog Dict; Wallace; Who Was Who Am 4.

MACFARLANE, JAMES (1819-1885). Geology. OCCUP: Lawyer (?). WORKS ABOUT: Am Journ Sci 130(1885):407; Sci 1st ser 6(1885):359.

McGEE, WILLIAM JOHN (1853-1912). Anthropology; Geology; Hydrology. OCCUP: U.S. government scientist. MSS: Ind NUCMC (62-4622). WORKS ABOUT: Am Men Sci 1-2 d3; Dict Am Biog; Elliott (s); Who Was Who Am 1. See also: Barr.

MacGOWAN, DANIEL JEROME (1815-1893). Geography (physical and human). OCCUP: Physician; Missionary. WORKS ABOUT: Kelly v2:132.

McGUIRE, JOSEPH DEAKINS (1842-1916). Archeology. OCCUP: Lawyer; Public official. WORKS ABOUT: Am Men Sci 1-2 d3; Dict Am Biog; Who Was Who Am 1.

McHATTON, HENRY (1856-1917). Medicine. OCCUP: Physician. WORKS ABOUT: Natl Cycl Am Biog 18; Sci 45(1917):458.

McILVAINE, CHARLES ("Tobe Hodge") (1840-1909). Mycology.
OCCUP: Engineer; Author. WORKS ABOUT: Adams O; Biog Dict
Synopsis Books; Wallace; Who Was Who Am 1.

McINTOSH, DONALD (-1915). Medicine. OCCUP:
Veterinarian; Professor. WORKS ABOUT: Sci 42(1915):373; Who
Was Who Am 3.

MACK, WINFRED BERDELL (1871?-1918). Bacteriology;
Veterinary science. OCCUP: Professor. MSS: Ind NUCMC (?).
WORKS ABOUT: Sci 47(1918):414.

McKAY, CHARLES LESLIE (-1883). Ichthyology. OCCUP: U.S.
government employee; Natural history collector. WORKS
ABOUT: Sci 1st ser 2(1883):635.

McKAY, JOHN SOPHRONUS (1850-1917). Mathematics; Physics.
OCCUP: Professor; Educator. WORKS ABOUT: Natl Cycl Am Biog
17; Sci 45(1917):289.

MACKENZIE, KENNETH ALEXANDER JAMES (1859-1920). Surgery.
OCCUP: Surgeon; Professor. WORKS ABOUT: Am Men Sci 2 d3;
Dict Am Med Biog; Who Was Who Am 1.

McKINNEY, ROBERT CHRISTIAN (1856?-1918). Topography. OCCUP:
U.S. government scientist. WORKS ABOUT: Sci 48(1918):341.

MACKINTOSH, JAMES BUCKTON (1856-1891). Chemistry. OCCUP:
Academic affiliate; Industrial chemist. MSS: Ind NUCMC (62-
28). WORKS ABOUT: Am Chem Soc Journ 13(1891):153-54; Am
Journ Sci 141(1891):444; Chem News 63(1891):224-25; Sch
Mines Q 12(1891):256-59.

McLANE, JAMES WOODS (1839-1912). Obstetrics. OCCUP:
Physician; Professor; Academic administrator. MSS: Ind
NUCMC (60-2122). WORKS ABOUT: Sci 36(1912):784; Who Was Who
Am 1.

MACLAURIN, RICHARD COCKBURN (1870-1920). Physics. OCCUP:
Professor; College president. MSS: Ind NUCMC. WORKS ABOUT:
Am Men Sci 2 d3; Dict Am Biog; Natl Cycl Am Biog 14; Who
Was Who Am 1. See also: Barr.

MACLAY, JAMES (1864-1919). Mathematics. OCCUP: Professor.
WORKS ABOUT: Am Men Sci 1-2 d3; Who Was Who Am 1.

MACLEAN, JOHN (1771-1814). Chemistry. OCCUP: Physician;
Professor. MSS: Ind NUCMC. WORKS ABOUT: Dict Am Biog; Dict
Sci Biog; Elliott; Who Was Who Am h. See also: Biog Geneal
Master Ind; Ireland Ind Sci.

MACLOSKIE, GEORGE (1834-1920). Biology. OCCUP: Clergyman;
Professor. WORKS ABOUT: Adams O; Allibone Supp; Am Men Sci
1-2 d3; Appleton's; Natl Cycl Am Biog 19; Wallace; Who Was
Who Am 1.

MACLURE, WILLIAM (1763-1840). Geology. OCCUP: Merchant; Natural history explorer; Benefactor. MSS: Ind NUCMC (69-1605). WORKS ABOUT: Dict Am Biog; Dict Sci Biog; Elliott; Natl Cycl Am Biog 13; Who Was Who Am h. See also: Barr; Biog Geneal Master Ind; Ireland Ind Sci; Pelletier.

McMAHON, BERNARD (1775-1816). Botany. OCCUP: Horticulturist. WORKS ABOUT: Dict Am Biog; Who Was Who Am h. See also: Biog Geneal Master Ind.

McMATH, ROBERT EMMETT (1833-1918). Civil engineering. OCCUP: Army engineer; Public official; Consultant. WORKS ABOUT: Am Men Sci 1-2 d3; Dict Am Biog; Natl Cycl Am Biog 26; Who Was Who Am 4.

McMURRAN, STOCKTON MOSBY (1887-1920). Plant pathology. OCCUP: U.S. government scientist. WORKS ABOUT: Natl Cycl Am Biog 20.

McMURRAY, WILLIAM JOSIAH (1842-1905). Hygiene. OCCUP: Physician; Public official. WORKS ABOUT: Natl Cycl Am Biog 8; Sci 22(1905):807; Wallace; Who Was Who Am 1.

McMURTRIE, WILLIAM (1851-1913). Chemistry. OCCUP: Industrial and U.S. government scientist. WORKS ABOUT: Am Men Sci 1-2 d3; Dict Am Biog; Elliott (s); Natl Cycl Am Biog 12; Who Was Who Am 1. See also: Barr; Biog Geneal Master Ind; Ireland Ind Sci; Pelletier.

MacNEVEN, WILLIAM JAMES (1763-1841). Chemistry. OCCUP: Physician; Professor. MSS: Ind NUCMC. WORKS ABOUT: Dict Am Biog; Natl Cycl Am Biog 9; Who Was Who Am h. See also: Biog Geneal Master Ind; Ireland Ind Sci.

MacNIDER, GEORGE MALLETT (1884-1917). Chemistry. OCCUP: State government scientist. WORKS ABOUT: Am Men Sci 2 d3.

MADISON, JAMES (1749-1812). Astronomy; Geology; Physics. OCCUP: Professor; College president; Clergyman. MSS: Ind NUCMC (67-184). WORKS ABOUT: Dict Am Biog; Elliott; Natl Cycl Am Biog 3, 7:216; Who Was Who Am h. See also: Biog Geneal Master Ind; Ireland Ind Sci.

MAGOWAN, CHARLES S. (-1907). Engineering. OCCUP: Professor. WORKS ABOUT: Sci 26(1907):767.

MAGRUDER, ERNEST P. (-1915). Medicine. OCCUP: Physician; Professor. WORKS ABOUT: Sci 41(1915):606.

MAHAN, DENNIS HART (1802-1871). Engineering. OCCUP: Army officer; Professor. MSS: Ind NUCMC (72-1351). WORKS ABOUT: Dict Am Biog; Natl Ac Scis Biog Mem 2; Natl Cycl Am Biog 10; Who Was Who Am h. See also: Biog Geneal Master Ind.

MAHLA, FREDERICK (fl. second half of 19th century). Chemistry; Toxicology. OCCUP: Professor. Scientific work: see Roy Soc Cat. WORKS ABOUT: Andreas v2:531, 538, 549, 554, 556, v3:864 (mentioned).

MAIN, WILLIAM (1844?-1918). Chemistry. OCCUP: Professor; Industrialist (?). WORKS ABOUT: Sci 48(1918):467.

MAINE, THOMAS (-1896). Engineering. OCCUP: Mechanical engineer. WORKS ABOUT: Sci 3(1896):839.

MAISCH, HENRY CHARLES CHRISTIAN (1862-1901). Chemistry. OCCUP: Industrial scientist. WORKS ABOUT: Who Was Who Am 1.

MAISCH, JOHN MICHAEL (1831-1893). Pharmacology. OCCUP: Pharmacist; Professor; Editor. WORKS ABOUT: Dict Am Biog; Natl Cycl Am Biog 5; Who Was Who Am h. See also: Biog Geneal Master Ind.

MAKUEN, GEORGE HUDSON (1855-1917). Medicine. OCCUP: Laryngologist; Professor. WORKS ABOUT: Am Men Sci 1-2 d3; Natl Cycl Am Biog 12, 19; Who Was Who Am 1.

MALL, FRANKLIN PAINE (1862-1917). Anatomy; Embryology; Physiology. OCCUP: Professor. MSS: Ind NUCMC (65-1952). WORKS ABOUT: Am Men Sci 1-2 d3; Dict Am Biog; Dict Am Med Biog; Dict Sci Biog; Natl Ac Scis Biog Mem 16; Natl Cycl Am Biog 14; Who Was Who Am 1. See also: Barr; Ireland Ind Sci.

MALLERY, GARRICK (1831-1894). Anthropology; Ethnology. OCCUP: Army officer; Research institution associate. WORKS ABOUT: Dict Am Biog; Natl Cycl Am Biog 7:506; Who Was Who Am h. See also: Barr; Biog Geneal Master Ind.

MALLET, JOHN WILLIAM (1832-1912). Chemistry. OCCUP: Professor. MSS: Ind NUCMC. WORKS ABOUT: Am Men Sci 1-2 d3; Dict Am Biog; Elliott (s); Natl Cycl Am Biog 13; Who Was Who Am 1. See also: Barr; Biog Geneal Master Ind; Ireland Ind Sci; Pelletier.

MALLORY, WILLIAM G. (-1918). Physics. OCCUP: Professor. WORKS ABOUT: Sci 48(1918):467-68.

MANCHESTER, JAMES EUGENE (1855-1913). Mathematics. OCCUP: Educator; Professor. WORKS ABOUT: Am Men Sci 1-2 d3; Sci 37(1913):330.

MANN, HORACE (1844-1868). Botany. OCCUP: Natural history collector; Curator. MSS: Ind NUCMC (60-1968). WORKS ABOUT: Am Journ Sci 97(1869):143; Boston Soc Nat Hist Proc 12(1869):152-55; Essex Inst Bull 1(1870):25-31, 41-50; Journ Bot 7(1869):168-70.

MANNING, CHARLES HENRY (1844-1919). Mechanical engineering. OCCUP: Naval engineer; Industrial manager. WORKS ABOUT: Am Men Sci 1 d3; Who Was Who Am 1.

MANNING, GEORGE LINCOLN (1865-1914). Physics. OCCUP: Professor. WORKS ABOUT: Am Men Sci 1 d3; Sci 40(1914):848.

MANROSS, NEWTON SPAULDING (1828?-1862). Chemistry; Geology. OCCUP: Mining engineer; Explorer. WORKS ABOUT: <u>Am Journ Sci</u> 84(1862):452; Appleton's; <u>Natl Cycl Am Biog</u> 11.

MANSFIELD, CHARLES M. (-1917). Photography. OCCUP: U.S. government scientist. WORKS ABOUT: <u>Sci</u> 46(1917):635.

MANSFIELD, JARED (1759-1830). Mathematics; Physics; Surveying. OCCUP: Educator; Professor; Army engineer. MSS: <u>Ind NUCMC</u> (63-262, 70-1346). WORKS ABOUT: <u>Dict Am Biog</u>; Elliott; <u>Natl Cycl Am Biog</u> 3; <u>Who Was Who Am</u> h. See also: Barr; <u>Biog Geneal Master Ind</u>.

MAPES, CHARLES VICTOR (1836-1916). Agricultural chemistry. OCCUP: Editor; Businessman. WORKS ABOUT: <u>Dict Am Biog</u>; Elliott; <u>Natl Cycl Am Biog</u> 3; <u>Who Was Who Am</u> 1. See also: Barr; <u>Biog Geneal Master Ind</u>.

MAPES, JAMES JAY (1806-1866). Agriculture; Chemistry. OCCUP: Businessman; Editor; Consultant (chemistry). WORKS ABOUT: <u>Dict Am Biog</u>; Elliott; <u>Natl Cycl Am Biog</u> 3; <u>Who Was Who Am</u> h. See also: <u>Biog Geneal Master Ind</u>; Ireland <u>Ind Sci</u>; Pelletier.

MARCH, ALDEN (1795-1869). Anatomy. OCCUP: Physician; Professor (anatomy and surgery). MSS: <u>Ind NUCMC</u>. WORKS ABOUT: <u>Dict Am Biog</u>; <u>Dict Am Med Biog</u>; <u>Natl Cycl Am Biog</u> 2; <u>Who Was Who Am</u> h. See also: <u>Biog Geneal Master Ind</u>.

MARCOU, JULES (1824-1898). Geology; Paleontology; Topography. OCCUP: Professor; Museum affiliate. MSS: <u>Ind NUCMC</u> (81-596). WORKS ABOUT: <u>Dict Am Biog</u>; <u>Dict Sci Biog</u>; Elliott; <u>Natl Cycl Am Biog</u> 25; <u>Who Was Who Am</u> h. See also: Barr; <u>Biog Geneal Master Ind</u>; Ireland <u>Ind Sci</u>.

MARCY, OLIVER (1820-1899). Geology. OCCUP: Professor; Museum curator. WORKS ABOUT: Elliott; <u>Natl Cycl Am Biog</u> 13; <u>Who Was Who Am</u> 1. See also: Barr; <u>Biog Geneal Master Ind</u>.

MARINDIN, HENRY LOUIS FRANCOIS (1843-1904). Hydrography. OCCUP: U.S. government scientist. WORKS ABOUT: <u>Sci</u> 19(1904):598; <u>Who Was Who Am</u> 1.

MARKOE, FRANCIS HARTMAN (1856-1907). Surgery. OCCUP: Surgeon (?); Professor. WORKS ABOUT: <u>Sci</u> 26(1907):390.

MARKOE, GEORGE F. H. (-1896). Chemistry. OCCUP: Professor; Consultant (chemistry) (?). WORKS ABOUT: <u>Sci</u> 4(1896):493-94.

MARKOE, THOMAS MASTERS (1819-1901). Surgery. OCCUP: Physician; Professor. WORKS ABOUT: Adams O; Allibone <u>Supp</u>; Appleton's; <u>Natl Cycl Am Biog</u> 11; <u>Sci</u> 14(1901):380; Wallace; <u>Who Was Who Am</u> 1.

MARKS, SOLON (1827-1914). Surgery. OCCUP: Surgeon; Professor. WORKS ABOUT: <u>Am Men Sci</u> 2 d3; <u>Natl Cycl Am Biog</u> 2; <u>Who Was Who Am</u> 1.

MARKS, WILLIAM DENNIS (1849-1914). Engineering. OCCUP: Engineer; Professor; Industrial engineer and officer. WORKS ABOUT: Adams O; Allibone <u>Supp</u>; <u>Am Men Sci</u> 1-2 d3; <u>Sci</u> 39(1914):206; <u>Twentieth Cent Biog Dict</u>; Wallace; <u>Who Was Who Am</u> 1.

MARSH, BENJAMIN V. (1818-   ). Astronomy. OCCUP: Businessman. WORKS ABOUT: Poggendorff v3.

MARSH, GEORGE PERKINS (1801-1882). Humanity and environment; Physical geography. OCCUP: Lawyer; Businessman; Public official. MSS: <u>Ind NUCMC</u> (81-315). WORKS ABOUT: <u>Dict Am Biog</u>; Elliott; Natl Ac Scis <u>Biog Mem</u> 6; <u>Natl Cycl Am Biog</u> 2; <u>Who Was Who Am</u> h. See also: Barr; <u>Biog Geneal Master Ind</u>; Pelletier.

MARSH, OTHNIEL CHARLES (1831-1899). Vertebrate paleontology. OCCUP: Professor; U.S. government scientist. MSS: <u>Ind NUCMC</u> (65-1076). WORKS ABOUT: <u>Dict Am Biog</u>; <u>Dict Sci Biog</u>; Elliott; Natl Ac Scis <u>Biog Mem</u> 20; <u>Natl Cycl Am Biog</u> 9; <u>Who Was Who Am</u> h. See also: Barr; <u>Biog Geneal Master Ind</u>; Ireland <u>Ind Sci</u>; Pelletier.

MARSHALL, HUMPHRY (1722-1801). Botany. OCCUP: Inherited wealth; Farmer. MSS: <u>Ind NUCMC</u>. WORKS ABOUT: <u>Dict Am Biog</u>; Elliott; <u>Who Was Who Am</u> h. See also: Barr; <u>Biog Geneal Master Ind</u>.

MARSHALL, JOHN POTTER (1824?-1901). Geology. OCCUP: Professor. WORKS ABOUT: <u>Sci</u> 13(1901):279.

MARTIN, ARTEMAS (1835-1918). Mathematics. OCCUP: Farmer; U.S. government scientist. WORKS ABOUT: <u>Am Men Sci</u> 1-2 d3; <u>Dict Am Biog</u>; <u>Natl Cycl Am Biog</u> 2; <u>Who Was Who Am</u> 1. See also: Barr; <u>Biog Geneal Master Ind</u>.

MARTIN, DOUGLAS S. (-1914). Electricity. OCCUP: Editor. WORKS ABOUT: <u>Sci</u> 40(1914):812.

MARTIN, GEORGE (1826-1886). Botany. OCCUP: Physician. WORKS ABOUT: <u>Journ Mycol</u> 2(1886):137-38; <u>Natl Cycl Am Biog</u> 15.

MARTIN, GEORGE A. (1831?-1904). Agriculture. OCCUP: Editor; Author. WORKS ABOUT: <u>Sci</u> 19(1904):743.

MARTIN, HENRY NEWELL (1848-1896). Physiology. OCCUP: Professor. WORKS ABOUT: <u>Dict Am Biog</u>; <u>Dict Am Med Biog</u>; <u>Dict Sci Biog</u>; Elliott; <u>Natl Cycl Am Biog</u> 12; <u>Who Was Who Am</u> h. See also: Barr; <u>Biog Geneal Master Ind</u>.

MARTIN, MARIA (1796-1863). Natural history; Nature painting. OCCUP: Artist. WORKS ABOUT: <u>Notable Am Wom</u>. See also: <u>Biog Geneal Master Ind</u>.

MARTINDALE, ISAAC C. (1842-1893). Botany; Entomology.
OCCUP: Banker. MSS: Ind NUCMC (66-61). WORKS ABOUT: Ent
News 4(1893):37-38; Gard and Forest 6(1893):36; Journ Bot
31(1893):159; Torrey Bot Club Bull 20(1893):98-100.

MARVIN, FRANK OLIN (1852-1915). Civil engineering. OCCUP:
Professor; Academic administrator. WORKS ABOUT: Am Men Sci
1-2 d3; Sci 41(1915):600-01; Who Was Who Am 1.

MARVIN, JAMES (1820-1901). Education; Mathematics. OCCUP:
Professor; College president; Clergyman. WORKS ABOUT: Natl
Cycl Am Biog 9; Sci 14(1901):158; Twentieth Cent Biog
Dict.

MARVINE, ARCHIBALD R. (1848-1876). Geology. OCCUP: U.S.
government scientist. WORKS ABOUT: Am Journ Sci
111(1876):424.

MARX, GEORGE (1838-1895). Entomology. OCCUP: Pharmacist;
Businessman; U.S. government scientist. WORKS ABOUT:
Elliott. See also: Ireland Ind Sci; Pelletier.

MASCHKE, HEINRICH (1853-1908). Mathematics. OCCUP:
Professor. WORKS ABOUT: Am Men Sci 1 d3; Dict Am Biog;
Elliott (s). See also: Barr; Ireland Ind Sci.

MASON, AMOS LAWRENCE (1842-1914). Medicine. OCCUP:
Physician; Professor. WORKS ABOUT: Am Men Sci 1-2 d3; Natl
Cycl Am Biog 2; Sci 39(1914):867; Twentieth Cent Biog
Dict.

MASON, CHARLES (1728?-1786). Astronomy; Geodesy. OCCUP:
Observatory affiliate; Surveyor. MSS: Ind NUCMC (68-1813,
68-1814). WORKS ABOUT: Dict Sci Biog; Natl Cycl Am Biog 10.
See also: Biog Geneal Master Ind; Pelletier.

MASON, EBENEZER PORTER (1819-1840). Astronomy. OCCUP:
Academic affiliate. WORKS ABOUT: Allibone; Am Journ Sci
40(1840-1841):407-08; Appleton's; Drake; Duyckinck; Natl
Cycl Am Biog 19.

MASON, JAMES WEIR (1836-1905). Mathematics. OCCUP:
Educator; Professor; Actuary. WORKS ABOUT: Elliott; Who Was
Who Am 1. See also: Barr.

MASON, OTIS TUFTON (1838-1908). Anthropology; Ethnology.
OCCUP: Educator; Museum curator. MSS: Ind NUCMC. WORKS
ABOUT: Am Men Sci 1 d3; Dict Am Biog; Natl Cycl Am Biog 10;
Who Was Who Am 1. See also: Barr; Biog Geneal Master Ind.

MASTERMAN, STILLMAN (1831-1863). Astronomy; Meteorology.
OCCUP: U.S. government science associate. WORKS ABOUT:
Elliott. See also: Barr; Biog Geneal Master Ind.

MATHER, COTTON (1662/63-1727/28). General science; Medicine. OCCUP: Clergyman. MSS: <u>Ind NUCMC</u>. WORKS ABOUT: <u>Dict Am Biog</u>; <u>Dict Am Med Biog</u>; Elliott; <u>Natl Cycl Am Biog</u> 4; <u>Who Was Who Am</u> h. See also: <u>Biog Geneal Master Ind</u>; Ireland <u>Ind Sci</u>; Pelletier.

MATHER, FRED (1833-1900). Pisciculture. OCCUP: Pisiculturist; State government scientist. WORKS ABOUT: <u>Dict Am Biog</u>; <u>Natl Cycl Am Biog</u> 13; <u>Who Was Who Am</u> 1. See also: Barr; <u>Biog Geneal Master Ind</u>.

MATHER, INCREASE (1639-1723). Astronomy; Promotion of science. OCCUP: Clergyman; College president. MSS: <u>Ind NUCMC</u>. WORKS ABOUT: <u>Dict Am Biog</u>; Elliott; <u>Natl Cycl Am Biog</u> 6; <u>Who Was Who Am</u> h. See also: <u>Biog Geneal Master Ind</u>.

MATHER, WILLIAM WILLIAMS (1804-1859). Geology. OCCUP: Army officer; Professor; State government scientist. MSS: <u>Ind NUCMC</u> (?). WORKS ABOUT: <u>Dict Am Biog</u>; <u>Dict Sci Biog</u>; Elliott; <u>Natl Cycl Am Biog</u> 8; <u>Who Was Who Am</u> h. See also: Barr; <u>Biog Geneal Master Ind</u>; Ireland <u>Ind Sci</u>; Pelletier.

MATHERS, GEORGE SCHRADER (1887?-1918). Medicine. OCCUP: Physician; Research institution associate. WORKS ABOUT: <u>Sci</u> 48(1918):443, 507-08.

MATHEWS, WILLIAM PHILLIP (1868-1918). Anatomy; Surgery. OCCUP: Surgeon; Professor. WORKS ABOUT: <u>Am Men Sci</u> 1-2 d3.

MATTHEWS, CHARLES PHILO (1867-1907). Electrical engineering. OCCUP: Professor. WORKS ABOUT: <u>Am Men Sci</u> 1 d3; <u>Sci</u> 26(1907):842-43.

MATTHEWS, WASHINGTON (1843-1905). Anthroplogy. OCCUP: Army surgeon. MSS: <u>Ind NUCMC</u> (69-1729). WORKS ABOUT: <u>Dict Am Biog</u>; <u>Natl Cycl Am Biog</u> 13; <u>Who Was Who Am</u> 1. See also: Barr; <u>Biog Geneal Master Ind</u>.

MAURY, MATTHEW FONTAINE (1806-1873). Meteorology; Oceanography; Physical geography. OCCUP: Naval officer; U.S. government scientist; Professor. MSS: <u>Ind NUCMC</u> (61-2970). WORKS ABOUT: <u>Dict Am Biog</u>; <u>Dict Sci Biog</u>; Elliott; <u>Natl Cycl Am Biog</u> 6; <u>Who Was Who Am</u> h. See also: Barr; <u>Biog Geneal Master Ind</u>; Ireland <u>Ind Sci</u>; Pelletier.

MAXIM, HIRAM STEVENS (1840-1916). Invention. OCCUP: Inventor; Engineer; Businessman. MSS: <u>Ind NUCMC</u>. WORKS ABOUT: <u>Dict Am Biog</u>; <u>Natl Cycl Am Biog</u> 6; <u>Who Was Who Am</u> 1. See also: Barr; <u>Biog Geneal Master Ind</u>; Ireland <u>Ind Sci</u>; Pelletier.

MAXON, ARTHUR RAY (1881?-1911). Mathematics. OCCUP: Academic instructor. WORKS ABOUT: <u>Sci</u> 33(1911):608.

MAXWELL, FRED BALDWIN (1862-1907). Biology. OCCUP: Teacher. WORKS ABOUT: <u>Am Men Sci</u> 1 d3.

MAXWELL, MARTHA (1831-1881). Natural history. OCCUP:
Collector; Taxidermist; Museum founder. WORKS ABOUT:
Benson.

MAYER, ALFRED MARSHALL (1836-1897). Physics. OCCUP:
Professor. MSS: Ind NUCMC (68-1740, 72-1059). WORKS ABOUT:
Dict Am Biog; Dict Sci Biog; Elliott; Natl Ac Scis Biog Mem
8; Natl Cycl Am Biog 13; Who Was Who Am h. See also: Barr;
Biog Geneal Master Ind; Ireland Ind Sci.

MAYER, FERDINAND F. (-ca.1866 to 1870). Chemistry;
Pharmacy. OCCUP: Professor; Pharmacist; Consultant
(analytical chemistry). Scientific work: see Roy Soc Cat.
WORKS ABOUT (mentioned in the following): Am Pharm Assoc
Proc 18(1870):22-23; Chymia 4(1953):172, 182-83; Druggists
Circ 10(1866):8 (advertisement); Sonnedecker; Wimmer p51,
53, 57, 60-62, 125.

MAYNARD, GEORGE WILLIAM (1839-1913). Chemistry; Mining
engineering. OCCUP: Engineer; Professor. WORKS ABOUT: Am
Men Sci 2-3 d4; Dict Am Biog; Who Was Who Am 1. See also:
Biog Geneal Master Ind; Pelletier.

MAYO, WILLIAM (ca.1684-1744). Surveying. OCCUP: Surveyor;
Public official. MSS: Ind NUCMC (?). WORKS ABOUT: Dict Am
Biog; Who Was Who Am h. See also: Biog Geneal Master Ind.

MAYO-SMITH, RICHMOND (1854-1901). Statistics. OCCUP:
Professor (political economy). MSS: Ind NUCMC (69-1413).
WORKS ABOUT: Dict Am Biog; Natl Ac Scis Biog Mem [Mem 17];
Natl Cycl Am Biog 29; Who Was Who Am 1. See also: Barr;
Biog Geneal Master Ind.

MEAD, SAMUEL B. (-1880). Botany. OCCUP: Not determined.
MSS: Ind NUCMC (?). WORKS ABOUT: Am Journ Sci
123(1882):333, 127(1884):242.

MEADE, WILLIAM (-1833). Mineralogy. OCCUP:
Physician (?). WORKS ABOUT: Elliott. See also: Barr; Biog
Geneal Master Ind.

MEARNS, EDGAR ALEXANDER (1856-1916). Natural history;
Botany; Zoology. OCCUP: Army surgeon; Museum associate;
Natural history explorer. MSS: Ind NUCMC (70-961). WORKS
ABOUT: Am Men Sci 1-2 d3; Dict Am Biog; Natl Cycl Am Biog
25; Who Was Who Am 1. See also: Biog Geneal Master Ind.

MEARS, JAMES EWING (1838-1919). Anatomy; Surgery. OCCUP:
Professor; Surgeon. WORKS ABOUT: Allibone Supp; Am Men Sci
1-2 d3; Indiana Authors 1917; Wallace.

MEARS, LEVERETT (1850-1917). Chemistry. OCCUP: Professor.
WORKS ABOUT: Am Men Sci 1-2 d3; Sci 45(1917):657; Who Was
Who Am 1.

MEASE, JAMES (1771-1846). General science; Medicine. OCCUP: Physician. MSS: Ind NUCMC. WORKS ABOUT: Dict Am Biog; Elliott; Who Was Who Am h. See also: Biog Geneal Master Ind; Ireland Ind Sci.

MEECH, LEVI WITTER (1821-1912). Mathematics; Actuarial science. OCCUP: U.S. government employee; Actuary. Scientific work: see Roy Soc Cat. WORKS ABOUT: Brigham v1:354 (mentioned); Brown Hist Cat 1904; Brown Hist Cat 1934; Poggendorff v3; U S Natl Lib Med Ind Cat; Wallbridge (mentioned).

MEEHAN, THOMAS (1826-1901). Botany; Horticulture. OCCUP: Horticulturist; Editor; State government scientist. MSS: Ind NUCMC (66-62). WORKS ABOUT: Dict Am Biog; Elliott; Natl Cycl Am Biog 11; Who Was Who Am 1. See also: Barr; Biog Geneal Master Ind.

MEEK, FIELDING BRADFORD (1817-1876). Geology; Paleontology. OCCUP: U.S. and state government scientist; Museum associate. MSS: Ind NUCMC (72-1244). WORKS ABOUT: Dict Am Biog; Dict Sci Biog; Elliott; Natl Ac Scis Biog Mem 4; Natl Cycl Am Biog 11; Who Was Who Am h. See also: Barr; Biog Geneal Master Ind; Ireland Ind Sci.

MEEK, SETH EUGENE (1859-1914). Zoology. OCCUP: Professor; Museum curator. WORKS ABOUT: Am Journ Sci 188(1914):370; Am Men Sci 1-2 d3; Sci 40(1914):95; Wallace; Who Was Who Am 1.

MEIER, EDWARD DANIEL (1841-1914). Engineering. OCCUP: Industrial engineer and manager; Businessman. MSS: Ind NUCMC (?). WORKS ABOUT: Natl Cycl Am Biog 15, 23; Sci 40(1914):930.

MEIGS, ARTHUR VINCENT (1850-1912). Medicine. OCCUP: Physician. WORKS ABOUT: Am Men Sci 1-2 d3; Dict Am Biog; Dict Am Med Biog; Natl Cycl Am Biog 15, 25; Who Was Who Am 1. See also: Barr; Biog Geneal Master Ind,

MEIGS, CHARLES DELUCENA (1792-1869). Medicine; Zoology. OCCUP: Physician; Professor (medicine). Scientific work: see Roy Soc Cat. MSS: Ind NUCMC. WORKS ABOUT: Dict Am Biog; Dict Am Med Biog; Natl Cycl Am Biog 6; Who Was Who Am h. See also: Biog Geneal Master Ind.

MEIGS, JAMES AITKEN (1829-1879). Craniology; Physiology. OCCUP: Physician; Professor (medicine). MSS: Ind NUCMC. WORKS ABOUT: Dict Am Biog; Natl Cycl Am Biog 8; Who Was Who Am h. See also: Biog Geneal Master Ind.

MEIGS, JOSIAH (1757-1822). Astronomy; Natural philosophy. OCCUP: Lawyer; Professor; Public official. MSS: Ind NUCMC. WORKS ABOUT: Dict Am Biog; Natl Cycl Am Biog 9; Who Was Who Am h. See also: Biog Geneal Master Ind.

MEIGS, MONTGOMERY CUNNINGHAM (1816-1892). Engineering.
OCCUP: Army officer; Engineer. MSS: Ind NUCMC (63-380, 80-247). WORKS ABOUT: Dict Am Biog; Natl Ac Scis Biog Mem 3; Natl Cycl Am Biog 4; Who Was Who Am h. See also: Biog Geneal Master Ind.

MELISH, JOHN (1771-1822). Geography. OCCUP: Businessman; Publisher; Map-maker. MSS: Ind NUCMC. WORKS ABOUT: Dict Am Biog; Who Was Who Am h. See also: Biog Geneal Master Ind.

MELL, PATRICK HUES (1850-1918). Botany; Geology. OCCUP: Professor; College president; State government scientist. MSS: Ind NUCMC (64-230). WORKS ABOUT: Am Men Sci 1-2 d3; Dict Am Biog; Natl Cycl Am Biog 15; Who Was Who Am 1. See also: Barr; Biog Geneal Master Ind.

MELSHEIMER, FRIEDRICH VALENTIN (1749-1814). Entomology. OCCUP: Clergyman. MSS: Ind NUCMC. WORKS ABOUT: Dict Am Biog; Elliott; Who Was Who Am h. See also: Barr; Biog Geneal Master Ind; Pelletier.

MELTZER, SAMUEL JAMES (1851-1920). Pathology; Pharmacology; Physiology. OCCUP: Physician; Research institution scientist. WORKS ABOUT: Am Men Sci 1-2 d3; Dict Am Biog; Dict Am Med Biog; Dict Sci Biog; Natl Ac Scis Biog Mem [Mem 21(9)]; Natl Cycl Am Biog 15; Who Was Who Am 1.

MELVILLE, GEORGE WALLACE (1841-1912). Mechanical engineering. OCCUP: Naval officer; Engineer. MSS: Ind NUCMC (73-899). WORKS ABOUT: Am Men Sci 1-2 d3; Dict Am Biog; Natl Cycl Am Biog 3; Who Was Who Am 1. See also: Barr; Biog Geneal Master Ind.

MELVIN, ALONZO DORUS (1862-1917). Veterinary science. OCCUP: U.S. government scientist. WORKS ABOUT: Am Men Sci 2 d3; Sci 46(1917):585; Who Was Who Am 1.

MENSCH, P. CALVIN (-1901). Biology. OCCUP: Professor. WORKS ABOUT: Sci 14(1901):228.

MERCUR, JAMES (1842-1896). Engineering. OCCUP: Army engineer; Professor. WORKS ABOUT: Adams O; Sci 3(1896):703; Twentieth Cent Biog Dict; Wallace.

MERRICK, JOHN VAUGHAN (1828-1906). Mechanical engineering. OCCUP: Mechanical engineer; Manufacturer. WORKS ABOUT: Franklin Inst Journ 161(1906):469-72; Natl Cycl Am Biog 13:334 (biography of father, Samuel V. Merrick); Who Was Who Am 1.

MERRILL, FREDERICK JAMES HAMILTON (1861-1916). Geology. OCCUP: State government scientist; Consultant (mining). MSS: Ind NUCMC. WORKS ABOUT: Am Men Sci 1-2 d3; Natl Cycl Am Biog 13; Sci 44(1916):814; Twentieth Cent Biog Dict; Who Was Who Am 1.

MERRILL, JAMES CUSHING (1853-1902). Ornithology. OCCUP: Army surgeon. WORKS ABOUT: <u>Dict Am Biog</u>; <u>Natl Cycl Am Biog</u> 15; <u>Who Was Who Am</u> 1. See also: Barr.

MERRILL, JOSHUA (1820-1904). Chemistry. OCCUP: Industrial scientist and officer; Businessman. WORKS ABOUT [birth dates differ]: <u>Dict Am Biog</u>; <u>Natl Cycl Am Biog</u> 13.

MERRILL, NATHAN FREDERICK (1849?-1915). Chemistry. OCCUP: Professor. WORKS ABOUT: <u>Sci</u> 42(1915):641.

MERRILL, WILLIAM STANLEY (1798-1880). Chemistry. OCCUP: Businessman. Scientific work: see Roy Soc <u>Cat</u>. WORKS ABOUT: Goss v3:829-30.

MESSENGER, HIRAM JOHN (1855-1913). Actuarial science; Mathematics. OCCUP: Industrial scientist; Professor. MSS: <u>Ind NUCMC</u> (?). WORKS ABOUT: <u>Am Men Sci</u> 1-2 d3; <u>Sci</u> 39(1914):63.

METCALF, WILLIAM (1838-1909). Engineering; Metallurgy. OCCUP: Engineer; Manufacturer. WORKS ABOUT: <u>Am Men Sci</u> 1-2 d3; <u>Dict Am Biog</u>; <u>Natl Cycl Am Biog</u> 12; <u>Who Was Who Am</u> 1. See also: <u>Biog Geneal Master Ind</u>.

METCALFE, SAMUEL LYTLER (1798-1856). Chemistry. OCCUP: Physician; Author; Researcher. WORKS ABOUT: <u>Dict Am Biog</u>; Elliott; <u>Natl Cycl Am Biog</u> 5; <u>Who Was Who Am</u> h. See also: <u>Biog Geneal Master Ind</u>.

MEYER, FRANK NICHOLAS (1875-1918). Botany. OCCUP: Horticulturist; Natural history explorer/ collector; U.S. government scientist. WORKS ABOUT: Jewett and McCausland p102-39; <u>Natl Cycl Am Biog</u> 20; Schapsmeier; <u>Sci</u> 47(1918):635, 48(1918):335-36.

MICHAEL, ELLIS LeROY (1881-1920). Zoology. OCCUP: Research institution scientist. WORKS ABOUT: <u>Am Men Sci</u> 2 d3.

MICHAEL, HELEN CECILIA DeSILVER ABBOTT (1857-1904). Chemistry; Plant chemistry. OCCUP: Laboratory associate; Author. WORKS ABOUT: Siegel and Finley; Wallace; <u>Who Was Who Am</u> 1.

MICHAELSON, HENRY (-1904). Forestry. OCCUP: Administrator; Author. WORKS ABOUT: <u>Sci</u> 19(1904):439.

MICHAUX, ANDRE (1746-1802). Botany; Silviculture. OCCUP: Natural history explorer/ collector; Planter; Government scientist (foreign). MSS: <u>Ind NUCMC</u>. WORKS ABOUT: <u>Dict Am Biog</u>; <u>Dict Sci Biog</u>; Elliott; <u>Who Was Who Am</u> h. See also: Barr; <u>Biog Geneal Master Ind</u>; Ireland <u>Ind Sci</u>; Pelletier.

MICHAUX, FRANCOIS ANDRE (1770-1855). Botany; Silviculture. OCCUP: Horticulturist; Government scientist (foreign); Natural history explorer/ collector. MSS: Ind NUCMC (60-2773). WORKS ABOUT: Dict Am Biog; Elliott; Who Was Who Am h. See also: Barr; Biog Geneal Master Ind; Ireland Ind Sci; Pelletier.

MICHEL, WILLIAM MIDDLETON (1822-1894). Medicine; Zoology. OCCUP: Physician; Professor. MSS: Ind NUCMC (70-1619). WORKS ABOUT: Dict Am Biog; Natl Cycl Am Biog 24; Who Was Who Am h. See also: Biog Geneal Master Ind.

MICHENER, EZRA (1794-1887). Natural history; Botany. OCCUP: Physician. WORKS ABOUT: Dict Am Biog; Elliott; Natl Cycl Am Biog 15; Who Was Who Am h. See also: Barr; Biog Geneal Master Ind.

MICHIE, PETER SMITH (1839-1901). Engineering; Physics. OCCUP: Army officer; Engineer; Professor. WORKS ABOUT: Dict Am Biog; Elliott; Who Was Who Am 1. See also: Barr; Biog Geneal Master Ind.

MIDDLETON, PETER (-1781). Chemistry. OCCUP: Physician; Professor. WORKS ABOUT: Dict Am Biog; Dict Am Med Biog; Who Was Who Am h. See also: Biog Geneal Master Ind; Pelletier.

MILES, MANLY (1826-1898). Agriculture; Natural history; Zoology. OCCUP: Physician; Professor; Researcher. WORKS ABOUT: Dict Am Biog; Elliott; Natl Cycl Am Biog 15. See also: Barr; Biog Geneal Master Ind.

MILLER, BLOOMFIELD JACKSON (1849-1905). Mathematics. OCCUP: Actuary; Industrial scientist and manager. WORKS ABOUT: Natl Cycl Am Biog 12; Who Was Who Am 1.

MILLER, EDMUND HOWD (1869-1906). Chemistry. OCCUP: Professor. WORKS ABOUT: Am Men Sci 1 d3; Natl Cycl Am Biog 20; Sci 24(1906):638; Wallace; Who Was Who Am 1.

MILLER, FRANK ELLSWORTH (1862-1919). Mathematics. OCCUP: Professor. WORKS ABOUT: Am Men Sci 1-2 d3.

MILLER, HARRIET MANN (also know as OLIVE THORNE MILLER) (1831-1918). Natural history; Nature writing. OCCUP: Author; Lecturer. WORKS ABOUT: Am Men Sci 1-2 d3; Dict Am Biog; Natl Cycl Am Biog 9; Notable Am Wom; Ogilvie; Siegel and Finley; Who Was Who Am 1. See also: Biog Geneal Master Ind; Ireland Ind Sci.

MILLER, JAMES BLAINE (1883-1915). Geodesy. OCCUP: U.S. government scientist. WORKS ABOUT: Sci 41(1915):859.

MILLER, JOSHUA (-1901). Archeology. OCCUP: Not determined. WORKS ABOUT: Sci 14(1901):191.

MILLER, SAMUEL ALMOND (1836-1897). Geology; Zoology. OCCUP: Lawyer. MSS: Ind NUCMC (?). WORKS ABOUT: Adams O; Allibone Supp; Geol Mag 5(1898):192; Ohio Authors; Ottawa Nat 11(1897):208; Wallace.

MILLER, WILLOUGHBY DAYTON (1853-1907). Bacteriology; Dentistry. OCCUP: Dentist; Professor. WORKS ABOUT: Dict Am Biog. See also: Barr; Biog Geneal Master Ind; Ireland Ind Sci.

MILLIKIN, BENJAMIN LOVE (1851-1916). Ophthalmology. OCCUP: Ophthalmologist; Professor; Academic administrator. WORKS ABOUT: Natl Cycl Am Biog 44; Sci 43(1916):95; Who Was Who Am 1.

MILLINGTON, JOHN (1779-1868). Natural philosophy; Chemistry; Mechanics; Civil engineering. OCCUP: Engineer; Professor. MSS: Ind NUCMC (67-186). WORKS ABOUT: Dict Am Biog; Dict Natl Biog. See also: Barr; Biog Geneal Master Ind; Pelletier.

MILLS, ADELBERT PHILO (1883-1918). Engineering. OCCUP: Professor. WORKS ABOUT: Natl Cycl Am Biog 18; Sci 48(1918):644; Wallace.

MILLS, HENRY (1813-1889). Microscopy; Study of sponges. OCCUP: Industrial affiliate. WORKS ABOUT: Am Soc Micr Proc 11(1889):152-53.

MILNE, WILLIAM JAMES (1843-1914). Mathematics. OCCUP: Professor (languages); College president. WORKS ABOUT: Sci 40(1914):408; Wallace; Who Was Who Am 4.

MINOT, CHARLES SEDGWICK (1852-1914). Biology; Anatomy; Embryology. OCCUP: Professor. MSS: Ind NUCMC (62-3755). WORKS ABOUT: Am Men Sci 1-2 d3; Dict Am Biog; Dict Am Med Biog; Dict Sci Biog; Elliott (s); Natl Ac Scis Biog Mem 9; Natl Cycl Am Biog 6; Who Was Who Am 1. See also: Barr; Biog Geneal Master Ind; Ireland Ind Sci.

MINOT, HENRY DAVIS (1859-1890). Ornithology. OCCUP: Businessman; Industrial officer. WORKS ABOUT: Adams O; Auk 8(1891):121; Wallace.

MINTO, WALTER (1753-1796). Mathematics. OCCUP: Educator; Professor. MSS: Ind NUCMC (?). WORKS ABOUT: Dict Am Biog; Elliott; Who Was Who Am h. See also: Biog Geneal Master Ind.

MITCHEL, ORMSBY MacKNIGHT (1809-1862). Astronomy. OCCUP: Professor; Observatory director; Lecturer. MSS: Ind NUCMC (71-1549). WORKS ABOUT: Dict Am Biog; Elliott; Natl Cycl Am Biog 3; Who Was Who Am h. See also: Barr; Biog Geneal Master Ind.

MITCHELL, ELISHA (1793-1857). Natural history; Botany; Geology. OCCUP: Professor; State government scientist. MSS: Ind NUCMC (64-580). WORKS ABOUT: Dict Am Biog; Dict Sci Biog; Elliott; Natl Cycl Am Biog 7:30; Who Was Who Am h. See also: Barr; Biog Geneal Master Ind; Ireland Ind Sci.

MITCHELL, HENRY (1830-1902). Hydrography. OCCUP: U.S. government scientist; Professor. WORKS ABOUT: Dict Am Biog; Elliott; Natl Ac Scis Biog Mem 20; Natl Cycl Am Biog 8; Who Was Who Am 1. See also: Barr; Biog Geneal Master Ind.

MITCHELL, IRVING NELSON (1850-1918). Biology. OCCUP: Educator; Professor. WORKS ABOUT: Am Men Sci 1-2 d3.

MITCHELL, JAMES ALFRED (1852-1902). Geology. OCCUP: Professor. WORKS ABOUT: Elliott; Who Was Who Am 1.

MITCHELL, JOHN (1711-1768). Natural history; Botany; Cartography; Chemistry. OCCUP: Physician; Author. MSS: Ind NUCMC. WORKS ABOUT: Dict Am Biog; Elliott; Who Was Who Am h. See also: Biog Geneal Master Ind; Ireland Ind Sci; Pelletier.

MITCHELL, JOHN KEARSLEY (1793-1858). Chemistry; Physiology. OCCUP: Physician; Professor (medicine). MSS: Ind NUCMC. WORKS ABOUT: Dict Am Biog; Dict Am Med Biog; Elliott; Natl Cycl Am Biog 9; Who Was Who Am h. See also: Biog Geneal Master Ind; Ireland Ind Sci; Pelletier.

MITCHELL, JOHN KEARSLEY (1859-1917). Neurology. OCCUP: Physician; Academic instructor. WORKS ABOUT: Am Men Sci 1-2 d3; Sci 45(1917):381.; Wallace; Who Was Who Am 1.

MITCHELL, MARIA (1818-1889). Astronomy. OCCUP: Librarian; U.S. government scientist; Professor. MSS: Ind NUCMC (66-1997). WORKS ABOUT: Dict Am Biog; Dict Sci Biog; Elliott; Natl Cycl Am Biog 5; Notable Am Wom; Ogilvie; Siegel and Finley; Who Was Who Am h. See also: Barr; Biog Geneal Master Ind; Ireland Ind Sci; Pelletier.

MITCHELL, OSCAR HOWARD (ca.1852-1889). Mathematics; Symbolic logic. OCCUP: Professor. WORKS ABOUT: Johns Hopkins Circ 8(1888-1889):110.

MITCHELL, SAMUEL AUGUSTUS (1792-1868). Geography. OCCUP: Publisher; Author; Map-maker. MSS: Ind NUCMC. WORKS ABOUT: Dict Am Biog; Who Was Who Am h. See also: Biog Geneal Master Ind.

MITCHELL, SILAS WEIR (1829-1914). Medicine; Neurology; Physiology. OCCUP: Physician; Author. MSS: Ind NUCMC. WORKS ABOUT: Am Men Sci 1-2 d3; Dict Am Biog; Dict Am Med Biog; Dict Sci Biog; Elliott (s); Natl Ac Scis Biog Mem 32; Natl Cycl Am Biog 9; Who Was Who Am 1. See also: Barr; Biog Geneal Master Ind; Ireland Ind Sci; Pelletier.

MITCHELL, THOMAS DUCHE (1791-1865). Chemistry; Medicine. OCCUP: Physician; Professor. WORKS ABOUT: Dict Am Biog; Who Was Who Am h. See also: Biog Geneal Master Ind; Pelletier.

MITCHELL, WILLIAM (1791-1869). Astronomy. OCCUP: Educator; Businessman; Public official. WORKS ABOUT: Dict Am Biog; Elliott; Natl Cycl Am Biog 11; Who Was Who Am h. See also: Barr; Biog Geneal Master Ind.

MITCHILL, SAMUEL LATHAM (1764-1831). Chemistry; Geology; Natural history; Zoology. OCCUP: Physician; Professor; Legislator. MSS: Ind NUCMC (61-1098). WORKS ABOUT: Dict Am Biog; Dict Am Med Biog; Elliott; Natl Cycl Am Biog 4; Who Was Who Am h. See also: Barr; Biog Geneal Master Ind; Pelletier.

MOHR, CHARLES THEODORE (1824-1901). Botany. OCCUP: Pharmacist; Researcher; U.S. government scientist. MSS: Ind NUCMC (84-1018). WORKS ABOUT: Dict Am Biog; Elliott; Natl Cycl Am Biog 26; Who Was Who Am 1. See also: Barr; Biog Geneal Master Ind.

MOLINEUX, ROLAND BURNHAM (1866-1917). Chemistry. OCCUP: Industrial scientist. WORKS ABOUT: Who Was Who Am 4.

MONELL, JOSEPH TARRIGAN (1857?-1915). Entomology. OCCUP: Mining engineer. WORKS ABOUT: Sci 42(1915):119, 217.

MONTGOMERY, EDMUND DUNCAN (1835-1911). Biology; Cell biology; Philosophy. OCCUP: Physician; Researcher; Author. MSS: Ind NUCMC (62-425). WORKS ABOUT: Am Men Sci 1-2 d3; Dict Am Biog; Dict Sci Biog; Who Was Who Am 1. See also: Biog Geneal Master Ind.

MONTGOMERY, HENRY (1848-   ). Anthropology; Geology. OCCUP: Professor; Museum curator. WORKS ABOUT: Am Men Sci 1-2 d3.

MONTGOMERY, JOHN JOSEPH (1858-1911). Invention; Aeronautics. OCCUP: Inventor; Professor. MSS: Ind NUCMC (78-2178). WORKS ABOUT: Cohen p90-93; Heinmuller p250; Maitland p34-35; Natl Cycl Am Biog 15.

MONTGOMERY, THOMAS HARRISON, JR. (1873-1912). Zoology. OCCUP: Professor; Museum administrator. WORKS ABOUT: Am Men Sci 1-2 d3; Dict Am Biog; Dict Sci Biog; Natl Cycl Am Biog 15; Who Was Who Am 1. See also: Barr; Biog Geneal Master Ind.

MOODY, MARY BLAIR (1837-1919). Languages; Medicine; Medical science. OCCUP: Physician. WORKS ABOUT: Am Men Sci 1-2 d3; Am Wom; Siegel and Finley.

MOOERS, EMMA WILSON DAVIDSON (-1911). Pathology. OCCUP: Hospital scientist; Museum curator. WORKS ABOUT: Sci 33(1911):925, 959.

MOORE, GIDEON EMMET (1842-1895). Chemistry. OCCUP:
Industrial scientist; Consultant. WORKS ABOUT: Am Chem and
Chem Eng; Am Chem Soc Journ 17(1895):659-63; Am Journ Sci
149(1895):430; Browne and Weeks p312, 319.

MOORE, JAMES EDWARD (1852-1918). Surgery. OCCUP: Surgeon;
Professor. WORKS ABOUT: Dict Am Biog; Natl Cycl Am Biog 6;
Who Was Who Am 1. See also: Biog Geneal Master Ind; Ireland
Ind Sci.

MOORE, JAMES W. (1844-1909). Physics. OCCUP: Professor.
WORKS ABOUT: Adams O; Appleton's; Natl Cycl Am Biog 11; Sci
29(1909):416; Twentieth Cent Biog Dict; Wallace; Who Was
Who Am 1.

MOORE, KATHLEEN CARTER (MRS. J. PERCY) (1866-1920).
Psychology. OCCUP: Teacher; Educator. WORKS ABOUT: Am Men
Sci 1-2 d3; Siegel and Finley.

MOORE, RICHARD B. (1815-1885). Natural history; Geology.
OCCUP: Tradesman. WORKS ABOUT: Cincinnati Soc Nat Hist
Journ 8(1886):67-69.

MORDECAI, ALFRED (1804-1887). Army ordnance; Meteorology.
OCCUP: Army officer; Engineer; Industrial officer.
Scientific work: see Roy Soc Cat. MSS: Ind NUCMC (83-
1068). WORKS ABOUT: Dict Am Biog; Natl Cycl Am Biog 10; Who
Was Who Am h. See also: Biog Geneal Master Ind.

MORE, THOMAS (fl. 1670-1724). Natural history. OCCUP: Not
determined; Natural history collector. WORKS ABOUT:
Elliott. See also: Pelletier.

MOREY, SAMUEL (1762-1843). Chemistry; Invention; Physics.
OCCUP: Businessman; Inventor. MSS: Ind NUCMC (84-77). WORKS
ABOUT: Dict Am Biog; Elliott; Natl Cycl Am Biog 11; Who Was
Who Am h. See also: Biog Geneal Master Ind.

MORFIT, CAMPBELL (1820-1897). Chemistry. OCCUP: Industrial
scientist; Businessman. MSS: Ind NUCMC. WORKS ABOUT: Dict
Am Biog; Elliott; Who Was Who Am h. See also: Biog Geneal
Master Ind.

MORGAN, ANDREW PRICE (1836-1906). Botany; Mycology. OCCUP:
Not determined. Am Men Sci 1 d3; Elliott (s).

MORGAN, JAMES DUDLEY (1862-1919). Medicine. OCCUP:
Physician; Professor. WORKS ABOUT: Am Men Sci 1-2 d3;
Appleton's Supp 1918-31; Natl Cycl Am Biog 22; Who Was Who
Am 1.

MORGAN, LEWIS HENRY (1818-1881). Anthropology; Ethnology.
OCCUP: Lawyer; Legislator; Researcher. MSS: Ind NUCMC (61-
1662, 74-287, 81-81). WORKS ABOUT: Dict Am Biog; Natl Ac
Scis Biog Mem 6; Natl Cycl Am Biog 6; Who Was Who Am h. See
also: Barr; Biog Geneal Master Ind; Ireland Ind Sci;
Pelletier.

MORISON, GEORGE SHATTUCK (1842-1903). Civil engineering. OCCUP: Engineer. WORKS ABOUT: <u>Dict Am Biog</u>; <u>Natl Cycl Am Biog</u> 10; <u>Who Was Who Am</u> 1. See also: Barr; <u>Biog Geneal Master Ind</u>.

MORONG, THOMAS (1827-1894). Botany. OCCUP: Clergyman; Curator; Collector. WORKS ABOUT: <u>Bot Gaz</u> 19(1894):225-28; <u>Journ Bot</u> 32(1894):319; Torrey Bot Club <u>Bull</u> 21(1894):239-44.

MORRILL, PARK (1860-1898). Meteorology. OCCUP: U.S. government scientist. WORKS ABOUT: U S Weather Bur <u>Mo Weather Rev</u> 26(1898):356-57.

MORRIS, EDWARD LYMAN (1870-1913). Botany. OCCUP: Educator; Museum curator. WORKS ABOUT: <u>Am Men Sci</u> 1-2 d3; <u>Sci</u> 38(1913):401, 476.

MORRIS, ELLWOOD (ca.1813-1872). Engineering. OCCUP: Civil engineer. Scientific work: see Roy Soc <u>Cat</u>. WORKS ABOUT: Am Phil Soc members; Am Phil Soc <u>Proc</u> 9(1862-1864):116, 12(1871-1872):437; Franklin Inst <u>Journ</u> 116(1883):391 (mentioned); Sinclair.

MORRIS, JOHN GOTTLIEB (1803-1895). Entomology. OCCUP: Clergyman; Lecturer. MSS: <u>Ind NUCMC</u> (?). WORKS ABOUT: <u>Dict Am Biog</u>; Elliott; <u>Who Was Who Am</u> h. See also: Barr; <u>Biog Geneal Master Ind</u>; Ireland <u>Ind Sci</u>; Pelletier.

MORRIS, JOHN LEWIS (1842-1905). Mechanics. OCCUP: Professor. MSS: <u>Ind NUCMC</u> (?). WORKS ABOUT: <u>Sci</u> 22(1905):726.

MORRIS, MARGARETTA HARE (1797-1867). Entomology. OCCUP: Engaged in independent study. WORKS ABOUT: Elliott; Siegel and Finley.

MORRIS, ORAN WILKINSON (1798-1877). Meteorology. OCCUP: Teacher; Educator. WORKS ABOUT: Torrey Bot Club <u>Bull</u> 6(1877):166-68.

MORRISON, HERBERT KNOWLES (1854-1885). Entomology. OCCUP: Natural history explorer/ collector. MSS: <u>Ind NUCMC</u> (?). WORKS ABOUT: <u>Ent Am</u> 1(1885-1886):100; Essig p709-10; <u>Psyche</u> 4(1890):287; <u>Sci</u> 1st ser 5(1885):532.

MORRISON, JOSEPH (1848-   ). Astronomy; Mathematics. OCCUP: Physician; U.S. government scientist; Professor. WORKS ABOUT: Adams O; <u>Am Men Sci</u> 1; <u>Who Was Who Am</u> 4.

MORROW, G. E. [GEORGE?] (1840?-1900). Agriculture. OCCUP: Professor. MSS: <u>Ind NUCMC</u> (?). WORKS ABOUT: <u>Sci</u> 11(1900):557.

MORROW, PRINCE ALBERT (1846-1913). Dermatology. OCCUP: Physician; Professor. WORKS ABOUT: <u>Dict Am Biog</u>; <u>Natl Cycl Am Biog</u> 21; <u>Who Was Who Am</u> 1. See also: Barr; <u>Biog Geneal Master Ind</u>.

MORSE, ELISHA WILSON (1866-1915). Biology. OCCUP: Academic instructor; U.S. government scientist. WORKS ABOUT: <u>Am Men Sci</u> 1-2; <u>Sci</u> 41(1915):677.

MORSE, HARMON NORTHROP (1848-1920). Chemistry; Organic chemistry. OCCUP: Museum administrator. WORKS ABOUT: <u>Am Men Sci</u> 1-2 d3; <u>Dict Am Biog</u>; Elliott (s); Natl Ac Scis <u>Biog Mem</u> [Mem 21(11)]; <u>Natl Cycl Am Biog</u> 16, 19; <u>Who Was Who Am</u> 1. See also: <u>Biog Geneal Master Ind</u>; Ireland <u>Ind Sci</u>.

MORSE, JEDIDIAH (1761-1826). Geography. OCCUP: Clergyman; Author. MSS: <u>Ind NUCMC</u> (60-2930, 81-1018). WORKS ABOUT: <u>Dict Am Biog</u>; <u>Dict Sci Biog</u>; <u>Natl Cycl Am Biog</u> 13; <u>Who Was Who Am</u> h. See also: <u>Biog Geneal Master Ind</u>.

MORSE, SAMUEL FINLEY BREESE (1791-1872). Invention; Telegraphy. OCCUP: Artist; Inventor. MSS: <u>Ind NUCMC</u> (66-1595, 66-1812, 77-246). WORKS ABOUT: <u>Dict Am Biog</u>; <u>Natl Cycl Am Biog</u> 4; <u>Who Was Who Am</u> h. See also: Barr; <u>Biog Geneal Master Ind</u>; Ireland <u>Ind Sci</u>; Pelletier.

MORTON, HENRY (1836-1902). Chemistry; Physics; Science education. OCCUP: Professor; College administrator. WORKS ABOUT: <u>Dict Am Biog</u>; Elliott; Natl Ac Scis <u>Biog Mem</u> 8; <u>Natl Cycl Am Biog</u> 11, 24; <u>Who Was Who Am</u> 1. See also: Barr; <u>Biog Geneal Master Ind</u>; Pelletier.

MORTON, SAMUEL GEORGE (1799-1851). Natural history; Anthropology; Craniology. OCCUP: Physician; Professor. MSS: <u>Ind NUCMC</u> (62-4974). WORKS ABOUT: <u>Dict Am Biog</u>; <u>Dict Am Med Biog</u>; <u>Dict Sci Biog</u>; Elliott; <u>Natl Cycl Am Biog</u> 10; <u>Who Was Who Am</u> h. See also: Barr; <u>Biog Geneal Master Ind</u>; Ireland <u>Ind Sci</u>; Pelletier.

MORTON, WILLIAM THOMAS GREEN (1819-1868). Medicine; Anaesthesiology. OCCUP: Dentist. MSS: <u>Ind NUCMC</u> (80-273). WORKS ABOUT: <u>Dict Am Biog</u>; <u>Dict Am Med Biog</u>; <u>Natl Cycl Am Biog</u> 8; <u>Who Was Who Am</u> h. See also: <u>Biog Geneal Master Ind</u>; Ireland <u>Ind Sci</u>; Pelletier.

MOSES, ALFRED JOSEPH (1859-1920). Mineralogy. OCCUP: Professor. WORKS ABOUT: Adams O; <u>Am Journ Sci</u> 199(1920):389; <u>Am Men Sci</u> 1-2 d3; <u>Natl Cycl Am Biog</u> 47; Wallace; <u>Who Was Who Am</u> 1.

MOSES, OTTO A. (1846-1906). Chemistry; Geology. OCCUP: Educator (?); State government scientist. WORKS ABOUT: <u>Sci</u> 23(1906):78.

MOSES, THOMAS FREEMAN (1836-1917). Archeology; Geology. OCCUP: Physician; Professor; College president. WORKS ABOUT: <u>Am Men Sci</u> 1-2 d3; Appleton's; <u>Twentieth Cent Biog Dict</u>; <u>Who Was Who Am</u> 1.

MOTT, HENRY AUGUSTUS, JR. (1852-1896). Chemistry. OCCUP: Industrial scientist; Professor; Consultant. WORKS ABOUT: Elliott; Natl Cycl Am Biog 3. See also: Biog Geneal Master Ind.

MOYES, HENRY (1750-1807). Chemistry. OCCUP: Lecturer. WORKS ABOUT: Am Chem and Chem Eng.

MUDGE, BENJAMIN FRANKLIN (1817-1879). Geology. OCCUP: Lawyer; State government scientist; Professor. WORKS ABOUT: Am Geol 23(1899):339-45; Am Journ Sci 119(1880):82; Appleton's; Kansas Ac Sci Trans 7(1881):7-11; Merrill p423-25; Twentieth Cent Biog Dict.

MUHLENBERG, GOTTHILF HENRY ERNEST (1753-1815). Botany. OCCUP: Clergyman. MSS: Ind NUCMC (60-2676, 61-1093). WORKS ABOUT: Dict Am Biog; Elliott; Natl Cycl Am Biog 9; Who Was Who Am h. See also: Barr; Biog Geneal Master Ind; Ireland Ind Sci.

MUIR, JOHN (1838-1914). Natural history; Geology. OCCUP: Explorer; Horticulturist; Activist (conservation). MSS: Ind NUCMC (61-2403, 62-397, 71-770, 72-779). WORKS ABOUT: Am Men Sci 1-2 d3; Dict Am Biog; Elliott (s); Natl Cycl Am Biog 9; Who Was Who Am 1. See also: Barr; Biog Geneal Master Ind; Ireland Ind Sci; Pelletier.

MUMFORD, JAMES GREGORY (1863-1914). Surgery. OCCUP: Physician. WORKS ABOUT: Am Men Sci 2 d3; Dict Am Biog; Dict Am Med Biog; Who Was Who Am 1. See also: Biog Geneal Master Ind.

MUNDE, PAUL FORTUNATUS (1846-1902). Gynecology. OCCUP: Physician; Editor; Professor. WORKS ABOUT: Dict Am Biog; Natl Cycl Am Biog 12; Who Was Who Am 1. See also: Barr; Biog Geneal Master Ind.

MUNRO, JOHN CUMMINGS (1858-1910). Surgery. OCCUP: Surgeon; Academic instructor. WORKS ABOUT: Am Men Sci 1-2 d3; Leonardo p297-98; Who Was Who Am 1.

MUNROE, GEORGE (-1896). Mathematics. OCCUP: Publisher; Academic instructor. WORKS ABOUT: Sci 3(1896):899.

MUNSELL, CHARLES EDWARD (1858-1918). Chemistry. OCCUP: City and state government scientist; Industrial scientist. WORKS ABOUT: Who Was Who Am 1.

MUNSON, ENEAS (1734-1826). Chemistry. OCCUP: Physician; Researcher. WORKS ABOUT: Am Chem and Chem Eng; Dict Am Med Biog.

MUNSON, THOMAS VOLNEY (1843-1913). Botany. OCCUP: Horticulturist. WORKS ABOUT: Am Men Sci 2 d3; Dict Am Biog; Natl Cycl Am Biog 18; Who Was Who Am 1. See also: Barr; Biog Geneal Master Ind.

MUNSTERBERG, HUGO (1863-1916). Psychology. OCCUP: Professor. MSS: Ind NUCMC. WORKS ABOUT: Am Men Sci 1-2 d3; Dict Am Biog; Dict Am Med Biog; Natl Cycl Am Biog 13; Who Was Who Am 1. See also: Barr; Biog Geneal Master Ind; Ireland Ind Sci.

MURPHY, EDGAR GARDNER (pseudonym KELVIN McKREADY) (1869-1913). Astronomy; Education. OCCUP: Clergyman; Author; Educator. MSS: Ind NUCMC (62-4664, 68-810). WORKS ABOUT: Dict Am Biog; Natl Cycl Am Biog 25; Who Was Who Am 1. See also: Barr; Biog Geneal Master Ind.

MURPHY, JOHN BENJAMIN (1857-1916). Surgery. OCCUP: Surgeon; Professor. WORKS ABOUT: Am Men Sci 2 d3; Dict Am Biog; Dict Am Med Biog; Natl Cycl Am Biog 13; Who Was Who Am 1. See also: Barr; Biog Geneal Master Ind; Ireland Ind Sci; Pelletier.

MURRAY, DAVID (1830-1905). Astronomy; Mathematics. OCCUP: Educator; Professor. MSS: Ind NUCMC (62-4521). WORKS ABOUT: Dict Am Biog; Who Was Who Am 1. See also: Barr; Biog Geneal Master Ind.

MURRAY, ROBERT DRAKE (1845-1903). Sanitation. OCCUP: Surgeon. WORKS ABOUT: Sci 18(1903):702, 734; Who Was Who Am 1.

MURTFELDT, MARY ESTHER (1848-1913). Botany; Entomology. OCCUP: U.S. and state government scientist; Writer. WORKS ABOUT: Am Men Sci 2; Ogilvie; Osborn Herb p165-66; Pop Sci Mo 74(1909):251-52; Wom Whos Who Am.

MUSSER, JOHN HERR (1856-1912). Medicine. OCCUP: Physician; Professor. MSS: Ind NUCMC. WORKS ABOUT: Am Men Sci 1-2 d3; Natl Cycl Am Biog 13; Sci 35(1912):571; Wallace; Who Was Who Am 1.

MYER, ALBERT JAMES (1829-1880). Meteorology. OCCUP: Army officer. MSS: Ind NUCMC (67-145, 77-1554). WORKS ABOUT: Dict Am Biog; Natl Cycl Am Biog 4, 24; Who Was Who Am h. See also: Barr; Biog Geneal Master Ind.

MYERS, JESSE J. (1876?-1914). Agriculture; Physiology. OCCUP: Professor. WORKS ABOUT: Am Journ Sci 118(1914):116; Sci 39(1914):825.

N

NASON, HENRY BRADFORD (1831-1895). Chemistry. OCCUP:
Professor. WORKS ABOUT: <u>Dict Am Biog</u>; Elliott; <u>Natl Cycl Am
Biog</u> 2; <u>Who Was Who Am</u> h. See also: Barr; <u>Biog Geneal
Master Ind</u>; Ireland <u>Ind Sci</u>; Pelletier.

NASON, WILLIAM ABBOT (1841-   ). Entomology. OCCUP:
Physician (?). WORKS ABOUT: <u>Am Men Sci</u> 2 d3.

NEF, JOHN ULRIC (1862-1915). Chemistry. OCCUP: Professor.
WORKS ABOUT: <u>Am Men Sci</u> 1-2 d3; <u>Dict Am Biog</u>; <u>Dict Sci
Biog</u>; Natl Ac Scis <u>Biog Mem</u> 34; <u>Natl Cycl Am Biog</u> 21; <u>Who
Was Who Am</u> 1. See also: Barr; <u>Biog Geneal Master Ind</u>;
Pelletier.

NEFF, PETER (1827-1903). Geology. OCCUP: Clergyman;
Inventor; Librarian. MSS: <u>Ind NUCMC</u> (75-1646). WORKS ABOUT:
<u>Natl Cycl Am Biog</u> 13:253; <u>Ohio Authors</u>.

NELSON, EDWARD THOMSON (-1897). Biology. OCCUP:
Professor. WORKS ABOUT: <u>Sci</u> 5(1897):442.

NELSON, HOWARD A. (-1915). Geology. OCCUP: Student.
WORKS ABOUT: <u>Sci</u> 42(1915):489.

NELSON, JULIUS (1858-1916). Biology. OCCUP: Professor;
State government scientist. WORKS ABOUT: <u>Am Men Sci</u> 1-2 d3;
<u>Dict Am Biog</u>; <u>Natl Cycl Am Biog</u> 18; <u>Who Was Who Am</u> 1. See
also: Barr.

NETTLETON, EDWIN S. (1831-1901). Agriculture; Engineering.
OCCUP: Engineer; State government engineer. WORKS ABOUT:
<u>Dict Am Biog</u>; <u>Who Was Who Am</u> h. See also: Barr.

NEUMOEGEN, BERTHOLD (1845-1895). Entomology. OCCUP: Banker;
Businessman. WORKS ABOUT: <u>Ent News</u> 6(1895):65-66.

NEWBERRY, JOHN STRONG (1822-1892). Geology; Paleontology. OCCUP: Physician; Professor; U.S. and state government scientist. MSS: Ind NUCMC. WORKS ABOUT: Dict Am Biog; Dict Sci Biog; Elliott; Natl Ac Scis Biog Mem 6; Natl Cycl Am Biog 9; Who Was Who Am h. See also: Barr; Biog Geneal Master Ind; Ireland Ind Sci.

NEWCOMB, SIMON (1835-1909). Astronomy. OCCUP: U.S. government scientist; Professor. MSS: Ind NUCMC (63-384). WORKS ABOUT: Am Men Sci 1 d3; Dict Am Biog; Dict Sci Biog; Elliott (s); Natl Ac Scis Biog Mem [Mem 17]; Natl Cycl Am Biog 7:17; Who Was Who Am 1. See also: Barr; Biog Geneal Master Ind; Ireland Ind Sci; Pelletier.

NEWCOMB, WESLEY (1808-1892). Conchology. OCCUP: Physician; Collector. WORKS ABOUT: Nautilus 5(1892):121-24; Sci 28(1908):243; Twentieth Cent Biog Dict; West Am Sci 4(1888):1-3.

NEWSON, HENRY BYRON (1860-1910). Mathematics. OCCUP: Professor. WORKS ABOUT: Am Men Sci 1 d3; Sci 31(1910):294-95; Wallace; Who Was Who Am 1.

NEWTON, HENRY (1845-1877). Geology; Mining engineering. OCCUP: U.S. and state government scientist; Academic affiliate. WORKS ABOUT: Elliott; Natl Cycl Am Biog 4. See also: Barr; Biog Geneal Master Ind.

NEWTON, HUBERT ANSON (1830-1896). Astronomy; Mathematics. OCCUP: Professor. MSS: Ind NUCMC. WORKS ABOUT: Dict Am Biog; Dict Sci Biog; Elliott; Natl Ac Scis Biog Mem 4; Natl Cycl Am Biog 9; Who Was Who Am h. See also: Barr; Biog Geneal Master Ind; Ireland Ind Sci.

NEWTON, JOHN (1823-1895). Engineering. OCCUP: Army officer; Engineer; Industrial officer. WORKS ABOUT: Dict Am Biog; Natl Ac Scis Biog Mem 4; Natl Cycl Am Biog 4; Who Was Who Am h. See also: Barr; Biog Geneal Master Ind.

NICHOLS, JAMES ROBINSON (1819-1888). Chemistry. OCCUP: Manufacturer; Editor. WORKS ABOUT: Dict Am Biog; Elliott; Natl Cycl Am Biog 5; Who Was Who Am h. See also: Biog Geneal Master Ind.

NICHOLS, OTHNIEL FOSTER (1845-1908). Civil engineering. OCCUP: Engineer; Industrial engineer and manager. WORKS ABOUT: Am Men Sci 1 d3; Natl Cycl Am Biog 9; Sci 27(1908):397; Twentieth Cent Biog Dict; Who Was Who Am 1.

NICHOLS, WILLIAM RIPLEY (1847-1886). Chemistry. OCCUP: Professor. WORKS ABOUT: Allibone Supp; Appleton's; Twentieth Cent Biog Dict; Wallace.

NICOLLET, JOSEPH NICOLAS (1786-1843). Astronomy; Exploration; Mathematics. OCCUP: Observatory affiliate; Explorer. MSS: Ind NUCMC. WORKS ABOUT: Dict Am Biog; Elliott; Who Was Who Am h. See also: Barr; Biog Geneal Master Ind; Pelletier.

NILES, WILLIAM HARMON (1838-1910). Geography; Geology.
OCCUP: Professor. WORKS ABOUT: Am Men Sci 1-2 d3;
Elliott (s); Natl Cycl Am Biog 12; Pop Sci Mo 55(1899):460;
Sci 32(1910):426; Twentieth Cent Biog Dict; Who Was Who
Am 1.

NIXON, HENRY BARBER (1857-1916). Mathematics. OCCUP:
Professor. WORKS ABOUT: Natl Cycl Am Biog 17; Sci
43(1916):492, 639.

NOBLE, ALFRED (1844-1914). Engineering. OCCUP: Engineer.
MSS: Ind NUCMC (80-1484). WORKS ABOUT: Dict Am Biog; Natl
Cycl Am Biog 9; Who Was Who Am 1. See also: Barr; Ireland
Ind Sci.

NOBLE, ALFRED J. (1858?-1916). Psychiatry. OCCUP: Medical
administrator. WORKS ABOUT: Sci 43(1916):131.

NORMAN, WESLEY W. (-1899). Biology. OCCUP: Professor.
WORKS ABOUT: Sci 10(1899):30.

NORRIS, WILLIAM FISHER (1839-1901). Ophthalmology. OCCUP:
Physician; Professor. WORKS ABOUT: Dict Am Biog; Natl Cycl
Am Biog 12, 18; Who Was Who Am 1. See also: Barr; Biog
Geneal Master Ind.

NORSWORTHY, NAOMI (1877-1916). Psychology. OCCUP:
Professor. WORKS ABOUT: Am Men Sci 1-2 d3; Dict Am Biog;
Siegel and Finley; Who Was Who Am 1. See also: Biog Geneal
Master Ind.

NORTH, ERASMUS DARWIN (1806-1858). Microscopy. OCCUP:
Physician; Academic instructor (elocution). WORKS ABOUT: Am
Journ Sci 76(1858):155; North p53-55, 98-99; Twentieth Cent
Biog Dict; Wallace.

NORTHROP, JOHN ISAIAH (1861-1891). Biology. OCCUP: Mining
engineer; Academic instructor. WORKS ABOUT: Auk
8(1891):400; New York Ac Scis Trans 11(1891-1892):9-12;
Osborn Hen; Sci 1st ser 18(1891):7, 33(1911):858.

NORTON, EDWARD (1823-1894). Natural history; Entomology.
OCCUP: Farmer; Businessman. WORKS ABOUT: Ent News
5(1894):161-63; Insect Life 6(1894):379.

NORTON, JOHN PITKIN (1822-1852). Agricultural chemistry.
OCCUP: Professor. MSS: Ind NUCMC (71-2029). WORKS ABOUT:
Dict Am Biog; Dict Sci Biog; Elliott; Natl Cycl Am Biog 8;
Who Was Who Am h. See also: Barr; Biog Geneal Master Ind;
Ireland Ind Sci; Pelletier.

NORTON, LEWIS MILLS (1855-1893). Chemistry. OCCUP:
Professor. WORKS ABOUT: Am Ac Arts Scis Proc 28(1893):348-
51; Am Chem and Chem Eng; Am Chem Soc Journ 15(1893):241-
44; Appleton's.

NORTON, SIDNEY AUGUSTUS (1835-1918). Chemistry. OCCUP: Professor. WORKS ABOUT: Adams O; Am Chem and Chem Eng; Am Men Sci 1-2 d3; Appleton's; Natl Cycl Am Biog 12; Sci 48(1918):597-98; Twentieth Cent Biog Dict; Wallace; Who Was Who Am 1.

NORTON, WILLIAM AUGUSTUS (1810-1883). Astronomy; Physics. OCCUP: Professor. WORKS ABOUT: Elliott; Natl Ac Scis Biog Mem 2; Natl Cycl Am Biog 9. See also: Barr; Biog Geneal Master Ind.

NORWOOD, JOSEPH GRANVILLE (1807-1895). Geology; Fossil study. OCCUP: Physician; State government scientist; Professor. MSS: Ind NUCMC (72-1529). WORKS ABOUT: Am Journ Sci 150(1895):79-80; Kelly and Burrage.

NOTT, JOSIAH CLARK (1804-1873). Ethnology; Medicine. OCCUP: Physician; Professor (medicine). MSS: Ind NUCMC. WORKS ABOUT: Dict Am Biog; Dict Am Med Biog; Natl Cycl Am Biog 19; Who Was Who Am h. See also: Biog Geneal Master Ind; Ireland Ind Sci.

NOYES, WILLIAM (1857-1915). Mental diseases. OCCUP: Physician; Academic instructor. WORKS ABOUT: Am Men Sci 1-2 d3; Sci 42(1915):688.

NUTTALL, THOMAS (1786-1859). Natural history; Botany; Ornithology. OCCUP: Natural history explorer/ collector; Academic instructor; Curator. MSS: Ind NUCMC. WORKS ABOUT: Dict Am Biog; Dict Sci Biog; Elliott; Natl Cycl Am Biog 8; Who Was Who Am h. See also: Barr; Biog Geneal Master Ind; Ireland Ind Sci; Pelletier.

NYSTROM, JOHN WILLIAM (1824-1885). Engineering. OCCUP: Civil engineer. WORKS ABOUT: Adams O; Allibone; Allibone Supp; Am Machinist 8(27 June 1885):6; Franklin Inst Journ 75(1863):276, 284-85 (autobiographical information); Rail Gaz 17(1885):318; Wallace.

O

OAKES, WILLIAM (1799-1848). Botany. OCCUP: Lawyer; Engaged in independent study. MSS: Ind NUCMC (62-3054). WORKS ABOUT: Am Journ Sci 57(1849):138-42.

OBER, FREDERICK ALBION (1849-1913). Ornithology. OCCUP: Natural history explorer/ collector; Businessman. WORKS ABOUT: Dict Am Biog; Elliott; Natl Cycl Am Biog 13, 54; Who Was Who Am 1. See also: Barr; Biog Geneal Master Ind.

O'CONNELL, JOSEPH J. (1866?-1916). Hygiene. OCCUP: Physician (?); Public official; Academic instructor. WORKS ABOUT: Sci 43(1916):65.

OGDEN, HERBERT GOUVERNEUR (1846-1906). Cartography; Topography. OCCUP: U.S. government scientist. WORKS ABOUT: Am Men Sci 1 d3; Dict Am Biog; Who Was Who Am 1.

OHLMACHER, ALBERT PHILIP (1865-1916). Pathology. OCCUP: Physician; Professor. WORKS ABOUT: Am Men Sci 1-2 d3; Dict Am Biog; Natl Cycl Am Biog 26; Who Was Who Am 1.

OHM, FREDERICK C. (1858?-1916). Geology. OCCUP: U.S. government scientist. WORKS ABOUT: Sci 43(1916):530.

OLDBERG, OSCAR (1846-1913). Pharmacy. OCCUP: Pharmacist; Professor; Academic administrator. MSS: Ind NUCMC (72-1787). WORKS ABOUT: Natl Cycl Am Biog 20; Sci 37(1913):408; Who Was Who Am 1. See also: Biog Geneal Master Ind.

OLIVER, ANDREW (1731-1799). Astronomy; Meteorology. OCCUP: Inherited wealth; Public official. MSS: Ind NUCMC. WORKS ABOUT: Dict Am Biog; Elliott; Who Was Who Am h. See also: Biog Geneal Master Ind.

OLIVER, CHARLES AUGUSTUS (1853-1911). Ophthalmology. OCCUP: Physician. WORKS ABOUT: Am Men Sci 1-2 d3; Dict Am Biog; Who Was Who Am 1. See also: Barr; Biog Geneal Master Ind.

OLIVER, JAMES EDWARD (1829-1895). Mathematics. OCCUP: U.S. government scientist; Professor. WORKS ABOUT: Elliott; Natl Ac Scis Biog Mem 4; Natl Cycl Am Biog 14. See also: Barr; Biog Geneal Master Ind.

OLIVER, MARSHAL (1843?-1900). Engineering. OCCUP: Professor. WORKS ABOUT: Sci 12(1900):893.

OLMSTED, DENISON (1791-1859). Astronomy; Geology; Physics. OCCUP: Professor; State government scientist. MSS: Ind NUCMC (74-1197). WORKS ABOUT: Dict Am Biog; Elliott; Natl Cycl Am Biog 8; Who Was Who Am h. See also: Barr; Biog Geneal Master Ind; Ireland Ind Sci; Pelletier.

OLMSTED, DENISON, JR. (1824-1846). Mineralogy. OCCUP: Academic affiliate; State government scientist. WORKS ABOUT: Am Journ Sci 52(1846):297.

OLNEY, STEPHEN THAYER (1812-1878). Botany. OCCUP: Merchant; Manufacturer. MSS: Ind NUCMC. WORKS ABOUT: Am Journ Sci 117(1879):179-80; Natl Cycl Am Biog 13.

OLSSON-SEFFER, PEHR HJALMAR (1873-1911). Botany. OCCUP: Academic instructor; Government scientist (foreign); Laboratory administrator. WORKS ABOUT: Am Men Sci 1-2 d3; Sci 33(1911):722.

OPPENHEIM, NATHAN (1865-1916). Medicine. OCCUP: Physician. WORKS ABOUT: Adams O; Am Men Sci 1-2 d3; Sci 43(1916):530; Wallace; Who Was Who Am 1.

ORD, GEORGE (1781-1866). Natural history; Zoology; Ornithology. OCCUP: Businessman. MSS: Ind NUCMC (61-769). WORKS ABOUT: Dict Am Biog; Elliott; Natl Cycl Am Biog 13; Who Was Who Am h. See also: Biog Geneal Master Ind.

ORDRONAUX, JOHN (1830-1908). Medicine. OCCUP: Lawyer; Academic instructor (medical jurisprudence); Public official. MSS: Ind NUCMC. WORKS ABOUT: Dict Am Biog; Natl Cycl Am Biog 12; Who Was Who Am 1. See also: Barr; Biog Geneal Master Ind.

ORDWAY, ALBERT (1843-1897). Natural history. OCCUP: Army officer; Businessman. Scientific work: see Roy Soc Cat. WORKS ABOUT: Appleton's Supp 1901; Boatner (listed); Powell p416; Wallace.

ORDWAY, JOHN MORSE (1823-1909). Chemistry. OCCUP: Industrial scientist and manager; Professor. MSS: Ind NUCMC. WORKS ABOUT: Am Chem and Chem Eng; Am Men Sci 1 d3; Appleton's; Natl Cycl Am Biog 7:259; Sci 30(1909):111; Twentieth Cent Biog Dict; Who Was Who Am 1.

O'REILLY, ROBERT MAITLAND (1845-1912). Medicine. OCCUP: Army officer; Surgeon. MSS: Ind NUCMC (61-2353). WORKS ABOUT: Dict Am Biog; Natl Cycl Am Biog 18; Who Was Who Am 1. See also: Barr.

ORTON, EDWARD FRANCIS BAXTER (1829-1899). Geology. OCCUP: Educator; Professor; State government scientist. MSS: Ind NUCMC. WORKS ABOUT: Dict Am Biog; Elliott; Natl Cycl Am Biog 24; Who Was Who Am 1. See also: Barr; Biog Geneal Master Ind; Ireland Ind Sci.

ORTON, JAMES (1830-1877). Exploration; Natural history; Zoology. OCCUP: Clergyman; Natural history explorer; Professor. MSS: Ind NUCMC. WORKS ABOUT: Dict Am Biog; Dict Sci Biog; Elliott; Natl Cycl Am Biog 11; Who Was Who Am h. See also: Barr; Biog Geneal Master Ind.

OSBORN, HENRY STAFFORD (1823-1894). Botany; Geography; Mineralogy. OCCUP: Clergyman; Professor; Map-maker. WORKS ABOUT: Dict Am Biog; Natl Cycl Am Biog 11; Who Was Who Am h. See also: Biog Geneal Master Ind.

OSLER, WILLIAM (1849-1919). Medicine. OCCUP: Physician; Professor. MSS: Ind NUCMC. WORKS ABOUT: Am Men Sci 1 d3; Dict Am Biog; Dict Am Med Biog; Natl Cycl Am Biog 12; Who Was Who Am 1. See also: Barr; Biog Geneal Master Ind; Ireland Ind Sci; Pelletier.

OSTEN SACKEN, CARL ROBERT ROMANOVICH VON DER (1828-1906). Entomology. OCCUP: Public official (foreign); Museum associate. MSS: Ind NUCMC. WORKS ABOUT: Dict Am Biog; Elliott. See also: Barr; Ireland Ind Sci; Pelletier.

OSTERBERG, MAX (1869-1904). Electrical engineering. OCCUP: Consultant. WORKS ABOUT: Wallace; Who Was Who Am 1.

OSWALD, FELIX LEOPOLD (1845-1906). Natural history. OCCUP: Journalist; Author. WORKS ABOUT: Adams O; Allibone Supp; Appleton's; Biog Dict Synopsis Books; Sci 24(1906):447; Wallace; Who Was Who Am 1.

OTT, ISAAC (1847-1916). Neurology; Physiology. OCCUP: Physician; Professor. WORKS ABOUT: Am Men Sci 1-2 d3; Dict Am Biog; Dict Sci Biog; Natl Cycl Am Biog 20; Who Was Who Am 1. See also: Biog Geneal Master Ind.

OVERMAN, FREDERICK (ca.1803-1852). Metallurgy. OCCUP: Consultant; Industrial engineer; Author. WORKS ABOUT: Dict Am Biog; Who Was Who Am h. See also: Biog Geneal Master Ind.

OVIATT, BOARDMAN LAMBERT (1863-1889). Zoology. OCCUP: Academic affiliate. WORKS ABOUT: Am Soc Micr Proc 11(1889):151-52.

OWEN, DAVID DALE (1807-1860). Geology. OCCUP: U.S. and state government scientist. MSS: Ind NUCMC. WORKS ABOUT: Dict Am Biog; Dict Sci Biog; Elliott; Natl Cycl Am Biog 8; Who Was Who Am h. See also: Barr; Biog Geneal Master Ind; Ireland Ind Sci.

OWEN, RICHARD (1810-1890). Geology; Paleontology. OCCUP:
Farmer; Professor; State government scientist. MSS: <u>Ind</u>
<u>NUCMC</u> (61-3063, 83-1288). WORKS ABOUT: Adams O; Allibone;
<u>Am Geol</u> 6(1890):135-45; <u>Am Journ Sci</u> 139(1890):414,
140(1890):96, 149(1895):247; <u>Am Meteorol Journ</u> 7(1890-
1891):47; Appleton's; Geol Soc Am <u>Bull</u> 2(1891):610,
5(1894):571-72; <u>Indiana Authors 1816</u>; <u>Leopoldina</u>
26(1890):111; <u>Natl Cycl Am Biog</u> 14; <u>Pop Sci Mo</u>
51(1897):145, 259-65; <u>Twentieth Cent Biog Dict</u>; Wallace.

P

PACKARD, ALPHEUS SPRING, JR. (1839-1905). Zoology; Entomology. OCCUP: Curator; Editor; Professor. MSS: <u>Ind NUCMC</u>. WORKS ABOUT: <u>Dict Am Biog</u>; <u>Dict Sci Biog</u>; Elliott; Natl Ac Scis <u>Biog Mem</u> 9; <u>Natl Cycl Am Biog</u> 3; <u>Who Was Who Am</u> 1. See also: Barr; <u>Biog Geneal Master Ind</u>; Ireland <u>Ind Sci</u>; Pelletier.

PACKARD, FREDERICK A. (1862?-1902). Medicine. OCCUP: Physician (?). WORKS ABOUT: <u>Sci</u> 16(1902):799.

PAGE, CHARLES GRAFTON (1812-1868). Physics. OCCUP: Physician; U.S. government scientist; Consultant (patent agent). WORKS ABOUT: <u>Dict Am Biog</u>; Elliott; <u>Natl Cycl Am Biog</u> 5; <u>Who Was Who Am</u> h. See also: Barr; <u>Biog Geneal Master Ind</u>; Pelletier.

PAGE, LOGAN WALLER (1870-1918). Civil engineering. OCCUP: U.S. and state government scientist; Administrator. WORKS ABOUT: <u>Am Men Sci</u> 2 d3; Wallace; <u>Who Was Who Am</u> 1.

PAINE, JOHN ALSOP (1840-1912). Archeology; Botany. OCCUP: Professor; Explorer; Curator. WORKS ABOUT: <u>Am Men Sci</u> 1-2 d3; <u>Dict Am Biog</u>; <u>Natl Cycl Am Biog</u> 13; <u>Who Was Who Am</u> 1. See also: Barr; <u>Biog Geneal Master Ind</u>.

PAINE, ROBERT TREAT (1803-1885). Astronomy. OCCUP: Lawyer; State government scientist; Independent wealth. WORKS ABOUT: Elliott.

PAINTER, JOSEPH H. (-1908). Botany. OCCUP: Museum affiliate. WORKS ABOUT: <u>Sci</u> 28(1908):921.

PALMER, ARTHUR WILLIAM (1861-1904). Chemistry. OCCUP: Professor. MSS: <u>Ind NUCMC</u> (71-1163). WORKS ABOUT: Elliott; <u>Who Was Who Am</u> 1. See also: Barr.

PALMER, EDWARD (1831-1911). Natural history; Anthropology. OCCUP: Physician; Natural history explorer; Collector. MSS: Ind NUCMC (60-1862). WORKS ABOUT: <u>Am Men Sci</u> 2; <u>Dict Sci Biog</u>.

PANCOAST, JOSEPH (1805-1882). Anatomy. OCCUP: Physician; Professor. WORKS ABOUT: <u>Dict Am Biog</u>; <u>Dict Am Med Biog</u>; <u>Natl Cycl Am Biog</u> 10; <u>Who Was Who Am</u> h. See also: <u>Biog Geneal Master Ind</u>; Ireland <u>Ind Sci</u>; Pelletier.

PANCOAST, SETH (1823-1889). Anatomy. OCCUP: Physician; Professor. WORKS ABOUT: <u>Dict Am Biog</u>; <u>Who Was Who Am</u> h. See also: <u>Biog Geneal Master Ind</u>.

PANCOAST, WILLIAM HENRY (1835-1897). Anatomy; Surgery. OCCUP: Surgeon; Professor. WORKS ABOUT: Appleton's; Leonardo p321-22; <u>Natl Cycl Am Biog</u> 10; Preston; <u>Sci</u> 5(1897):103; <u>Twentieth Cent Biog Dict</u>.

PARK, AUSTIN F. (1825-1893). Natural history; Ornithology. OCCUP: Civil engineer; Instrument-maker. WORKS ABOUT: <u>Auk</u> 10(1893):384-85.

PARK, ROSWELL (1852-1914). Surgery. OCCUP: Surgeon; Professor. MSS: <u>Ind NUCMC</u> (77-1876). WORKS ABOUT: <u>Am Men Sci</u> 1-2 d3; <u>Dict Am Biog</u>; <u>Dict Am Med Biog</u>; <u>Natl Cycl Am Biog</u> 8; <u>Who Was Who Am</u> 1. See also: <u>Biog Geneal Master Ind</u>; Ireland <u>Ind Sci</u>.

PARKER, CHARLES F. (1820-1883). Natural history; Botany. OCCUP: Businessman; Curator. WORKS ABOUT: Ac Nat Scis Philadelphia <u>Proc</u> (1883):260-65; <u>Am Journ Sci</u> 127(1884):243.

PARKER, HENRY WEBSTER (ca.1822-1903). Chemistry; Natural history. OCCUP: Clergyman; Professor; Editor. WORKS ABOUT: Adams O; Allibone; Appleton's; <u>Carolyn Sherwin Bailey Collection</u>; Drake; Duyckinck; Osborn Herb p157-58; <u>Twentieth Cent Biog Dict</u>; Wallace.

PARKHURST, HENRY MARTYN (1825-1908). Astronomy. OCCUP: Clerk. <u>Am Men Sci</u> 1 d3.

PARKMAN, FRANCIS (1823-1893). Horticulture. OCCUP: Author (history); Professor. MSS: <u>Ind NUCMC</u>. WORKS ABOUT: <u>Dict Am Biog</u>; <u>Natl Cycl Am Biog</u> 1; <u>Who Was Who Am</u> h. See also: <u>Biog Geneal Master Ind</u>.

PARKMAN, THEODORE (1837-1862). Chemistry. OCCUP: Student; Soldier. WORKS ABOUT: <u>Am Journ Sci</u> 85(1863):155-56.

PARLOA, MARIA (1843-1909). Nutrition; Home economics. OCCUP: Teacher; Lecturer; Author. MSS: <u>Ind NUCMC</u>. WORKS ABOUT: <u>Notable Am Wom</u>; <u>Who Was Who Am</u> 1. See also: Barr; <u>Biog Geneal Master Ind</u>.

PARMLY, CHARLES HOWARD (1868-1917). Physics; Electrical engineering. OCCUP: Professor. WORKS ABOUT: Natl Cycl Am Biog 17; Sci 46(1917):261.

PARRISH, CELESTIA SUSANNAH (1853-1918). Philosophy of science; Psychology. OCCUP: Teacher; Professor (psychology, education); Public official. WORKS ABOUT: Am Men Sci 1; Dict Am Biog; Notable Am Wom; Siegel and Finley; Who Was Who Am 1. See also: Biog Geneal Master Ind.

PARRY, CHARLES CHRISTOPHER (1823-1890). Botany. OCCUP: Physician; U.S. government scientist; Natural history collector. MSS: Ind NUCMC (62-2804). WORKS ABOUT: Dict Am Biog; Elliott; Natl Cycl Am Biog 13; Who Was Who Am h. See also: Barr; Biog Geneal Master Ind; Ireland Ind Sci.

PATERSON, JOHN (1801-1883). Mathematics. OCCUP: Skilled laborer; State government scientist. WORKS ABOUT: Appleton's.

PATRICK, GEORGE EDWARD (1851-1916). Chemistry. OCCUP: Professor; U.S. and state government scientist. WORKS ABOUT: Sci 43(1916):639; Who Was Who Am 1.

PATTEN, JOSEPH HURLBUT (1801-1877). Instrumentation; Technology. OCCUP: Lawyer. Scientific work: see Roy Soc Cat. WORKS ABOUT: Baldwin T p133 (mentioned); Brown Hist Cat 1904.

PATTERSON, CARLILE POLLOCK (1816-1881). Hydrographic surveying; Geodesy. OCCUP: Naval officer; U.S. government scientist. MSS: Ind NUCMC (?). WORKS ABOUT: Appleton's; Lanman; Natl Cycl Am Biog 4; Sci 44(1916):47.

PATTERSON, ROBERT (1743-1824). Mathematics. OCCUP: Educator; Professor; Public official. MSS: Ind NUCMC. WORKS ABOUT: Dict Am Biog; Elliott; Natl Cycl Am Biog 1, 26; Who Was Who Am h. See also: Biog Geneal Master Ind; Ireland Ind Sci; Pelletier.

PATTERSON, ROBERT MASKELL (1787-1854). Chemistry; Natural philosophy. OCCUP: Professor; Public official; Businessman. MSS: Ind NUCMC. WORKS ABOUT: Elliott; Natl Cycl Am Biog 1, 26. See also: Biog Geneal Master Ind; Pelletier.

PATTISON, GRANVILLE SHARP (1791 or 1792-1851). Anatomy. OCCUP: Professor. WORKS ABOUT: Dict Am Biog; Dict Am Med Biog; Natl Cycl Am Biog 6; Who Was Who Am h. See also: Biog Geneal Master Ind.

PATTON, WILLIAM HAMPTON (1853-1918). Entomology. OCCUP: U.S. government scientist. MSS: Ind NUCMC (?). WORKS ABOUT: Ent News 32(1921):33-41; Osborn Herb p221-22.

PAUL WILHELM, DUKE OF WURTTEMBERG (1797-1860). Natural history; Exploration. OCCUP: Inherited wealth; Natural history explorer/ collector. MSS: Ind NUCMC. WORKS ABOUT [form of name may differ]: Appleton's; Groce and Wallace; Who Was Who Am Hist [as Wurttemberg].

PAULMIER, FREDERICK CLARK (1873-1906). Zoology. OCCUP: Museum affiliate. WORKS ABOUT: Am Men Sci 1 d3; Sci 23(1906):556-57.

PAVY, OCTAVE (1844-1884). Natural history. OCCUP: Physician; Explorer. WORKS ABOUT: Dict Am Biog; Natl Cycl Am Biog 7:534; Who Was Who Am h. See also: Biog Geneal Master Ind.

PEABODY, GEORGE (1795-1869). Philanthropy. OCCUP: Businessman; Benefactor. MSS: Ind NUCMC (73-459). WORKS ABOUT: Dict Am Biog; Natl Cycl Am Biog 5; Who Was Who Am h. See also: Barr; Biog Geneal Master Ind; Pelletier.

PEABODY, GEORGE LIVINGSTON (1850-1914). Medicine. OCCUP: Physician; Professor. WORKS ABOUT: Am Men Sci 1-2 d3; Natl Cycl Am Biog 24; Sci 40(1914):667; Who Was Who Am 1.

PEABODY, SELIM HOBART (1829-1903). Entomology; Physics. OCCUP: Educator; Professor; College president. MSS: Ind NUCMC (65-165). WORKS ABOUT: Dict Am Biog; Natl Cycl Am Biog 1; Who Was Who Am 1. See also: Biog Geneal Master Ind.

PEALE, ALBERT CHARLES (1849-1914). Geology; Paleobotany. OCCUP: U.S. government scientist; Museum affiliate. WORKS ABOUT: Am Journ Sci 189(1915):230; Am Men Sci 2 d3; Natl Cycl Am Biog 21; Sci 40(1914):888; Who Was Who Am 1.

PEALE, CHARLES WILLSON (1741-1827). Museum administration; Natural history. OCCUP: Artist; Museum founder. MSS: Ind NUCMC. WORKS ABOUT: Dict Am Biog; Dict Sci Biog; Elliott; Natl Cycl Am Biog 6; Who Was Who Am h. See also: Barr; Biog Geneal Master Ind; Ireland Ind Sci; Pelletier.

PEALE, REMBRANDT (1778-1860). Natural history; Technology. OCCUP: Artist. MSS: Ind NUCMC. WORKS ABOUT: Dict Am Biog; Dict Sci Biog; Natl Cycl Am Biog 5; Who Was Who Am h. See also: Biog Geneal Master Ind.

PEALE, TITIAN RAMSAY (1799-1885). Natural history; Zoology. OCCUP: Museum administrator; U.S. government scientist; Natural history collector. MSS: Ind NUCMC (67-626). WORKS ABOUT: Dict Am Biog; Dict Sci Biog; Elliott; Natl Cycl Am Biog 21; Who Was Who Am h. See also: Barr; Biog Geneal Master Ind; Ireland Ind Sci.

PEARSE, JOHN BARNARD SWETT (1842-1914). Metallurgy. OCCUP: Industrial scientist and manager; State government scientist; Consultant. WORKS ABOUT: Dict Am Biog; Natl Cycl Am Biog 26; Who Was Who Am 4. See also: Biog Geneal Master Ind.

PEARSON, HERBERT WILLIAM (1850-1916). Astronomy; Geology. OCCUP: Businessman. WORKS ABOUT: Natl Cycl Am Biog 18.

PEARSON, LEONARD (1868-1909). Veterinary medicine. OCCUP: Veterinarian; Professor; State government scientist. MSS: Ind NUCMC (?). WORKS ABOUT: Am Men Sci 1 d3; Dict Am Biog; Who Was Who Am 1. See also: Barr.

PEARY, ROBERT EDWIN (1856-1920). Arctic exploration. OCCUP: Naval officer; Explorer. MSS: Ind NUCMC (66-349, 76-726, 84-84, 84-85). WORKS ABOUT: Am Men Sci 1-2 d3; Dict Am Biog; Natl Cycl Am Biog 14, 37; Who Was Who Am 1. See also: Barr; Biog Geneal Master Ind; Ireland Ind Sci; Pelletier.

PEASE, W. HARPER (-1872). Zoology. OCCUP: Not determined; Collector; Researcher. WORKS ABOUT: Am Journ Sci 103(1872):320.

PECK, CHARLES HORTON (1833-1917). Botany; Mycology. OCCUP: State government scientist. MSS: Ind NUCMC. WORKS ABOUT: Am Men Sci 1-2 d3; Dict Am Biog; Elliott (s); Natl Cycl Am Biog 13; Who Was Who Am 3. See also: Barr; Biog Geneal Master Ind.

PECK, JAMES INGRAHAM (1863-1898). Biology. OCCUP: Professor; Laboratory administrator. WORKS ABOUT: Nature 59(1898-1899):154-55; Sci 8(1898):667, 744, 783; Woods Hole Biol Lectures (1898):339-41.

PECK, LOUIS W. (1851?-1898). Physics. OCCUP: Professor. WORKS ABOUT: Am Journ Sci 157(1899):248.

PECK, WILLIAM DANDRIDGE (1763-1822). Natural history; Botany; Entomology. OCCUP: Farmer (?); Professor; Curator. MSS: Ind NUCMC (65-1260). WORKS ABOUT: Dict Am Biog; Elliott; Who Was Who Am h. See also: Biog Geneal Master Ind; Ireland Ind Sci; Pelletier.

PECK, WILLIAM GUY (1820-1892). Mathematics. OCCUP: Army officer; Professor. WORKS ABOUT: Elliott; Natl Cycl Am Biog 5; Who Was Who Am h. See also: Barr; Biog Geneal Master Ind.

PECKHAM, GEORGE WILLIAMS (1845-1914). Biology; Entomology; Psychology. OCCUP: Educator; Librarian. WORKS ABOUT: Am Men Sci 1-2 d3; Dict Am Biog; Natl Cycl Am Biog 12; Who Was Who Am 1. See also: Barr; Biog Geneal Master Ind; Pelletier.

PECKHAM, STEPHEN FARNUM (1839-1918). Chemistry. OCCUP: Professor; City and state government scientist. MSS: Ind NUCMC. WORKS ABOUT: Am Men Sci 1-2 d3; Dict Am Biog; Natl Cycl Am Biog 9; Who Was Who Am 1. See also: Barr; Biog Geneal Master Ind.

PEEBLES, ALVIN ROY (1884-1917). Medicine. OCCUP: Physician; Professor. WORKS ABOUT: Who Was Who Am 1.

PEET, STEPHEN DENISON (1831-1914). Anthropology; Archeology. OCCUP: Clergyman; Editor and publisher. WORKS ABOUT: Am Men Sci 1; Dict Am Biog; Who Was Who Am 4. See also: Barr; Biog Geneal Master Ind.

PEIRCE, BENJAMIN (1809-1880). Astronomy; Mathematics. OCCUP: Professor; U.S. government scientist. MSS: Ind NUCMC (65-1262). WORKS ABOUT: Dict Am Biog; Dict Sci Biog; Elliott; Natl Cycl Am Biog 8; Who Was Who Am h. See also: Barr; Biog Geneal Master Ind; Ireland Ind Sci; Pelletier.

PEIRCE, BENJAMIN OSGOOD (1854-1914). Mathematics; Physics. OCCUP: Professor. MSS: Ind NUCMC (76-1993). WORKS ABOUT: Am Men Sci 1-2 d3; Dict Am Biog; Dict Sci Biog; Elliott (s); Natl Ac Scis Biog Mem 8; Natl Cycl Am Biog 20; Who Was Who Am 1. See also: Barr; Biog Geneal Master Ind; Ireland Ind Sci; Pelletier.

PEIRCE, CHARLES SANDERS (1839-1914). Geodesy; History of science; Logic; Mathematics; Philosophy of science; Psychology. OCCUP: U.S. government scientist; Academic instructor; Engaged in independent study. MSS: Ind NUCMC. WORKS ABOUT: Am Men Sci 1-2 d3; Dict Am Biog; Dict Sci Biog; Elliott (s); Natl Cycl Am Biog 8; Who Was Who Am 1. See also: Biog Geneal Master Ind; Ireland Ind Sci; Pelletier.

PEIRCE, GEORGE (1883-1919). Physiological chemistry. OCCUP: Physician; Academic instructor; Industrial scientist. WORKS ABOUT: Natl Cycl Am Biog 24.

PEIRCE, JAMES MILLS (1834-1906). Mathematics. OCCUP: Professor; Academic administrator. MSS: Ind NUCMC. WORKS ABOUT: Am Men Sci 1 d3; Dict Am Biog; Elliott (s); Natl Cycl Am Biog 10; Who Was Who Am 1. See also: Barr; Biog Geneal Master Ind.

PEIRCE, WILLIAM (ca.1590-1641). Almanac making. OCCUP: Shipmaster. WORKS ABOUT: Dict Am Biog; Who Was Who Am h. See also: Biog Geneal Master Ind.

PEMBERTON, HENRY (1826-1911). Chemistry. OCCUP: Industrial scientist and manager. MSS: Ind NUCMC (?). Wallace; Who Was Who Am 1.

PEMBERTON, HENRY, JR. (1855-1913). Chemistry. OCCUP: Industrial scientist and manager. WORKS ABOUT: Natl Cycl Am Biog 15; Wallace.

PENDLETON, EDMUND MONROE (1815-1884). Chemistry. OCCUP: Physician; Planter; Manufacturer. WORKS ABOUT: Dict Am Biog; Who Was Who Am h. See also: Biog Geneal Master Ind.

PENFIELD, SAMUEL LEWIS (1856-1906). Mineralogy. OCCUP: Professor. WORKS ABOUT: Am Journ Sci 172(1906):264, 353-67; Am Men Sci 1 d3; Elliott (s): Natl Ac Scis Biog Mem 6; Sci 24(1906):221, 252-53; Wallace; Who Was Who Am 1.

PENGER, CHRISTIAN (1850?-1902). Surgery. OCCUP: Surgeon (?); Professor. WORKS ABOUT: Sci 15(1902):438.

PENHALLOW, DAVID PEARCE (1854-1910). Botany. OCCUP: Professor. WORKS ABOUT: Am Journ Sci 180(1910):431; Am Men Sci 1-2 d3; Macmillan Dict Canadian Biog; Natl Cycl Am Biog 20; Nature 85(1910):16; Pop Sci Mo 76(1910):204; Sci 32(1910):626; Wallace; Who Was Who Am 1.

PENROSE, RICHARD ALEXANDER FULLERTON (1827-1908). Surgery. OCCUP: Physician; Professor. MSS: Ind NUCMC. WORKS ABOUT: Appleton's; Natl Cycl Am Biog 2; Sci 29(1909):73; Who Was Who Am 1.

PENROSE, THOMAS NEALL (1835?-1902). Medicine. OCCUP: Naval physician. WORKS ABOUT: Sci 15(1902):318.

PEPPER, WILLIAM (1843-1898). Medicine. OCCUP: Physician; Professor; Academic administrator. MSS: Ind NUCMC (68-459). WORKS ABOUT: Dict Am Biog; Dict Am Med Biog; Natl Cycl Am Biog 1; Who Was Who Am h. See also: Barr; Biog Geneal Master Ind; Ireland Ind Sci.

PERCIVAL, JAMES GATES (1795-1856). Geology. OCCUP: Author; State government scientist. MSS: Ind NUCMC. WORKS ABOUT: Dict Am Biog; Elliott; Natl Cycl Am Biog 8; Who Was Who Am h. See also: Barr; Biog Geneal Master Ind; Ireland Ind Sci.

PERGANDE, THEODORE (1840-1916). Entomology. OCCUP: U.S. and state government scientist. WORKS ABOUT: Essig p733-34; Mallis; Osborn Herb p180, 348; Sci 43(1916):492.

PERKINS, GEORGE ROBERTS (1812-1876). Mathematics. OCCUP: Educator; Professor; State government engineer. WORKS ABOUT: Adams O; Allibone; Appleton's; Drake; Twentieth Cent Biog Dict; Wallace.

PERKINS, JACOB (1766-1849). Invention; Steam pressure. OCCUP: Manufacturer; Printer. MSS: Ind NUCMC (62-3294). WORKS ABOUT: Dict Am Biog; Natl Cycl Am Biog 10; Who Was Who Am h. See also: Biog Geneal Master Ind.

PERRINE, FREDERIC AUTEN COMBS (1862-1908). Electrical engineering. OCCUP: Industrial engineer and manager; Professor. WORKS ABOUT: Am Men Sci 1 d3; Dict Am Biog; Natl Cycl Am Biog 19; Who Was Who Am 1. See also: Barr; Biog Geneal Master Ind.

PERRINE, HENRY (1797-1840). Botany. OCCUP: Physician; Public official; Natural history explorer. WORKS ABOUT: Dict Am Biog; Elliott; Who Was Who Am h.

PERRY, JOHN BUCKLEY (1825-1872). Geology. OCCUP: Clergyman; Museum affiliate; Academic instructor. MSS: Ind NUCMC (66-960). WORKS ABOUT: Am Journ Sci 104(1872):424; Harvard Hist Register.

PERRY, THOMAS HOBART (1813-1849). Geology; Mathematics. OCCUP: Professor. Scientific work: see Roy Soc <u>Cat</u>. WORKS ABOUT: Colby <u>Obit Rec</u> p27-28; Pennsylvania <u>Gen Cat</u>.

PERSONS, AUGUSTUS ARCHILUS (1866-1917). Chemistry. OCCUP: Professor; Consultant. WORKS ABOUT: <u>Am Men Sci</u> 1-2 d3; <u>Who Was Who Am</u> 1.

PETER, ROBERT (1805-1894). Chemistry. OCCUP: Professor; Physician; State government scientist. MSS: <u>Ind NUCMC</u> (61-961). WORKS ABOUT: <u>Dict Am Biog</u>; <u>Dict Am Med Biog</u>; Elliott; <u>Natl Cycl Am Biog</u> 4; <u>Who Was Who Am</u> h. See also: <u>Biog Geneal Master Ind</u>.

PETERS, CHRISTIAN HENRY FREDERICK (1813-1890). Astronomy. OCCUP: Government scientist; Professor; Observatory director. MSS: <u>Ind NUCMC</u>. WORKS ABOUT: <u>Dict Am Biog</u>; <u>Dict Sci Biog</u>; Elliott; <u>Natl Cycl Am Biog</u> 13; <u>Who Was Who Am</u> h. See also: Barr; <u>Biog Geneal Master Ind</u>.

PETERS, EDWARD DYER (1849-1917). Metallurgy. OCCUP: Mining engineer; Professor. WORKS ABOUT: <u>Dict Am Biog</u>; <u>Who Was Who Am</u> 1. See also: Barr; <u>Biog Geneal Master Ind</u>.

PETTEE, WILLIAM HENRY (1838-1904). Geology. OCCUP: Professor. MSS: <u>Ind NUCMC</u>. WORKS ABOUT: Appleton's; <u>Sci</u> 19(1904):870, 20(1904):58-60; <u>Who Was Who Am</u> 1.

PETTEGREW, DAVID LYMAN (-1914). Mathematics. OCCUP: Not determined. WORKS ABOUT: <u>Am Men Sci</u> 1-2 d3.

PETTIT, JAMES HARVEY (1876-1914). Soil chemistry. OCCUP: Professor; State government scientist. WORKS ABOUT: <u>Am Men Sci</u> 2 d3; <u>Sci</u> 41(1915):61.

PFIZER, CHARLES (1823-1906). Chemistry. OCCUP: Manufacturer. WORKS ABOUT: <u>Am Chem and Chem Eng</u>.

PHELPS, ABEL MIX (1851-1902). Orthopedics. OCCUP: Surgeon; Professor. WORKS ABOUT: <u>Natl Cycl Am Biog</u> 12; <u>Sci</u> 16(1902):638.

PHELPS, ALMIRA HART LINCOLN (1793-1884). Botany; Writing of science textbooks. OCCUP: Teacher; Educator. MSS: <u>Ind NUCMC</u> (?). WORKS ABOUT: <u>Dict Am Biog</u>; Elliott; <u>Natl Cycl Am Biog</u> 11; <u>Notable Am Wom</u>; Ogilvie; Siegel and Finley; <u>Who Was Who Am</u> h. See also: <u>Biog Geneal Master Ind</u>; Pelletier.

PHETTEPLACE, THURSTON MASON (1877-1913). Mechanical engineering. OCCUP: Professor. WORKS ABOUT: <u>Am Men Sci</u> 2 d3.

PHILLIPS, ANDREW WHEELER (1844-1915). Mathematics. OCCUP: Professor; Academic administrator. WORKS ABOUT: <u>Am Journ Sci</u> 189(1915):326; <u>Am Men Sci</u> 1-2 d3; <u>Who Was Who Am</u> 1; <u>Who Was Who North Am Authors</u>.

PHILLIPS, FRANCIS CLIFFORD (1850-1920). Chemistry. OCCUP: Professor. WORKS ABOUT: Am Men Sci 1-2 d3; Dict Am Biog; Elliott (s); Natl Cycl Am Biog 18; Who Was Who Am 1. See also: Biog Geneal Master Ind; Pelletier.

PHILLIPS, FRANK JAY (1881-1911). Forestry. OCCUP: U.S. government scientist; Professor. WORKS ABOUT: Am Men Sci 2 d3; Sci 33(1911):297.

PHILLIPS, WILLIAM BATTLE (1857-1918). Geology; Mining engineering. OCCUP: Industrial and state government scientist; Professor; Mining engineer. MSS: Ind NUCMC. WORKS ABOUT: Am Journ Sci 196(1918):692; Who Was Who Am 1.

PHYSICK, PHILIP SYNG (1768-1837). Surgery. OCCUP: Surgeon; Professor. MSS: Ind NUCMC. WORKS ABOUT: Dict Am Biog; Dict Am Med Biog; Natl Cycl Am Biog 6; Who Was Who Am h. See also: Barr; Biog Geneal Master Ind; Ireland Ind Sci; Pelletier.

PICKERING, CHARLES (1805-1878). Natural history. OCCUP: Physician; U.S. government scientist; Researcher. MSS: Ind NUCMC. WORKS ABOUT: Dict Am Biog; Elliott; Natl Cycl Am Biog 13; Who Was Who Am h. See also: Barr; Biog Geneal Master Ind; Ireland Ind Sci.

PICKERING, EDWARD CHARLES (1846-1919). Astronomy. OCCUP: Professor; Observatory director. MSS: Ind NUCMC (65-1263). WORKS ABOUT: Am Men Sci 1-2 d3; Dict Am Biog; Dict Sci Biog; Elliott (s); Natl Ac Scis Biog Mem 15; Natl Cycl Am Biog 6; Who Was Who Am 1. See also: Barr; Biog Geneal Master Ind; Ireland Ind Sci; Pelletier.

PIERCE, ARTHUR HENRY (1867-1914). Psychology. OCCUP: Professor. MSS: Ind NUCMC. WORKS ABOUT: Am Men Sci 1-2 d3; Sci 39(1914):323, 456-57; Wallace; Who Was Who Am 1.

PIERCE, JOHN (-1897). Chemistry. OCCUP: Professor. WORKS ABOUT: Sci 5(1897):442.

PIERCE, JOSIAH, JR. (1861-1902). Civil engineering. OCCUP: Professor; Industrial and U.S. government engineer; Consultant. WORKS ABOUT: Natl Cycl Am Biog 5; Sci 16(1902):239; Who Was Who Am 1.

PIERCE, NEWTON BARRIS (1856-1916). Botany; Vegetable pathology. OCCUP: Consultant (lumber inspector); U.S. government scientist. WORKS ABOUT: Am Men Sci 1-2 d3; Sci 44(1916):814; Wallace; Who Was Who Am 1.

PIGGOT, AARON SNOWDEN (1822-1869). Chemistry; Metallurgy. OCCUP: Physician; Professor; Consultant. WORKS ABOUT: Allibone; Am Chem and Chem Eng.

PIGGOT, CAMERON (1856?-1911). Chemistry. OCCUP: Professor. Scientific work: See Roy Soc Cat v 17 (under co-author H. N. Morse). WORKS ABOUT: Johns Hopkins Half Cent Dir; Sci 33(1911):723.

PIKE, NICHOLAS (1743-1819). Mathematics. OCCUP: Educator; Public official. MSS: <u>Ind NUCMC</u> (73-28). WORKS ABOUT: <u>Dict Am Biog</u>; Elliott; <u>Natl Cycl Am Biog</u> 20; <u>Who Was Who Am</u> h. See also: <u>Biog Geneal Master Ind</u>.

PIKE, NICHOLAS (1818-1905). Biology. OCCUP: Businessman; Public official (diplomatic service); Collector. WORKS ABOUT: <u>Natl Cycl Am Biog</u> 24; <u>Sci</u> 21(1905):677.

PIKE, WILLIAM ABBOT (-1895). Engineering. OCCUP: Professor. WORKS ABOUT: <u>Sci</u> 2(1895):623.

PILLING, JAMES CONSTANTINE (1846-1895). Ethnology. OCCUP: U.S. government scientist. MSS: <u>Ind NUCMC</u>. WORKS ABOUT: <u>Dict Am Biog</u>; <u>Natl Cycl Am Biog</u> 15; <u>Who Was Who Am</u> h. See also: Barr; <u>Biog Geneal Master Ind</u>.

PILLSBURY, JOHN ELLIOTT (1846-1919). Geography; Oceanography. OCCUP: Naval officer. WORKS ABOUT: <u>Am Men Sci</u> 2 d3; <u>Dict Am Biog</u>; <u>Natl Cycl Am Biog</u> 20; <u>Who Was Who Am</u> 1.

PILLSBURY, JOHN HENRY (1846-1910). Biology. OCCUP: Clergyman; Professor; Educator. WORKS ABOUT: <u>Am Men Sci</u> 1-2 d3; Wallace; <u>Who Was Who Am</u> 1.

PINCKNEY, ELIZABETH LUCAS (1722?-1793). Botany; Agriculture. OCCUP: Planter. MSS: <u>Ind NUCMC</u>. WORKS ABOUT: <u>Dict Am Biog</u>; <u>Notable Am Wom</u>; <u>Who Was Who Am</u> h. See also: <u>Biog Geneal Master Ind</u>.

PIRSSON, LOUIS VALENTINE (1860-1919). Geology. OCCUP: Professor. MSS: <u>Ind NUCMC</u>. WORKS ABOUT: <u>Am Men Sci</u> 1-2 d3; <u>Dict Am Biog</u>; Elliott (s); Natl Ac Scis <u>Biog Mem</u> 34; <u>Natl Cycl Am Biog</u> 10, 28; <u>Who Was Who Am</u> 1. See also: Barr; <u>Biog Geneal Master Ind</u>.

PITCHER, ZINA (1797-1872). Natural history. OCCUP: Physician; Army surgeon; Public official. MSS: <u>Ind NUCMC</u>. WORKS ABOUT: <u>Dict Am Biog</u>; <u>Dict Am Med Biog</u>; Elliott; <u>Natl Cycl Am Biog</u> 12; <u>Who Was Who Am</u> h. See also: <u>Biog Geneal Master Ind</u>.

PITKIN, CHARLES ALFRED (1853?-1916). Mathematics; Physics. OCCUP: Teacher. WORKS ABOUT: <u>Sci</u> 44(1916):850.

PITMAN, BENN (1822-1910). Phonetics; Phonography. OCCUP: Teacher; Lecturer; Publisher. MSS: <u>Ind NUCMC</u>. WORKS ABOUT: <u>Am Men Sci</u> 1-2 d3; <u>Dict Am Biog</u>; <u>Natl Cycl Am Biog</u> 4; <u>Who Was Who Am</u> 1. See also: Barr; <u>Biog Geneal Master Ind</u>.

PLATT, FRANKLIN (1844-1900). Geology. OCCUP: State government scientist; Consultant; Industrial officer. WORKS ABOUT: Elliott; <u>Natl Cycl Am Biog</u> 5; <u>Who Was Who Am</u> 1. See also: <u>Biog Geneal Master Ind</u>.

PLUMMER, FREDERICK G. (1844?-1913). Geography. OCCUP: U.S. government scientist. WORKS ABOUT: Sci 38(1913):300.

PLUMMER, JOHN THOMAS (1807-1865). Chemistry; Natural history. OCCUP: Physician. MSS: Ind NUCMC (?). WORKS ABOUT: Elliott. See also: Barr.

PLYMPTON, GEORGE WASHINGTON (1827-1907). Physics. OCCUP: Engineer; Professor; Editor. WORKS ABOUT: Adams O; Allibone; Allibone Supp; Am Men Sci 1 d3; Appleton's; Natl Cycl Am Biog 9; Preston; Sci 26(1907):423; Twentieth Cent Biog Dict; Wallace; Who Was Who Am 1.

POHLMAN, JULIUS (1848-1910). Geology; Physiology. OCCUP: Museum administrator; Physician; Professor. WORKS ABOUT: Am Men Sci 1-2 d3; Who Was Who Am 1.

POINIER, PORTER (1853-1876). Physics. OCCUP: Academic affiliate. WORKS ABOUT: Am Journ Sci 112(1876):164.

POLK, WILLIAM MECKLENBERG (1844-1918). Gynecology. OCCUP: Physician; Professor; Academic administrator. MSS: Ind NUCMC. WORKS ABOUT: Am Men Sci 1-2 d3; Dict Am Biog; Natl Cycl Am Biog 2, 26; Who Was Who Am 1. See also: Barr; Biog Geneal Master Ind.

POND, GEORGE GILBERT (1861-1920). Chemistry. OCCUP: Professor; Academic administrator. WORKS ABOUT: Am Chem and Chem Eng; Am Men Sci 1-2 d3; Natl Cycl Am Biog 20; Wallace; Who Was Who Am 1.

POOLE, HERMAN (1849-1906). Chemistry; Metallurgy. OCCUP: Consultant (?). WORKS ABOUT: Adams O; Wallace; Who Was Who Am 4.

POPE, FRANKLIN LEONARD (1840-1895). Electricity; Invention. OCCUP: Industrial affiliate; Consultant; Editor. WORKS ABOUT: Dict Am Biog; Natl Cycl Am Biog 7:414; Who Was Who Am h. See also: Barr; Biog Geneal Master Ind.

POPENOE, EDWIN ALONZO (1853-1913). Entomology. OCCUP: Professor; State government scientist. WORKS ABOUT: Am Men Sci 2 d3.

PORCHER, FRANCIS PEYRE (1825-1895). Botany. OCCUP: Physician; Professor (medicine). MSS: Ind NUCMC (70-1621). WORKS ABOUT: Dict Am Biog; Dict Am Med Biog; Elliott; Who Was Who Am h. See also: Barr; Biog Geneal Master Ind.

PORTER, ALBERT BROWN (1864-1909). Physics. OCCUP: Professor; Consultant; Businessman. WORKS ABOUT: Am Men Sci 1 d3; Sci 29(1909):696, 962-63.

PORTER, ANDREW (1743-1813). Surveying. OCCUP: Educator; Surveyor; Farmer. MSS: Ind NUCMC (61-433). WORKS ABOUT: Dict Am Biog; Natl Cycl Am Biog 1; Who Was Who Am h. See also: Biog Geneal Master Ind.

PORTER, CHARLES BURNHAM (1840-1909). Medicine. OCCUP: Surgeon; Professor. WORKS ABOUT: Leonardo p344-45; Sci 29(1909):893; Who Was Who Am 1.

PORTER, JOHN ADDISON (1822-1866). Chemistry. OCCUP: Professor; Academic administrator. MSS: Ind NUCMC (?). WORKS ABOUT: Dict Am Biog; Elliott; Who Was Who Am h. See also: Barr; Biog Geneal Master Ind.

PORTER, JOHN TALBOT (1825-1910). Engineering. OCCUP: Engineer (?). WORKS ABOUT: Sci 32(1910):339.

PORTER, THOMAS CONRAD (1822-1901). Botany. OCCUP: Clergyman; Professor. MSS: Ind NUCMC (66-353). WORKS ABOUT: Dict Am Biog; Elliott; Natl Cycl Am Biog 11; Who Was Who Am 1. See also: Barr; Biog Geneal Master Ind; Ireland Ind Sci.

PORY, JOHN (1572-1635). Geography. OCCUP: Public official; Explorer. WORKS ABOUT: Dict Am Biog; Natl Cycl Am Biog 8; Who Was Who Am h. See also: Biog Geneal Master Ind.

POST, GEORGE EDWARD (1838-1909). Biology. OCCUP: Clergyman; Physician; Professor (medicine). WORKS ABOUT: Dict Am Biog; Natl Cycl Am Biog 13; Who Was Who Am 1. See also: Barr; Biog Geneal Master Ind.

POTAMIAN, BROTHER (MICHAEL FRANCIS O'REILLY) (1847-1917). Physics. OCCUP: Educator; Professor. WORKS ABOUT: Am Men Sci 1-2 d3; Dict Am Biog. See also: Biog Geneal Master Ind.

POUND, THOMAS (ca.1650-1703). Cartography. OCCUP: Naval officer. WORKS ABOUT: Dict Am Biog; Who Was Who Am h.

POURTALES, LOUIS FRANCOIS de (1823 or 1824-1880). Natural history; Zoology; Oceanography. OCCUP: U.S. government scientist; Museum curator; Inherited wealth. WORKS ABOUT: Dict Am Biog; Dict Sci Biog; Elliott; Natl Ac Scis Biog Mem 5; Who Was Who Am h. See also: Barr; Biog Geneal Master Ind.

POWALKY, KARL RUDOLPH (1817-1881). Astronomy. OCCUP: Observatory affiliate; Computer. WORKS ABOUT: Dict Sci Biog.

POWELL, JOHN WESLEY (1834-1902). Natural history; Geology; Ethnology. OCCUP: Academic instructor; U.S. government scientist. MSS: Ind NUCMC (62-4848, 68-1270). WORKS ABOUT: Dict Am Biog; Dict Sci Biog; Elliott; Natl Ac Scis Biog Mem 8; Natl Cycl Am Biog 3; Who Was Who Am 1. See also: Barr; Biog Geneal Master Ind; Ireland Ind Sci; Pelletier.

PRANG, LOUIS (1824-1909). Study of color. OCCUP: Printer; Publisher. MSS: Ind NUCMC (?). WORKS ABOUT: Am Men Sci 1 d3; Dict Am Biog; Natl Cycl Am Biog 11; Who Was Who Am 1. See also: Biog Geneal Master Ind.

PRATT, WILLIAM HENRY (1822-1893). General science. OCCUP: Teacher; Educator. WORKS ABOUT: Minnesota Ac Nat Scis Bull 4(1896):37-39.

PRENTISS, ALBERT NELSON (1836-1896). Botany. OCCUP: Professor. MSS: Ind NUCMC (?). WORKS ABOUT: Am Nat 30(1896):1043-44; Bot Gaz 21(1896):283-89; Natl Cycl Am Biog 4; Sci 4(1896):267, 523-24; Twentieth Cent Biog Dict; Wallace.

PRENTISS, CHARLES WILLIAM (1874-1915). Zoology; Anatomy. OCCUP: Professor. WORKS ABOUT: Am Men Sci 1-2 d3; Sci 42(1915):178-79; Wallace.

PRENTISS, DANIEL WEBSTER (1843-1899). Medicine; Ornithology. OCCUP: Physician; Professor (medicine). WORKS ABOUT: Auk 17(1900):91-92; Natl Cycl Am Biog 3; Sci 11(1900):159; Who Was Who Am 1.

PRENTISS, ROBERT WOODWORTH (1857-1913). Mathematics. OCCUP: U.S. government scientist; Professor. WORKS ABOUT: Am Men Sci 1-2 d3; Sci 37(1913):556.

PRESCOTT, ALBERT BENJAMIN (1832-1905). Chemistry. OCCUP: Professor; Academic administrator. MSS: Ind NUCMC. WORKS ABOUT: Dict Am Biog; Elliott; Natl Cycl Am Biog 13; Who Was Who Am 1. See also: Barr; Biog Geneal Master Ind; Ireland Ind Sci; Pelletier.

PRESTON, ERASMUS DARWIN (1851-1906). Geodesy. OCCUP: U.S. government scientist. WORKS ABOUT: Am Men Sci 1 d3; Who Was Who Am 1.

PRICE, MARSHALL LANGTON (1878-1915). Public health. OCCUP: Physician; Public official. WORKS ABOUT: Am Men Sci 2 d3; Who Was Who Am 1.

PRICE, OVERTON WESTFELDT (1873-1914). Forestry. OCCUP: U.S. government scientist. WORKS ABOUT: Am Men Sci 1-2 d3; Sci 40(1914):446; Wallace; Who Was Who Am 1.

PRICE, SADIE F. (-1903). Botany. OCCUP: Teacher; Natural history explorer. WORKS ABOUT: Who Was Who Am 1.

PRIESTLEY, JOSEPH (1733-1804). Natural philosophy; Chemistry; Electricity; Theology. OCCUP: Clergyman; Teacher; Librarian. MSS: Ind NUCMC (62-4316, 66-472). WORKS ABOUT: Dict Am Biog; Dict Sci Biog; Elliott; Natl Cycl Am Biog 6; Who Was Who Am h. See also: Barr; Biog Geneal Master Ind; Ireland Ind Sci; Pelletier.

PRIME, FREDERICK (1846-1915). Geology. OCCUP: Professor; Industrial officer; State government scientist. WORKS ABOUT: Am Men Sci 1-2 d3; Appleton's; Natl Cycl Am Biog 4; Sci 42(1915):118; Who Was Who Am 1.

PRIME, TEMPLE (1832-1903). Conchology. OCCUP: Inherited wealth (?); Businessman. WORKS ABOUT: Elliott. See also: Biog Geneal Master Ind.

PRINCE, JOHN (1751-1836). Mechanics; Instrumentation. OCCUP: Clergyman. WORKS ABOUT: Allibone; Am Journ Sci 31(1837):201-22; Appleton's; Drake; Natl Cycl Am Biog 7:345.

PRINGLE, CYRUS GUERNSEY (1838-1911). Botany. OCCUP: Natural history collector; Herbarium curator. MSS: Ind NUCMC. WORKS ABOUT: Am Men Sci 1-2 d3; Dict Am Biog; Natl Cycl Am Biog 23; Who Was Who Am 1. See also: Barr.

PRITCHETT, CARR WALLER (1823-1910). Astronomy. OCCUP: Clergyman; Educator; Observatory director. WORKS ABOUT: Sci 31(1910):454; Twentieth Cent Biog Dict; Who Was Who Am 1.

PROCTER, JOHN ROBERT (1844-1903). Geology. OCCUP: State government scientist; Public official. WORKS ABOUT: Dict Am Biog; Elliott. See also: Biog Geneal Master Ind.

PROCTER, WILLIAM (1817-1874). Chemistry; Pharmacy. OCCUP: Pharmacist; Professor; Editor. WORKS ABOUT: Dict Am Biog; Dict Am Med Biog; Natl Cycl Am Biog 5; Who Was Who Am h.

PROSSER, CHARLES SMITH (1860-1916). Geology. OCCUP: Professor. MSS: Ind NUCMC. WORKS ABOUT: Am Men Sci 1-2 d3; Dict Am Biog; Elliott (s); Natl Cycl Am Biog 12; Who Was Who Am 1. See also: Barr; Biog Geneal Master Ind.

PROSSER, THOMAS (-1870). Engineering; Studies of steam. OCCUP: Engineer; Architect; Manufacturer. WORKS ABOUT: Am Soc Civil Eng Trans 36(1896):564-65.

PROUT, HIRAM AUGUSTUS (1808-1862). Geology; Paleontology. OCCUP: Physician. WORKS ABOUT: Ac Sci Saint Louis Bull 1(1935):59-61; Am Journ Sci 83(1862):453.

PUGH, EVAN (1828-1864). Chemistry. OCCUP: Professor; College president. WORKS ABOUT: Dict Am Biog; Elliott; Natl Cycl Am Biog 11; Who Was Who Am h. See also: Barr; Biog Geneal Master Ind; Ireland Ind Sci; Pelletier.

PULSIFER, WILLIAM HENRY (1831-1905). Chemistry. OCCUP: Manufacturer. WORKS ABOUT: Allibone Supp; Sci 21(1905):758; Wallace.

PURDIE, HENRY AUGUSTUS (1840-1911). Ornithology. OCCUP: State government employee. WORKS ABOUT: Auk 28(1911):387, 29(1912):1-15.

PURDUE, ALBERT HOMER (1861-1917). Geology. OCCUP: Professor; State government scientist. MSS: Ind NUCMC (76-1656). WORKS ABOUT: Am Men Sci 1-2 d3; Sci 47(1918):67; Who Was Who Am 1.

PURINTON, GEORGE DANA (1856-1897). Biology; Chemistry. OCCUP: Professor; Industrial and state government scientist; Physician. WORKS ABOUT: <u>Natl Cycl Am Biog</u> 8; <u>Twentieth Cent Biog Dict</u>.

PURSH, FREDERICK (1774-1820). Botany. OCCUP: Horticulturist; Natural history explorer/ collector. MSS: <u>Ind NUCMC</u>. WORKS ABOUT: <u>Dict Am Biog</u>; <u>Dict Sci Biog</u>; Elliott; <u>Who Was Who Am</u> h. See also: <u>Biog Geneal Master Ind</u>.

PURYEAR, BENNET (1826-1914). Agriculture; Chemistry. OCCUP: Professor. WORKS ABOUT: <u>Dict Am Biog</u>; Elliott; <u>Natl Cycl Am Biog</u> 11. See also: <u>Biog Geneal Master Ind</u>.

PUTNAM, FREDERIC WARD (1839-1915). Natural history; Zoology; Anthropology; Archeology. OCCUP: Professor; Museum curator. MSS: <u>Ind NUCMC</u> (65-1267). WORKS ABOUT: <u>Am Men Sci</u> 1-2 d3; <u>Dict Am Biog</u>; <u>Dict Sci Biog</u>; Elliott (s); Natl Ac Scis <u>Biog Mem</u> 16; <u>Natl Cycl Am Biog</u> 3, 23; <u>Who Was Who Am</u> 1. See also: <u>Biog Geneal Master Ind</u>.

PUTNAM, JAMES JACKSON (1846-1918). Neurology. OCCUP: Physician; Professor. WORKS ABOUT: <u>Am Men Sci</u> 2 d3; <u>Dict Am Biog</u>; <u>Natl Cycl Am Biog</u> 18; <u>Who Was Who Am</u> 1. See also: Barr; <u>Biog Geneal Master Ind</u>; Ireland <u>Ind Sci</u>; Pelletier.

PUTNAM, JOSEPH DUNCAN (1855-1881). Entomology. OCCUP: Independent wealth. WORKS ABOUT: <u>Canadian Ent</u> 13(1881):256; Osborn Herb p158-59, 348; <u>Papilio</u> 1(1881):223-24; <u>Pop Sci Mo</u> 51(1897):87-95; <u>Psyche</u> 3(1886):312.

PUTNAM, MARY LOUISE DUNCAN (1832-1903). General science. OCCUP: Benefactor. MSS: <u>Ind NUCMC</u> (?). WORKS ABOUT: <u>Pop Sci Mo</u> 51(1897):88-95; <u>Sci</u> 17(1903):399, 632-33.

PUTNEY, FRED SILVER (1881-1918). Biology; Animal nutrition. OCCUP: Professor. WORKS ABOUT: <u>Sci</u> 48(1918):486-87.

Q

QUANTZ, JOHN O. (1868-1903). Psychology. OCCUP: Professor.
WORKS ABOUT: <u>Sci</u> 17(1903):238-39.

QUARLES, JAMES ADDISON (1837-1907). Philosophy;
Psychology. OCCUP: Clergyman; Professor. WORKS ABOUT: <u>Am
Men Sci</u> 1 d3; Knight; <u>Sci</u> 25(1907):678; <u>Who Was Who Am</u> 1.

QUINBY, ISAAC FERDINAND (1821-1891). Natural philosophy;
Mathematics. OCCUP: Professor; Army officer. WORKS ABOUT:
<u>Dict Am Biog</u>; <u>Who Was Who Am</u> h. See also: <u>Biog Geneal
Master Ind</u>.

R

RADELFINGER, FRANK GUSTAVE (ca.1870-1904). Mathematics. OCCUP: Professor; Lawyer. WORKS ABOUT: <u>Sci</u> 20(1904):319.

RAFINESQUE, CONSTANTINE SAMUEL (1783-1840). Natural history; Archeology; Botany; Zoology. OCCUP: Businessman; Professor. MSS: <u>Ind NUCMC</u> (61-766, 66-354, 80-2206). WORKS ABOUT: <u>Dict Am Biog</u>; <u>Dict Sci Biog</u>; Elliott; <u>Natl Cycl Am Biog</u> 8; <u>Who Was Who Am</u> h. See also: Barr; <u>Biog Geneal Master Ind</u>; Ireland <u>Ind Sci</u>; Pelletier.

RAFTER, GEORGE W. (1851-1907). Civil engineering. OCCUP: Engineer; State government engineer. WORKS ABOUT: <u>Dict Am Biog</u>; <u>Natl Cycl Am Biog</u> 12. See also: <u>Biog Geneal Master Ind</u>.

RAGSDALE, GEORGE HENRY (-1895). Ornithology. OCCUP: Editor. WORKS ABOUT: <u>Auk</u> 12(1895):316.

RAHT, AUGUST WILHELM (1843-1916). Metallurgy. OCCUP: Industrial scientist and manager. MSS: <u>Ind NUCMC</u> (?). WORKS ABOUT: <u>Dict Am Biog</u>.

RAINS, GEORGE WASHINGTON (1817-1898). Chemistry. OCCUP: Army officer; Industrial officer; Professor. MSS: <u>Ind NUCMC</u> (64-623). WORKS ABOUT: <u>Dict Am Biog</u>; <u>Who Was Who Am</u> h. See also: Barr; <u>Biog Geneal Master Ind</u>; Pelletier.

RALPH, WILLIAM L. (1850-1907). Ornithology; Oology. OCCUP: Museum curator. WORKS ABOUT: <u>Sci</u> 26(1907):126.

RAMSAY, ALEXANDER (1754?-1824). Anatomy. OCCUP: Teacher; Lecturer; Physician. WORKS ABOUT: <u>Dict Am Biog</u>; <u>Dict Am Med Biog</u>; Elliott; <u>Who Was Who Am</u> h. See also: <u>Biog Geneal Master Ind</u>.

RAND, THEODORE DEHON (1836-1903). Geology. OCCUP: Lawyer. WORKS ABOUT: Appleton's; <u>Who Was Who Am</u> 1.

RANDALL, JOHN WITT (1813-1892). Natural history; Zoology.
OCCUP: Inherited wealth (?). WORKS ABOUT: Adams O;
Allibone; Appleton's; Burke and Howe; Chamberlain v2:221-
22; Drake; Shaw supp 1 (1972); Wallace.

RANDOLPH, ISHAM (1848-1920). Civil engineering. OCCUP:
Industrial and city government engineer; Consultant. WORKS
ABOUT: Am Men Sci 2 d3; Dict Am Biog; Natl Cycl Am Biog 19;
Who Was Who Am 1. See also: Biog Geneal Master Ind.

RANDOLPH, NATHANIEL ARCHER (1858-1887). Medicine;
Physiology. OCCUP: Physician; Professor. WORKS ABOUT: Am
Phil Soc Proc 26(1889):359-65; Appleton's.

RATHBUN, RICHARD (1852-1918). Natural history; Zoology.
OCCUP: U.S. government scientist; Museum administrator.
MSS: Ind NUCMC (72-1252). WORKS ABOUT: Am Men Sci 1-2 d3;
Dict Am Biog; Elliott (s); Natl Cycl Am Biog 13; Who Was
Who Am 1. See also: Barr; Biog Geneal Master Ind.

RAU, CHARLES (1826-1887). Archeology. OCCUP: Teacher;
Museum curator. MSS: Ind NUCMC. WORKS ABOUT: Dict Am Biog;
Natl Cycl Am Biog 2; Who Was Who Am h. See also: Barr; Biog
Geneal Master Ind.

RAVENEL, EDMUND (1797-1871). Natural history; Zoology;
Conchology. OCCUP: Physician; Professor; Planter. MSS: Ind
NUCMC. WORKS ABOUT: Dict Am Biog; Elliott; Who Was Who
Am h.

RAVENEL, HENRY WILLIAM (1814-1887). Botany. OCCUP: Planter;
Businessman; Editor. MSS: Ind NUCMC (62-2034, 74-997).
WORKS ABOUT: Dict Am Biog; Elliott; Natl Cycl Am Biog 10;
Who Was Who Am h. See also: Barr; Biog Geneal Master Ind.

RAVENEL, ST. JULIEN (1819-1882). Agricultural chemistry.
OCCUP: Physician; Industrial scientist. MSS: Ind NUCMC.
WORKS ABOUT: Dict Am Biog; Elliott; Natl Cycl Am Biog 10;
Who Was Who Am h. See also: Barr; Biog Geneal Master Ind.

RAYMOND, ROSSITER WORTHINGTON (1840-1918). Metallurgy.
OCCUP: Mining engineer; Editor; U.S. government scientist.
MSS: Ind NUCMC. WORKS ABOUT: Am Men Sci 1-2 d3; Dict Am
Biog; Natl Cycl Am Biog 8; Who Was Who Am 1. See also:
Barr; Biog Geneal Master Ind.

READ, MATTHEW CANFIELD (1823-1902). Geology. OCCUP: Lawyer;
Public official; Government scientist. WORKS ABOUT: Ohio
Authors.

READ, MOTTE ALSTON (1872-1920). Paleontology; Physiography.
OCCUP: Academic instructor; Engaged in independent study.
MSS: Ind NUCMC (82-1649). WORKS ABOUT: Am Men Sci 1-2 d3;
Harvard Class of 1893 Rep.

REDFIELD, JOHN HOWARD (1815-1895). Botany; Zoology (conchology). OCCUP: Businessman. MSS: Ind NUCMC (66-73). WORKS ABOUT: Ac Nat Scis Philadelphia Proc (1895):292-301; Allibone; Am Journ Sci 149(1895):485; Appleton's; Bot Gaz 20(1895):175-76; Journ Bot 33(1895):313-14; Sci 1(1895):470-71, 3(1896):281; Torrey Bot Club Bull 22(1895):162-71; Wallace.

REDFIELD, WILLIAM C. (1789-1857). Meteorology; Paleontology. OCCUP: Tradesman; Businessman. WORKS ABOUT: Dict Am Biog; Dict Sci Biog; Elliott; Natl Cycl Am Biog 7:354; Who Was Who Am h. See also: Barr; Biog Geneal Master Ind; Pelletier.

REED, EVA M. (-1901). Botany. OCCUP: Librarian. WORKS ABOUT: Sci 14(1901):158.

REED, JOHN OREN (1856-1916). Physics. OCCUP: Professor; Academic administrator. WORKS ABOUT: Am Men Sci 1-2 d3; Sci 16(1902):1039, 43(1916):131, 46(1917):207-08; Wallace; Who Was Who Am 1.

REED, STEPHEN (1801?-1877). Geology. OCCUP: Not determined; Physician (?). WORKS ABOUT: Am Journ Sci 114(1877):168.

REED, WALTER (1851-1902). Medicine; Bacteriology. OCCUP: Army physician; Professor. MSS: Ind NUCMC (66-1489, 71-1950). WORKS ABOUT: Dict Am Biog; Dict Am Med Biog; Dict Sci Biog; Natl Cycl Am Biog 13, 33; Who Was Who Am h. See also: Barr; Biog Geneal Master Ind; Ireland Ind Sci; Pelletier.

REED, WILLIAM HARLOW (1848-1915). Paleontology. OCCUP: Museum curator; Academic instructor. WORKS ABOUT: Am Men Sci 2 d3; Sci 41(1915):722.

REES, JOHN KROM (1851-1907). Astronomy; Geodesy. OCCUP: Professor; Observatory director. WORKS ABOUT: Am Men Sci 1 d3; Dict Am Biog; Elliott (s); Natl Cycl Am Biog 11; Who Was Who Am 1. See also: Barr; Biog Geneal Master Ind.

REESE, JACOB (1825-1907). Metallurgy. OCCUP: Industrial engineer and manager; Manufacturer; Inventor. WORKS ABOUT: Am Men Sci 1 d3; Dict Am Biog.

REID, DAVID BOSWELL (1805-1863). Chemistry; Ventilation engineering. OCCUP: Educator; Engineer; Professor. WORKS ABOUT: Dict Am Biog; Who Was Who Am h. See also: Biog Geneal Master Ind.

REID, JAMES DOUGLAS (1819?-1901). Telegraphy. OCCUP: Not determined; Editor (?). WORKS ABOUT: Sci 13(1901):755.

REID, JAMES L. (1844-1910). Experimentation with corn. OCCUP: Farmer; Businessman. WORKS ABOUT: Dict Am Biog; Who Was Who Am h (add), 4. See also: Biog Geneal Master Ind; Ireland Ind Sci.

REMINGTON, JOSEPH PRICE (1847-1918). Chemistry; Pharmacy. OCCUP: Pharmacist; Professor; Academic administrator. MSS: Ind NUCMC (?). WORKS ABOUT: Am Men Sci 1-2 d3; Dict Am Biog; Natl Cycl Am Biog 5, 33; Who Was Who Am 1. See also: Barr; Biog Geneal Master Ind.

RENSSELAER. See VAN RENSSELAER.

RENWICK, EDWARD SABINE (1823-1912). Engineering. OCCUP: Engineer; Consultant (patent expert); Inventor. MSS: Ind NUCMC. WORKS ABOUT: Dict Am Biog; Natl Cycl Am Biog 11; Who Was Who Am 1. See also: Barr; Biog Geneal Master Ind.

RENWICK, JAMES, SR. (1792-1863). Physics; Engineering. OCCUP: Professor; Engineer. MSS: Ind NUCMC. WORKS ABOUT: Dict Am Biog; Elliott; Natl Cycl Am Biog 11; Who Was Who Am h. See also: Barr; Biog Geneal Master Ind.

REX, GEORGE ABRAHAM (1845-1895). Botany. OCCUP: Physician. MSS: Ind NUCMC (66-75). WORKS ABOUT: Am Journ Sci 149(1895):328; Kelly and Burrage.

REYNOLDS, C. LESLIE (1858?-1913). Botany. OCCUP: Museum (botanic garden) administrator. WORKS ABOUT: Sci 38(1913):300.

REYNOLDS, EDWIN (1831-1909). Engineering. OCCUP: Industrial engineer and manager; Inventor. MSS: Ind NUCMC (?). WORKS ABOUT: Am Men Sci 1-2 d3; Dict Am Biog; Natl Cycl Am Biog 2; Who Was Who Am 1. See also: Biog Geneal Master Ind.

REYNOLDS, ELMER ROBERT (1846-1907). Botany; Ethnology. OCCUP: Teacher; U.S. government employee. WORKS ABOUT: Adams O; Allibone Supp; Appleton's; Sci 26(1907):423; Twentieth Cent Biog Dict; Who Was Who Am 1.

REYNOLDS, JOSEPH JONES (1822-1899). Mechanics; Natural philosophy. OCCUP: Army officer; Professor; Businessman. MSS: Ind NUCMC. WORKS ABOUT: Dict Am Biog; Natl Cycl Am Biog 9; Who Was Who Am h, 1. See also: Barr; Biog Geneal Master Ind.

RHEES, WILLIAM JONES (1830-1907). Bibliography; Statistics. OCCUP: Research institution administrator. MSS: Ind NUCMC (61-3269, 72-1253). WORKS ABOUT: Am Men Sci 1 d3; Dict Am Biog; Natl Cycl Am Biog 12; Who Was Who Am 1. See also: Barr; Biog Geneal Master Ind.

RICE, CHARLES (1841-1901). Chemistry; Pharmacology. OCCUP: Pharmacist; City government scientist; Hospital scientist. WORKS ABOUT: Dict Am Biog; Elliott; Who Was Who Am h. See also: Barr.

RICE, GEORGE STAPLES (1849-1920). Civil engineering. OCCUP: Engineer; City government engineer; Academic instructor. WORKS ABOUT: Am Men Sci 2 d3; Natl Cycl Am Biog 12; Who Was Who Am 1.

RICE, JOHN MINOT (1833-1901). Mathematics. OCCUP: Professor. WORKS ABOUT: Allibone <u>Supp</u>; <u>Sci</u> 13(1901):438.

RICHARDS, CHARLES BRINCKERHOFF (1833-1919). Mechanical engineering. OCCUP: Industrial engineer; Professor. WORKS ABOUT: <u>Am Men Sci</u> 1-2 d3; <u>Dict Am Biog</u>; <u>Natl Cycl Am Biog</u> 25; <u>Who Was Who Am</u> 1. See also: Barr.

RICHARDS, ELLEN HENRIETTA SWALLOW (1842-1911). Chemistry; Sanitary chemistry; Home economics. OCCUP: Academic instructor. MSS: <u>Ind NUCMC</u> (77-1818). WORKS ABOUT: <u>Am Men Sci</u> 1-2 d3; <u>Dict Am Biog</u>; Elliott (s); <u>Natl Cycl Am Biog</u> 7:343; <u>Notable Am Wom</u>; Ogilvie; Siegel and Finley; <u>Who Was Who Am</u> 1. See also: Barr; <u>Biog Geneal Master Ind</u>; Ireland <u>Ind Sci</u>; Pelletier.

RICHARDS, EUGENE LAMB (1838-1912). Mathematics. OCCUP: Professor. WORKS ABOUT: <u>Am Men Sci</u> 1-2 d3; <u>Sci</u> 36(1912):211; Wallace; <u>Who Was Who Am</u> 1.

RICHARDSON, MAURICE HOWE (1851-1912). Surgery. OCCUP: Surgeon; Professor. WORKS ABOUT: <u>Dict Am Biog</u>; <u>Dict Am Med Biog</u>; <u>Natl Cycl Am Biog</u> 13, 18; <u>Who Was Who Am</u> 1. See also: Barr; Ireland <u>Ind Sci</u>.

RICKETTS, HOWARD TAYLOR (1871-1910). Pathology. OCCUP: Academic instructor. MSS: <u>Ind NUCMC</u> (64-153). WORKS ABOUT: <u>Am Men Sci</u> 1 d3; <u>Dict Am Biog</u> supp 1; <u>Dict Am Med Biog</u>; <u>Dict Sci Biog</u>; <u>Natl Cycl Am Biog</u> 34. See also: Barr; <u>Biog Geneal Master Ind</u>; Ireland <u>Ind Sci</u>; Pelletier.

RICKETTS, PIERRE DE PEYSTER (1848?-1918). Chemistry; Metallurgy. OCCUP: Professor; Consultant. WORKS ABOUT: <u>Am Journ Sci</u> 197(1919):84; <u>Am Men Sci</u> 1-2 d3; <u>Sci</u> 48(1918):571; Wallace; <u>Who Was Who Am</u> 1.

RIDDELL, JOHN LEONARD (1807-1865). Botany; Invention; Microscopy. OCCUP: Professor; Physician; U.S. government scientist. WORKS ABOUT: <u>Dict Am Biog</u>; Elliott; <u>Natl Cycl Am Biog</u> 21; <u>Who Was Who Am</u> h. See also: Barr; <u>Biog Geneal Master Ind</u>; Pelletier.

RIDGWAY, AUDUBON WHELOCK (1877-1901). Ornithology. OCCUP: Museum affiliate. WORKS ABOUT: <u>Sci</u> 13(1901):599.

RILEY, CHARLES VALENTINE (1843-1895). Entomology. OCCUP: U.S. and state government scientist. MSS: <u>Ind NUCMC</u> (74-978). WORKS ABOUT: <u>Dict Am Biog</u>; Elliott; <u>Natl Cycl Am Biog</u> 9; <u>Who Was Who Am</u> h. See also: Barr; <u>Biog Geneal Master Ind</u>; Ireland <u>Ind Sci</u>; Pelletier.

RISING, WILLARD BRADLEY (1839-1910). Chemistry. OCCUP: Professor. MSS: <u>Ind NUCMC</u> (?). WORKS ABOUT: <u>Am Chem and Chem Eng</u>; <u>Am Men Sci</u> 1 d3; Appleton's; Elliott (s); <u>Natl Cycl Am Biog</u> 12, 25; <u>Sci</u> 31(1910):261; <u>Who Was Who Am</u> 1.

RISLEY, SAMUEL DOTY (1845-1920). Ophthalmology. OCCUP:
Physician; Professor. WORKS ABOUT: Am Men Sci 1-2 d3;
Appleton's; Natl Cycl Am Biog 7:513; Ohio Authors; Who Was
Who Am 1.

RISSER, JONATHAN (1869-1917). Zoology. OCCUP: Professor.
WORKS ABOUT: Am Men Sci 1-2 d3; Sci 45(1917):334.

RITTENHOUSE, DAVID (1732-1796). Natural philosophy;
Astronomy; Mathematics; Technology; Telescope
manufacturing. OCCUP: Instrument maker; Surveyor; Public
official. MSS: Ind NUCMC (61-501, 61-782, 76-937). WORKS
ABOUT: Dict Am Biog; Dict Sci Biog; Elliott; Natl Cycl Am
Biog 1; Who Was Who Am h. See also: Barr; Biog Geneal
Master Ind; Ireland Ind Sci; Pelletier.

RITTER, WILLIAM FRANCIS MCKNIGHT (1846?-1917). Astronomy.
OCCUP: U.S. government scientist. WORKS ABOUT: Sci
46(1917):614.

RIVERS, JOHN JAMES (1824-1913). Biology; Entomology. OCCUP:
Independent wealth; Museum curator. WORKS ABOUT: Essig
p746-47; Sci 39(1914):170, 319-20.

ROBBINS, JAMES WATSON (1801-1879). Biology; Botany. OCCUP:
Physician. WORKS ABOUT: Am Journ Sci 117(1879):180,
119(1880):77.

ROBERTS, CHARLOTTE FITCH (1859-1917). Chemistry. OCCUP:
Professor. WORKS ABOUT: Am Men Sci 2 d3; Natl Cycl Am Biog
19; Sci 47(1918):140; Wallace; Who Was Who Am 1; Wom Whos
Who Am.

ROBIE, THOMAS (1688/89-1729). Astronomy; Medicine;
Meteorology. OCCUP: Publisher (almanac); Academic
instructor; Physician. MSS: Ind NUCMC. WORKS ABOUT:
Elliott. See also: Biog Geneal Master Ind; Pelletier.

ROBINSON, CHARLES BUDD, JR. (1871-1913). Botany. OCCUP:
Teacher; Government scientist (foreign). WORKS ABOUT: Am
Journ Sci 187(1914):208; Am Men Sci 2 d3; Sci 39(1914):20.

ROBINSON, COLEMAN TOWNSEND (1838-1872). Entomology. OCCUP:
Independent wealth (?). WORKS ABOUT: Natl Cycl Am Biog 21.

ROBINSON, FRANKLIN CLEMENT (1852-1910). Chemistry;
Mineralogy. OCCUP: Professor; State government scientist.
WORKS ABOUT: Am Journ Sci 180(1910):96; Am Men Sci 1 d3;
Natl Cycl Am Biog 14; Sci 31(1910):854; Wallace; Who Was
Who Am 1.

ROBINSON, FREDERICK BYRON (1855-1910). Anatomy; Gynecology.
OCCUP: Surgeon; Professor. WORKS ABOUT: Dict Am Biog. See
also: Biog Geneal Master Ind.

ROBINSON, HORATIO NELSON (1806-1867). Mathematics. OCCUP: Teacher; Educator; Author. WORKS ABOUT: Adams O; Allibone; Appleton's; Drake; Natl Cycl Am Biog 2; Twentieth Cent Biog Dict; Wallace.

ROBINSON, OTIS HALL (1835-1912). Astronomy; Physics. OCCUP: Professor; Librarian. WORKS ABOUT: Am Men Sci 1-2 d3; Dict Am Lib Biog.

ROBINSON, PHILIP ELY (-1920). Physics. OCCUP: Professor. WORKS ABOUT: Am Men Sci 2 d3.

ROBINSON, STILLMAN WILLIAMS (1838-1910). Mechanical engineering. OCCUP: Professor; Inventor; Consultant. MSS: Ind NUCMC. WORKS ABOUT: Am Men Sci 1-2 d3; Dict Am Biog; Natl Cycl Am Biog 10, 35; Who Was Who Am 1. See also: Biog Geneal Master Ind.

ROCK, MILES (1840-1901). Astronomy. OCCUP: U.S. government scientist; Government scientist (foreign); Observatory affiliate. WORKS ABOUT: Sci 13(1901):237, 278-79, 978-80.

ROCKHILL, WILLIAM WOODVILLE (1854-1914). Geography; Ethnology. OCCUP: Public official (diplomatic service). MSS: Ind NUCMC (81-627). WORKS ABOUT: Am Men Sci 2 d3; Dict Am Biog; Natl Cycl Am Biog 8; Who Was Who Am 1. See also: Barr; Biog Geneal Master Ind.

ROCKWELL, ALFRED PERKINS (1834-1903). Mining engineering. OCCUP: Professor; Industrial officer. MSS: Ind NUCMC. WORKS ABOUT: Adams O; Natl Cycl Am Biog 21; Wallace; Who Was Who Am 1.

ROCKWOOD, CHARLES GREENE, JR. (1843-1913). Mathematics; Seismology. OCCUP: Professor. WORKS ABOUT: Am Journ Sci 186(1913):576; Am Men Sci 1-2 d3; Appleton's; Davison p144-46; Natl Cycl Am Biog 16; Sci 38(1913):47; Twentieth Cent Biog Dict; Who Was Who Am 1.

RODGERS, JOHN (1812-1882). Exploration. OCCUP: Naval officer. MSS: Ind NUCMC (69-2048). WORKS ABOUT: Dict Am Biog; Natl Ac Scis Biog Mem 6; Natl Cycl Am Biog 5, 25; Who Was Who Am h. See also: Biog Geneal Master Ind.

RODMAN, WILLIAM LOUIS (1858-1916). Surgery. OCCUP: Surgeon; Professor. WORKS ABOUT: Dict Am Med Biog; Leonardo p366-67; Natl Cycl Am Biog 27; Sci 43(1916):384; Wallace; Who Was Who Am 1.

ROEBER, EUGENE FRANZ (1867-1917). Chemistry; Electrochemistry. OCCUP: Electrical engineer; Editor. WORKS ABOUT: Am Chem and Chem Eng; Natl Cycl Am Biog 17; Who Was Who Am 1.

ROEBLING, JOHN AUGUSTUS (1806-1869). Civil engineering. OCCUP: Engineer; Manufacturer. MSS: Ind NUCMC. WORKS ABOUT: Dict Am Biog; Natl Cycl Am Biog 4; Who Was Who Am h. See also: Biog Geneal Master Ind; Ireland Ind Sci; Pelletier.

ROEMER, (KARL) FERDINAND (1818-1891). Geology. OCCUP:
Natural history explorer; Professor; Museum curator. MSS:
Ind NUCMC. WORKS ABOUT: Dict Am Biog; Dict Sci Biog; Who
Was Who Am h. See also: Barr; Ireland Ind Sci.

ROGERS, ARTHUR CURTIS (1856-1917). Medicine. OCCUP:
Physician; Administrator. MSS: Ind NUCMC (82-1300). WORKS
ABOUT: Am Men Sci 2 d3; Natl Cycl Am Biog 17; Who Was Who
Am 1.

ROGERS, ELLIOT FOLGER (-1895). Chemistry. OCCUP:
Academic instructor. WORKS ABOUT: Sci 2(1895):483.

ROGERS, FAIRMAN (1833-1900). Civil engineering. OCCUP:
Professor. WORKS ABOUT: Adams O; Allibone Supp; Appleton's;
Natl Ac Scis Biog Mem 6; Natl Cycl Am Biog 11; Preston; Sci
12(1900):347; Twentieth Cent Biog Dict; Wallace; Who Was
Who Am 1.

ROGERS, HENRY DARWIN (1808-1866). Geology. OCCUP:
Professor; State government scientist. MSS: Ind NUCMC.
WORKS ABOUT: Dict Am Biog; Dict Sci Biog; Elliott; Natl
Cycl Am Biog 1, 7:543; Who Was Who Am h. See also: Barr;
Biog Geneal Master Ind; Ireland Ind Sci; Pelletier.

ROGERS, JAMES BLYTHE (1802-1852). Chemistry. OCCUP:
Industrial and state government scientist; Professor. MSS:
Ind NUCMC. WORKS ABOUT: Dict Am Biog; Elliott; Natl Cycl Am
Biog 8; Who Was Who Am h. See also: Barr; Biog Geneal
Master Ind; Ireland Ind Sci; Pelletier.

ROGERS, PATRICK KERR (1776-1828). Chemistry. OCCUP:
Physician; Lecturer; Professor. WORKS ABOUT: Am Chem and
Chem Eng.

ROGERS, ROBERT EMPIE (1813-1884). Chemistry. OCCUP: State
government scientist; Professor. MSS: Ind NUCMC. WORKS
ABOUT: Dict Am Biog; Elliott; Natl Ac Scis Biog Mem 5; Natl
Cycl Am Biog 7:518; Who Was Who Am h. See also: Barr; Biog
Geneal Master Ind; Ireland Ind Sci; Pelletier.

ROGERS, WILLIAM AUGUSTUS (1832-1898). Astronomy;
Mathematics; Physics. OCCUP: Professor; Observatory
affiliate. MSS: Ind NUCMC. WORKS ABOUT: Dict Am Biog;
Elliott; Natl Ac Scis Biog Mem 4, 6; Natl Cycl Am Biog 9;
Who Was Who Am h. See also: Barr; Biog Geneal Master Ind.

ROGERS, WILLIAM BARTON (1804-1882). Geology; Science
education. OCCUP: Professor; State government scientist;
College president. MSS: Ind NUCMC (61-2418, 80-2097). WORKS
ABOUT: Dict Am Biog; Dict Sci Biog; Elliott; Natl Ac Scis
Biog Mem 3; Natl Cycl Am Biog 7:410; Who Was Who Am h. See
also: Barr; Biog Geneal Master Ind; Ireland Ind Sci;
Pelletier.

ROHE, GEORGE HENRY (1851-1899). Public health. OCCUP: Physician; Administrator. WORKS ABOUT: Dict Am Biog; Natl Cycl Am Biog 7:275; Who Was Who Am 1. See also: Barr; Biog Geneal Master Ind.

ROMANS, BERNARD (ca.1720-ca.1784). Cartography; Civil engineering; Natural history. OCCUP: Engineer; Surveyor; Public official. MSS: Ind NUCMC. WORKS ABOUT: Dict Am Biog; Elliott; Natl Cycl Am Biog 7:176; Who Was Who Am h. See also: Biog Geneal Master Ind.

ROMINGER, CARL LUDWIG (1820-1907). Geology. OCCUP: Physician; State government scientist. MSS: Ind NUCMC (65-520). WORKS ABOUT: Am Men Sci 1 d3; Merrill p444-45; Sci 25(1907):718; Who Was Who Am 1.

ROOD, OGDEN NICHOLAS (1831-1902). Physics. OCCUP: Professor. MSS: Ind NUCMC (66-1601). WORKS ABOUT: Dict Am Biog; Dict Sci Biog; Elliott; Natl Ac Scis Biog Mem 6; Natl Cycl Am Biog 13; Who Was Who Am 1. See also: Barr; Biog Geneal Master Ind.

ROOSA, DANIEL BENNETT ST. JOHN (1838-1908). Medicine. OCCUP: Physician; Professor; Academic administrator. MSS: Ind NUCMC (?). WORKS ABOUT: Dict Am Biog; Natl Cycl Am Biog 9; Who Was Who Am 1. See also: Barr; Biog Geneal Master Ind.

ROOSEVELT, ROBERT BARNWELL (1829-1906). Conservation. OCCUP: Lawyer; Businessman; Public official. MSS: Ind NUCMC (63-392). WORKS ABOUT: Dict Am Biog; Natl Cycl Am Biog 3; Who Was Who Am 1. See also: Biog Geneal Master Ind.

ROOSEVELT, THEODORE (1858-1919). Natural history; Conservation. OCCUP: Public official; President. MSS: Ind NUCMC (60-59, 61-194, 64-812, 70-503, 73-905, 76-973, 78-714, 82-451, 81-629). WORKS ABOUT: Cutright; Dict Am Biog; Natl Cycl Am Biog 9, 11, 14; Who Was Who Am h (add), 1, 4 (add). See also: Barr; Biog Geneal Master Ind; Ireland Ind Sci.

ROOT, EDWIN W. (1840?-1870). Chemistry. OCCUP: Professor. WORKS ABOUT: Am Journ Sci 101(1871):75.

ROOT, ELIHU (1845-1880). Mathematics; Physics. OCCUP: Professor. WORKS ABOUT: Amherst Biog Rec 1973; Sci 1st ser 3(1884):467.

ROOT, OREN (ca.1803-1885). Natural history; Geology; Mathematics. OCCUP: Educator; Professor. Scientific work: see Roy Soc Cat. WORKS ABOUT: Jessup [biography of his son; see its index for page references].

ROOT, OREN (1838-1907). Mathematics. OCCUP: Professor; Clergyman. WORKS ABOUT: Am Men Sci 1 d3; Sci 26(1907):326; Twentieth Cent Biog Dict; Wallace; Who Was Who Am 1.

ROPER, EDWARD WARREN (1858-1898). Conchology. OCCUP: Editor. WORKS ABOUT: Nautilus 12(1898-1899):129-31.

ROSS, HENRY SCHUYLER (-1902). Engineering. OCCUP: Naval engineer (?). WORKS ABOUT: Sci 16(1902):678.

ROSSE, IRVING COLLINS (1859?-1901). Medicine. OCCUP: Physician (?); Professor; Explorer. WORKS ABOUT: Sci 13(1901):796.

ROTCH, ABBOTT LAWRENCE (1861-1912). Meteorology. OCCUP: Independent wealth; Observatory director; Professor. MSS: Ind NUCMC (84-1940). WORKS ABOUT: Am Men Sci 1-2 d3; Dict Am Biog; Natl Cycl Am Biog 15; Who Was Who Am 1. See also: Barr; Biog Geneal Master Ind.

ROTCH, THOMAS MORGAN (1849-1914). Pediatrics. OCCUP: Physician; Professor. WORKS ABOUT: Am Men Sci 1-2 d3; Dict Am Biog; Dict Am Med Biog; Natl Cycl Am Biog 19; Who Was Who Am 1. See also: Barr; Biog Geneal Master Ind; Ireland Ind Sci.

ROTHER, REINHOLD FRIEDRICH WILHELM (1843-1889). Pharmacy. OCCUP: Pharmacist. WORKS ABOUT: Am Journ Pharm 61(1889):639-40.

ROTHWELL, RICHARD PENNEFATHER (1836-1901). Mining. OCCUP: Mining engineer; Editor. WORKS ABOUT: Dict Am Biog; Natl Cycl Am Biog 10; Who Was Who Am 1. See also: Barr; Biog Geneal Master Ind.

ROTZELL, WILLETT ENOS (1871-1913). Natural history; Botany; Zoology. OCCUP: Physician; Professor; Editor. WORKS ABOUT: Am Men Sci 1; Natl Cycl Am Biog 12; Wallace; Who Was Who Am 1.

ROWE, RICHARD BURTON (1872?-1902). Geology. OCCUP: U.S. government scientist. WORKS ABOUT: Sci 15(1902):1037.

ROWLAND, HENRY AUGUSTUS (1848-1901). Physics. OCCUP: Professor. MSS: Ind NUCMC (60-1646). WORKS ABOUT: Dict Am Biog; Dict Sci Biog; Elliott; Natl Ac Scis Biog Mem 5; Natl Cycl Am Biog 11; Who Was Who Am 1. See also: Barr; Biog Geneal Master Ind; Ireland Ind Sci; Pelletier.

ROYCE, JOSIAH (1855-1916). Philosophy; Psychology. OCCUP: Professor. MSS: Ind NUCMC (65-1272). WORKS ABOUT: Am Men Sci 1-2 d3; Dict Am Biog; Natl Ac Scis Biog Mem 33; Natl Cycl Am Biog 11, 25; Who Was Who Am 1. See also: Barr; Biog Geneal Master Ind.

RUFFIN, EDMUND (1794-1865). Agriculture; Agricultural chemistry. OCCUP: Farmer; Public official; Editor and publisher. MSS: Ind NUCMC (64-1098, 68-2067, 69-549). WORKS ABOUT: Dict Am Biog; Elliott; Natl Cycl Am Biog 5; Who Was Who Am h. See also: Biog Geneal Master Ind; Pelletier.

RUFFNER, WILLIAM HENRY (1824-1908). Geology. OCCUP: Clergyman; Educator; Consultant. MSS: Ind NUCMC. WORKS ABOUT: Dict Am Biog; Natl Cycl Am Biog 12; Who Was Who Am 1. See also: Biog Geneal Master Ind.

RUGGLES, DANIEL (1810-1897). Geology; Study of lake tides. OCCUP: Army officer; Businessman; Farmer. MSS: Ind NUCMC (60-3247, 62-1034). Scientific work: see Roy Soc Cat. WORKS ABOUT: Appleton's; Cullum v1; Drake; Twentieth Cent Biog Dict.

RUMFORD, COUNT. See THOMPSON, BENJAMIN.

RUMSEY, JAMES (1743-1792). Invention. OCCUP: Businessman; Inventor. MSS: Ind NUCMC (62-1035). WORKS ABOUT: Dict Am Biog; Natl Cycl Am Biog 5; Who Was Who Am h. See also: Biog Geneal Master Ind; Ireland Ind Sci; Pelletier.

RUNKLE, JOHN DANIEL (1822-1902). Mathematics; Engineering education. OCCUP: U.S. government scientist; Professor; College president. MSS: Ind NUCMC. WORKS ABOUT: Dict Am Biog; Elliott; Natl Cycl Am Biog 6; Who Was Who Am 1. See also: Barr; Biog Geneal Master Ind; Ireland Ind Sci.

RUSCHENBERGER, WILLIAM SAMUEL WAITHMAN (1807-1895). Natural history. OCCUP: Naval surgeon. MSS: Ind NUCMC (66-356). WORKS ABOUT: Ac Nat Scis Philadelphia Proc 47(1895):452-62; Adams O; Allibone; Allibone Supp; Am Nat 29(1895):405; Am Phil Soc Proc 34(1895):361-64; Appleton's; Drake; Duyckinck; Major v2:769-70; Natl Cycl Am Biog 13; Nautilus 9(1895-1896):31-32; Sci 1(1895):417; Twentieth Cent Biog Dict; Wallace.

RUSH, BENJAMIN (1746-1813). Chemistry; Medicine; Psychiatry. OCCUP: Physician; Professor; Public official. MSS: Ind NUCMC (61-774, 61-2703, 73-710, 76-173). WORKS ABOUT: Dict Am Biog; Dict Am Med Biog; Dict Sci Biog; Elliott; Natl Cycl Am Biog 3; Who Was Who Am h. See also: Barr; Biog Geneal Master Ind; Ireland Ind Sci; Pelletier.

RUSH, JAMES (1786-1869). Psychology. OCCUP: Physician; Engaged in independent study. MSS: Ind NUCMC (61-2704). WORKS ABOUT: Dict Am Biog; Elliott; Natl Cycl Am Biog 6; Who Was Who Am h. See also: Biog Geneal Master Ind.

RUSSELL, FRANK (1868-1903). Anthropology. OCCUP: Academic instructor; U.S. government scientist. WORKS ABOUT: Adams O; Natl Cycl Am Biog 12; Oxford Canadian Hist; Sci 18(1903):637; Wallace; Who Was Who Am 1.

RUSSELL, ISRAEL COOK (1852-1906). Geology; Physical geography. OCCUP: U.S. government scientist; Professor. MSS: Ind NUCMC. WORKS ABOUT: Am Men Sci 1 d3; Dict Am Biog; Elliott (s); Natl Cycl Am Biog 10; Who Was Who Am 1. See also: Barr; Biog Geneal Master Ind.

RUSSELL, JOHN LEWIS (1808-1873). Natural history; Botany. OCCUP: Clergyman; Lecturer; Engaged in independent study. WORKS ABOUT: Am Ac Arts Scis <u>Proc</u> 9(1873-1875):321; <u>Am Journ Sci</u> 107(1874):240; Essex Inst <u>Bull</u> 5(1874):103-04, 12(1881):92.

RUSSELL, LINUS ELI (1848-1917). Medicine. OCCUP: Surgeon; Professor. WORKS ABOUT: <u>Natl Cycl Am Biog</u> 16; <u>Sci</u> 46(1917):135.

RUSSELL, THOMAS H. (1853?-1916). Surgery. OCCUP: Surgeon; Professor. WORKS ABOUT: <u>Sci</u> 43(1916):207.

RUST, WILLIAM P. (1826?-1891). Geology. OCCUP: U.S. government scientist; Collector. WORKS ABOUT: <u>Am Journ Sci</u> 143(1892):167-68.

RUTHERFURD, LEWIS MORRIS (1816-1892). Astrophysics. OCCUP: Lawyer; Engaged in independent study. MSS: <u>Ind NUCMC</u>. WORKS ABOUT: <u>Dict Am Biog</u>; <u>Dict Sci Biog</u>; Elliott; Natl Ac Scis <u>Biog Mem</u> 3; <u>Natl Cycl Am Biog</u> 6; <u>Who Was Who Am</u> h. See also: Barr; <u>Biog Geneal Master Ind</u>; Ireland <u>Ind Sci</u>.

RUTTER, CLOUDSLEY (1867-1903). Zoology. OCCUP: U.S. government scientist. WORKS ABOUT: <u>Sci</u> 18(1903):767.

RYDER, JOHN ADAM (1852-1895). Embryology. OCCUP: U.S. government scientist; Professor. MSS: <u>Ind NUCMC</u> (?). WORKS ABOUT: Ac Nat Scis Philadelphia <u>Proc</u> (1896):222-56; <u>Am Mo Micr Journ</u> 16(1895):205; <u>Am Nat</u> 29(1895):401-05; Am Phil Soc <u>Proc (Memorial Vol)</u>:1-8; <u>Leopoldina</u> 31(1895):103; <u>Natl Cycl Am Biog</u> 16; <u>Nautilus</u> 9(1895-1896):29-31; Roy Micr Soc <u>Journ</u> (1895):594; <u>Sci</u> 1(1895):417, 613, 2(1895):334-36, 3(1896):164, 41(1915):748-49; Smithsonian <u>Rep</u> (1896):673-87.

S

SABINE, WALLACE CLEMENT WARE (1868-1919). Physics. OCCUP: Professor; Academic administrator. MSS: <u>Ind NUCMC</u>. WORKS ABOUT: <u>Am Men Sci</u> 1-2 d3; <u>Dict Am Biog</u>; <u>Dict Sci Biog</u>; Natl Ac Scis <u>Biog Mem</u> [<u>Mem</u> 21(13)]; <u>Natl Cycl Am Biog</u> 15, 27; <u>Who Was Who Am</u> 1, 2. See also: Barr; <u>Biog Geneal Master Ind</u>; Pelletier.

SACHS, THEODORE BERNARD (1868-1916). Medicine. OCCUP: Physician; Administrator. WORKS ABOUT: <u>Dict Am Biog</u>. See also: Barr.

SAFFORD, JAMES MERRILL (1822-1907). Chemistry; Geology; Natural history. OCCUP: Professor; State government scientist. MSS: <u>Ind NUCMC</u> (?). WORKS ABOUT: <u>Am Men Sci</u> 1 d3; <u>Dict Am Biog</u>; Elliott (s); <u>Natl Cycl Am Biog</u> 8; <u>Who Was Who Am</u> 1. See also: Barr; <u>Biog Geneal Master Ind</u>; Ireland <u>Ind Sci</u>.

SAFFORD, TRUMAN HENRY (1836-1901). Astronomy; Mathematics. OCCUP: Observatory director; Professor. MSS: <u>Ind NUCMC</u>. WORKS ABOUT: <u>Dict Am Biog</u>; Elliott; <u>Natl Cycl Am Biog</u> 13; <u>Who Was Who Am</u> 1. See also: Barr; <u>Biog Geneal Master Ind</u>.

SALISBURY, JAMES HENRY (1823-1905). Chemistry; Medicine; Microscopy; Plant pathology. OCCUP: State government scientist; Physician. WORKS ABOUT: <u>Dict Am Biog</u>; Elliott; <u>Natl Cycl Am Biog</u> 8; <u>Who Was Who Am</u> 1. See also: <u>Biog Geneal Master Ind</u>.

SALMON, DANIEL ELMER (1850-1914). Pathology; Veterinary medicine. OCCUP: Veterinarian; U.S. government scientist; Academic administrator. WORKS ABOUT: <u>Am Men Sci</u> 1-2 d3; <u>Dict Am Biog</u>; <u>Dict Am Med Biog</u>; <u>Who Was Who Am</u> 1. See also: Barr; <u>Biog Geneal Master Ind</u>; Ireland <u>Ind Sci</u>.

SAMPSON, FRANCIS ASBURY (1842-1918). Geology; Conchology. OCCUP: Lawyer; Banker; Librarian. Scientific work: see Roy Soc <u>Cat</u>. WORKS ABOUT: Knight; <u>Who Was Who Am</u> 1.

SAMPSON, WILLIAM THOMAS (1840-1902). Astronomy. OCCUP: Naval officer. Scientific work: see Roy Soc <u>Cat</u>. MSS: <u>Ind NUCMC</u>. WORKS ABOUT: <u>Dict Am Biog</u>; <u>Natl Cycl Am Biog</u> 9; <u>Who Was Who Am</u> 1. See also: <u>Biog Geneal Master Ind</u>.

SAMUELS, EDWARD AUGUSTUS (1836-1908). Nature study; Zoology; Ornithology. OCCUP: State government scientist; Publisher. WORKS ABOUT: <u>Dict Am Biog</u>; Elliott; <u>Who Was Who Am</u> 1. See also: Barr; <u>Biog Geneal Master Ind</u>.

SANBORN, FRANCIS GREGORY (1838-1884). Conchology; Entomology. OCCUP: Museum curator; State government scientist; Academic instructor. WORKS ABOUT: <u>Canadian Ent</u> 16(1884):103-05; Essex Inst <u>Bull</u> 17(1886):61.

SANFORD, E. W. (-1918). Medicine. OCCUP: Academic instructor; Researcher. WORKS ABOUT: <u>Sci</u> 48(1918):137.

SANFORD, LEONARD JACOB (1833-1896). Anatomy. OCCUP: Physician; Professor. WORKS ABOUT: Am Med Assoc <u>Journ</u> 27(1896):1360; <u>Sci</u> 4(1896):915.

SANGER, CHARLES ROBERT (1860-1912). Chemistry. OCCUP: Professor. MSS: <u>Ind NUCMC</u> (76-2007). WORKS ABOUT: <u>Am Men Sci</u> 1-2 d3; <u>Dict Am Biog</u>; <u>Natl Cycl Am Biog</u> 28; <u>Who Was Who Am</u> 1. See also: Barr; Pelletier.

SARGENT, WINTHROP (1753-1820). Botany; Meteorology. OCCUP: Administrator; Public official. MSS: <u>Ind NUCMC</u> (60-1245, 60-2437, 66-1688, 68-238). WORKS ABOUT: <u>Dict Am Biog</u>; <u>Natl Cycl Am Biog</u> 13; <u>Who Was Who Am</u> h. See also: <u>Biog Geneal Master Ind</u>.

SARTWELL, HENRY PARKER (1792-1867). Botany. OCCUP: Physician. MSS: <u>Ind NUCMC</u>. WORKS ABOUT: <u>Dict Am Biog</u>; Elliott; <u>Who Was Who Am</u> h. See also: Barr; <u>Biog Geneal Master Ind</u>.

SAUGRAIN DE VIGNI, ANTOINE FRANCOIS (1763-1820). Instrument making; Natural history; Physical science. OCCUP: Physician; Landowner. WORKS ABOUT: <u>Dict Am Biog</u>; <u>Dict Am Med Biog</u>; Elliott; <u>Who Was Who Am</u> h. See also: Pelletier.

SAUNDERS, WILLIAM (1822-1900). Horticulture. OCCUP: Architect (landscape); U.S. government scientist. MSS: <u>Ind NUCMC</u> (79-932). WORKS ABOUT: <u>Dict Am Biog</u>; <u>Natl Cycl Am Biog</u> 10; <u>Who Was Who Am</u> h. See also: Barr; <u>Biog Geneal Master Ind</u>; Ireland <u>Ind Sci</u>.

SAVAGE, THOMAS STAUGHTON (1804-1880). Natural history; Zoology. OCCUP: Clergyman. MSS: <u>Ind NUCMC</u>. WORKS ABOUT: <u>Dict Am Biog</u>; Elliott; <u>Natl Cycl Am Biog</u> 26; <u>Who Was Who Am</u> h.

SAWYER, EDWIN FORREST (1849-   ). Astronomy. OCCUP: Banker. WORKS ABOUT: <u>Am Men Sci</u> 1-2; <u>Natl Cycl Am Biog</u> 8.

SAWYER, HARRIS EASTMAN (1868-1911). Chemistry. OCCUP: Industrial and U.S. government scientist. WORKS ABOUT: Am Men Sci 1-2 d3; Sci 34(1911):49.

SAXTON, JOSEPH (1799-1873). Scientific instrumentation. OCCUP: Craftsman; U.S. government scientist. WORKS ABOUT: Dict Am Biog; Dict Sci Biog; Elliott; Natl Ac Scis Biog Mem 1; Natl Cycl Am Biog 9; Who Was Who Am h. See also: Biog Geneal Master Ind.

SAY, LUCY WAY SISTARE (1801-1886). Science illustration. OCCUP: Artist. MSS: Ind NUCMC (66-80). WORKS ABOUT: Groce and Wallace; Ogilvie; Who Was Who Am h.

SAY, THOMAS (1787-1834). Conchology; Entomology. OCCUP: Natural history explorer; Professor; Curator. MSS: Ind NUCMC (61-779, 66-81). WORKS ABOUT: Dict Am Biog; Dict Sci Biog; Elliott; Natl Cycl Am Biog 6; Who Was Who Am h. See also: Barr; Biog Geneal Master Ind; Ireland Ind Sci; Pelletier.

SAYRE, HERBERT ARMISTEAD (1866-1916). Physics. OCCUP: Professor. WORKS ABOUT: Am Men Sci 1-2 d3; Who Was Who Am 1.

SAYRE, LEWIS ALBERT (1820-1900). Surgery. OCCUP: Surgeon; Professor. MSS: Ind NUCMC (68-794). WORKS ABOUT: Dict Am Biog; Dict Am Med Biog; Natl Cycl Am Biog 2; Who Was Who Am 1. See also: Barr; Biog Geneal Master Ind; Ireland Ind Sci.

SCAMMON, CHARLES M. (1825-1911). Natural history. OCCUP: Shipmaster; U.S. government employee. MSS: Ind NUCMC (65-1171). WORKS ABOUT: Allibone Supp; Appleton's; Hanley; Sci 33(1911):887; Wallace.

SCANNELL, JAMES J. (-1915). Bacteriology. OCCUP: Laboratory administrator; Public official. WORKS ABOUT: Sci 41(1915):322.

SCHAEFFER, GEORGE CHRISTIAN (1815-1873). Chemistry. OCCUP: Professor; Librarian. Scientific work: see Roy Soc Cat. WORKS ABOUT: Poggendorff v3.

SCHAFFER, CHARLES (1838-1903). Botany; Geology. OCCUP: Physician. MSS: Ind NUCMC (?). WORKS ABOUT: Kelly and Burrage; Who Was Who Am 1.

SCHANCK, JOHN STILLWELL (1817-1898). Chemistry. OCCUP: Physician; Professor; Museum curator. WORKS ABOUT: Appleton's; Sci 8(1898):902; Twentieth Cent Biog Dict.

SCHENCK, CHARLES CARROLL (1871-1914). Physics. OCCUP: Academic instructor. WORKS ABOUT: Am Men Sci 1-2 d3.

SCHLOTTERBECK, JULIUS OTTO (1865-1917). Pharmacognosy; Pharmacology. OCCUP: Professor; Academic administrator. WORKS ABOUT: Am Men Sci 1-2 d3; Sci 46(1917):34; Who Was Who Am 1.

SCHMITT, COOPER DAVIS (1859-1910). Mathematics. OCCUP: Professor; Academic administrator. WORKS ABOUT: Am Men Sci 1-2 d3; Who Was Who Am 3.

SCHOFIELD, JOHN McALLISTER (1831-1906). Astronomy. OCCUP: Army officer; Professor. MSS: Ind NUCMC (63-397). WORKS ABOUT: Dict Am Biog; Natl Cycl Am Biog 4; Who Was Who Am 1. See also: Biog Geneal Master Ind.

SCHONEY, LAZARUS (1838-1914). Pathology. OCCUP: Surgeon; Professor. WORKS ABOUT: Am Men Sci 1-2 d3; Sci 39(1914):323.

SCHOOLCRAFT, HENRY ROWE (1793-1864). Ethnology; Geology. OCCUP: U.S. government employee; Explorer. MSS: Ind NUCMC (61-1757, 62-4596). WORKS ABOUT: Dict Am Biog; Dict Sci Biog; Elliott; Natl Cycl Am Biog 5; Who Was Who Am h. See also: Barr; Biog Geneal Master Ind; Ireland Ind Sci; Pelletier.

SCHOPF, JOHANN DAVID (1752-1800). Natural history; Geology; Travels; Zoology. OCCUP: Physician; Explorer. WORKS ABOUT: Dict Am Biog; Elliott; Who Was Who Am h. See also: Ireland Ind Sci.

SCHOTT, ARTHUR CARL VICTOR (1814-1875). Geology; Natural history. OCCUP: Mining manager; U.S. government scientist. MSS: Ind NUCMC. WORKS ABOUT: Elliott. See also: Biog Geneal Master Ind.

SCHOTT, CHARLES ANTHONY (1826-1901). Geodesy; Geography; Geophysics; Terrestrial magnetism. OCCUP: U.S. government scientist. WORKS ABOUT: Dict Am Biog; Dict Sci Biog; Elliott; Natl Ac Scis Biog Mem 8; Who Was Who Am 1. See also: Barr; Biog Geneal Master Ind.

SCHUBERT, ERNST (1813-1873). Astronomy. OCCUP: Observatory affiliate; U.S. government scientist. WORKS ABOUT: Elliott.

SCHULTE, EDWARD DELAVAN NELSON (1877-1920). Electrical engineering. OCCUP: Professor. WORKS ABOUT: Am Men Sci 2 d3.

SCHWATKA, FREDERICK (1849-1892). Geography; Exploration. OCCUP: Army officer; Explorer. MSS: Ind NUCMC (?). WORKS ABOUT: Dict Am Biog; Natl Cycl Am Biog 3; Who Was Who Am h. See also: Biog Geneal Master Ind.

SCHWEINITZ, EMIL ALEXANDER DE (1864?-1904). Bacteriology; Biochemistry. OCCUP: U.S. government scientist; Professor. WORKS ABOUT [birth dates may differ]: Dict Am Biog; Elliott; Who Was Who Am 1. See also: Barr; Biog Geneal Master Ind.

SCHWEINITZ, LEWIS DAVID VON (1780-1834). Botany; Mycology. OCCUP: Clergyman; Educator. MSS: Ind NUCMC (66-82). WORKS ABOUT: Dict Am Biog; Elliott; Natl Cycl Am Biog 8; Who Was Who Am h. See also: Barr; Biog Geneal Master Ind; Ireland Ind Sci.

SCHWEITZER, HUGO (1861-1917). Chemistry. OCCUP: Consultant; Industrial scientist. WORKS ABOUT: Am Men Sci 1-2 d3; Sci 47(1918):19.

SCHWEITZER, (JOHANN) PAUL (1840-1911). Chemistry; Agricultural chemistry. OCCUP: Professor; State government scientist. WORKS ABOUT: Am Men Sci 1-2 d3; Sci 34(1911):210; Who Was Who Am 1.

SCOTT, WILLIAM EARL DODGE (1852-1910). Natural history; Ornithology. OCCUP: Curator. WORKS ABOUT: Adams O; Am Men Sci 1 d3; Natl Cycl Am Biog 13; Sci 32(1910):303; Wallace; Who Was Who Am 1.

SCOTT, WINFIELD GERMAIN (1854-1919). Chemistry. OCCUP: Industrial scientist; Consultant. WORKS ABOUT: Natl Cycl Am Biog 19; Wallace.

SCOVELL, JOSIAH THOMAS (1841-1915). Botany; Chemistry; Mineralogy. OCCUP: Professor; Teacher. WORKS ABOUT: Am Men Sci 1-2 d3; Indiana Authors 1816; Natl Cycl Am Biog 16.

SCOVELL, MELVILLE AMASA (1855-1912). Chemistry. OCCUP: Professor; State government scientist. WORKS ABOUT: Am Men Sci 1-2 d3; Dict Am Biog; Natl Cycl Am Biog 15; Who Was Who Am 1. See also: Barr; Biog Geneal Master Ind; Pelletier.

SCUDDER, SAMUEL HUBBARD (1837-1911). Natural history; Entomology. OCCUP: Administrator; Librarian; U.S. government scientist. MSS: Ind NUCMC. WORKS ABOUT: Am Men Sci 1-2 d3; Dict Am Biog; Dict Sci Biog; Elliott (s); Natl Ac Scis Biog Mem [Mem 17]; Natl Cycl Am Biog 3, 24; Who Was Who Am 1. See also: Barr; Biog Geneal Master Ind; Ireland Ind Sci; Pelletier.

SEABURY, GEORGE JOHN (1844-1909). Chemistry; Pharmacy. OCCUP: Manufacturer; Pharmacist. WORKS ABOUT: Dict Am Biog. See also: Biog Geneal Master Ind.

SEAMAN, WILLIAM HENRY (1837-1910). Chemistry. OCCUP: Professor; U.S. government scientist. MSS: Ind NUCMC (?). WORKS ABOUT: Am Men Sci 1 d3; Elliott (s); Natl Cycl Am Biog 14; Sci 31(1910):985; Who Was Who Am 1.

SEARLE, ARTHUR (1837-1920). Astronomy. OCCUP: Professor. MSS: Ind NUCMC (65-1274). WORKS ABOUT: Am Men Sci 1-2 d3; Dict Am Biog; Elliott (s); Natl Cycl Am Biog 18, 19; Who Was Who Am 1. See also: Biog Geneal Master Ind.

SEARLE, GEORGE MARY (1839-1918). Astronomy. OCCUP: Observatory affiliate; Clergyman; Professor. MSS: Ind NUCMC (?). WORKS ABOUT: Adams O; Allibone Supp; Am Lit Yearbook; Am Men Sci 1-2 d3; Appleton's; Sci 48(1918):66-67; Twentieth Cent Biog Dict; Wallace; Who Was Who Am 1.

SEATON, HENRY ELIASON (1869-1893). Botany. OCCUP: Academic instructor; Herbarium curator. WORKS ABOUT: Am Journ Sci 145(1893):526-27.

SEAVER, JAY WEBBER (1855-1915). Hygiene. OCCUP: Physician; Professor. WORKS ABOUT: Sci 41(1915):722; Wallace.

SEELY, HENRY MARTYN (1828-1917). Geology. OCCUP: Professor. WORKS ABOUT: Am Men Sci 1-2 d3; Sci 45(1917):458; Twentieth Cent Biog Dict; Who Was Who Am 1.

SELFRIDGE, THOMAS E. (1882-1908). Aeronautics. OCCUP: Army officer. WORKS ABOUT: Heinmuller p269; Howard F; Sci 28(1908):404.

SELLERS, COLEMAN (1827-1907). Engineering. OCCUP: Industrial engineer; Consultant; Professor. MSS: Ind NUCMC (?). WORKS ABOUT: Dict Am Biog; Natl Cycl Am Biog 11; Who Was Who Am 1. See also: Barr; Biog Geneal Master Ind.

SELLERS, WILLIAM (1824-1905). Mechanical engineering. OCCUP: Manufacturer; Inventor. WORKS ABOUT: Dict Am Biog; Natl Cycl Am Biog 7:185; Who Was Who Am 1. See also: Barr; Biog Geneal Master Ind.

SELOUS, PERCY S. (-1900). Ornithology. OCCUP: Not determined. WORKS ABOUT: Sci 12(1900):158.

SENN, NICHOLAS (1844-1908). Surgery. OCCUP: Physician; Professor. WORKS ABOUT: Dict Am Biog; Dict Am Med Biog; Natl Cycl Am Biog 6; Who Was Who Am 1. See also: Barr; Biog Geneal Master Ind; Ireland Ind Sci.

SENNETT, GEORGE BURRITT (1840-1900). Ornithology. OCCUP: Manufacturer. MSS: Ind NUCMC (70-1969). WORKS ABOUT: Dict Am Biog; Elliott; Natl Cycl Am Biog 14; Who Was Who Am h. See also: Barr.

SENSENEY, EDGAR MOORE (1856?-1916). Medicine. OCCUP: Professor; Physician (?). WORKS ABOUT: Sci 43(1916):639.

SESTINI, BENEDICT (1816-1890). Astronomy; Mathematics. OCCUP: Clergyman; Observatory affiliate; Academic instructor. WORKS ABOUT: Dict Am Biog; Elliott; Who Was Who Am h. See also: Biog Geneal Master Ind.

SEVERSON, B. O. [BURNS OSCAR?] (-1918). Zoology; Animal breeding. OCCUP: Professor. WORKS ABOUT: Natl Union Cat; Sci 48(1918):644.

SEWALL, STEPHEN (1734-1804). Magnetism. OCCUP: Professor (languages). MSS: Ind NUCMC (65-1275). WORKS ABOUT: Dict Am Biog; Natl Cycl Am Biog 6; Who Was Who Am h. See also: Biog Geneal Master Ind.

SEYBERT, ADAM (1773-1825). Chemistry; Mineralogy. OCCUP: Physician; Pharmacist; Legislator. MSS: Ind NUCMC (76-944). WORKS ABOUT: Dict Am Biog; Natl Cycl Am Biog 4; Who Was Who Am h. See also: Biog Geneal Master Ind; Ireland Ind Sci.

SEYBERT, HENRY (1801-1883). Mineralogy. OCCUP: Inherited wealth; Benefactor. MSS: Ind NUCMC. WORKS ABOUT: Dict Am Biog; Elliott; Who Was Who Am h. See also: Barr; Pelletier.

SHALER, NATHANIEL SOUTHGATE (1841-1906). Geology. OCCUP: Professor; Academic administrator; U.S. government scientist. MSS: Ind NUCMC. WORKS ABOUT: Am Men Sci 1 d3; Dict Am Biog; Dict Sci Biog; Elliott (s); Natl Cycl Am Biog 9; Who Was Who Am 1. See also: Barr; Biog Geneal Master Ind; Ireland Ind Sci.

SHAPLEIGH, WALDRON (1848-1901). Chemistry. OCCUP: Industrial scientist. WORKS ABOUT: Sci 14(1901):463; Who Was Who Am 1.

SHARP, BENJAMIN (1858-1915). Zoology. OCCUP: Professor; Natural history collector. WORKS ABOUT: Am Journ Sci 189(1915):326; Am Men Sci 2 d3; Sci 41(1915):204; Who Was Who Am 1.

SHARSWOOD, WILLIAM (1836-1905). Chemistry. OCCUP: Not determined; Author. Scientific work: see Roy Soc Cat. MSS: Ind NUCMC (61-783). WORKS ABOUT: Allibone; Appleton's; Drake; Pennsylvania Biog Cat (listed); Pennsylvania Gen Cat; Wallace.

SHATTUCK, LEMUEL (1793-1859). Statistics; Public health. OCCUP: Merchant; Public official. MSS: Ind NUCMC (68-244). WORKS ABOUT: Dict Am Biog; Dict Am Med Biog; Who Was Who Am h. See also: Biog Geneal Master Ind; Ireland Ind Sci; Pelletier.

SHATTUCK, LYDIA WHITE (1822-1889). Natural history; Botany. OCCUP: Teacher. WORKS ABOUT: Elliott; Notable Am Wom; Siegel and Finley. See also: Barr; Biog Geneal Master Ind.

SHATTUCK, SAMUEL WALKER (1841-1915). Mathematics. OCCUP: Professor. WORKS ABOUT: Am Men Sci 1-2 d3; Natl Cycl Am Biog 18; Sci 41(1915):284; Who Was Who Am 1.

SHAW, BENJAMIN SHURTLEFF (1827-1893). Chemistry; Medicine. OCCUP: Physician. Scientific work: see Roy Soc <u>Cat</u>. WORKS ABOUT: <u>Boston Med Surg Journ</u> 128(1893):478; Harrington v2:910 and v3:1484 (listed); Harvard <u>Quinquennial Cat</u>.

SHAW, CHARLES HUGH (1875-1910). Botany. OCCUP: Professor. WORKS ABOUT: <u>Am Men Sci</u> 1; <u>Sci</u> 32(1910):239-40, 33(1911):449-50.

SHAW, EDWARD RICHARD (1850-1903). Pedagogy; Science teaching. OCCUP: Educator; Professor. WORKS ABOUT: <u>Dict Am Biog</u>; <u>Who Was Who Am</u> 1. See also: Barr; <u>Biog Geneal Master Ind</u>.

SHAW, HENRY (1800-1889). Botany; Founder of Missouri Botanical Garden. OCCUP: Businessman; Benefactor. MSS: <u>Ind NUCMC</u> (84-232). WORKS ABOUT: <u>Dict Am Biog</u>; <u>Natl Cycl Am Biog</u> 9; <u>Who Was Who Am</u> h. See also: Barr; <u>Biog Geneal Master Ind</u>.

SHAW, JOSEPH P. (-1895). Chemistry. OCCUP: Professor. WORKS ABOUT: <u>Sci</u> 2(1895):46.

SHAW, THOMAS (1843-1919?). Agriculture. OCCUP: Professor. WORKS ABOUT: Adams O; <u>Am Men Sci</u> 2 d3; Wallace; <u>Who Was Who Am</u> 4.

SHEAFER, PETER WENRICH (1819-1891). Geology; Mining engineering. OCCUP: Mining engineer; Mining manager; State government scientist. MSS: <u>Ind NUCMC</u> (60-3141). WORKS ABOUT: Am Phil Soc <u>Proc</u> 29(1891):39-42; Appleton's.

SHECUT, JOHN LINNAEUS EDWARD WHITRIDGE (1770-1836). Botany. OCCUP: Physician. WORKS ABOUT: <u>Dict Am Biog</u>; Elliott; <u>Who Was Who Am</u> h. See also: <u>Biog Geneal Master Ind</u>.

SHELDON, RALPH EDWARD (1883-1918). Anatomy. OCCUP: Professor. WORKS ABOUT: <u>Am Men Sci</u> 2 d3; <u>Who Was Who Am</u> 1.

SHELDON, SAMUEL (1862-1920). Physics; Electrical engineering. OCCUP: Professor; Consultant. MSS: <u>Ind NUCMC</u>. WORKS ABOUT: <u>Am Lit Yearbook</u>; <u>Am Men Sci</u> 1-2 d3; <u>Natl Cycl Am Biog</u> 14; Wallace; <u>Who Was Who Am</u> 1.

SHEPARD, CHARLES UPHAM (1804-1886). Mineralogy. OCCUP: Professor. MSS: <u>Ind NUCMC</u> (75-20). WORKS ABOUT: <u>Dict Am Biog</u>; Elliott; <u>Natl Cycl Am Biog</u> 5; <u>Who Was Who Am</u> h. See also: Barr; <u>Biog Geneal Master Ind</u>; Ireland <u>Ind Sci</u>; Pelletier.

SHEPARD, CHARLES UPHAM, JR. (1842-1915). Chemistry. OCCUP: Professor; U.S. and state government scientist; Consultant (?). WORKS ABOUT: <u>Am Men Sci</u> 2 d3; Appleton's; <u>Sci</u> 21(1905):878.

SHEPARD, JAMES HENRY (1850-1918). Chemistry. OCCUP: Educator; Professor; State government scientist. WORKS ABOUT: Dict Am Biog; Elliott; Natl Cycl Am Biog 17; Who Was Who Am 1. See also: Biog Geneal Master Ind.

SHEPARD, WILLIAM A. (-1895). Chemistry. OCCUP: Professor. WORKS ABOUT: Sci 1(1895):668.

SHEPHERD, FORREST (1800-1888). Geography; Geology. OCCUP: Natural history explorer; Professor; Consultant. Scientific work: see Roy Soc Cat. WORKS ABOUT: Waite p309-11, 320, 334, 343, 490 (mentioned); Yale Obit Rec (1889):490.

SHERMAN, FRANK ASBURY (1841-1915). Mathematics. OCCUP: Professor. WORKS ABOUT: Am Men Sci 1-2 d3; Natl Cycl Am Biog 9; Sci 41(1915):422; Who Was Who Am 1.

SHERMAN, JOHN (1613-1685). Mathematics. OCCUP: Clergyman; Teacher. WORKS ABOUT: Dict Am Biog; Elliott; Natl Cycl Am Biog 7:75; Who Was Who Am h. See also: Biog Geneal Master Ind.

SHIELDS, CHARLES WOODRUFF (1825-1904). Science and religion. OCCUP: Clergyman; Professor. WORKS ABOUT: Dict Am Biog; Natl Cycl Am Biog 13; Who Was Who Am 1. See also: Barr; Biog Geneal Master Ind.

SHIMER, HENRY (1828-1895). Entomology. OCCUP: Physician; State government scientist. WORKS ABOUT: Ent News 6(1895):305-06.

SHORT, CHARLES WILKINS (1794-1863). Botany. OCCUP: Physician; Professor. MSS: Ind NUCMC (61-781, 62-3397, 64-1103, 84-2407). WORKS ABOUT: Dict Am Biog; Elliott; Natl Cycl Am Biog 4; Who Was Who Am h. See also: Barr; Biog Geneal Master Ind.

SHORT, SIDNEY HOWE (1858-1902). Electrical engineering; Invention. OCCUP: Professor; Industrial scientist; Manufacturer. WORKS ABOUT: Dict Am Biog; Natl Cycl Am Biog 13; Who Was Who Am h. See also: Barr.

SHRADY, GEORGE FREDERICK (1837-1907). Medicine. OCCUP: Physician; Editor. WORKS ABOUT: Dict Am Biog; Dict Am Med Biog; Natl Cycl Am Biog 7:271; Who Was Who Am 1. See also: Barr; Biog Geneal Master Ind.

SHUMARD, BENJAMIN FRANKLIN (1820-1869). Geology; Paleontology. OCCUP: Physician; U.S. and state government scientist; Professor (medicine). MSS: Ind NUCMC (?). WORKS ABOUT: Elliott; Natl Cycl Am Biog 8. See also: Barr.

SIAS, SOLOMON (1829-   ). Geology. OCCUP: Educator. WORKS ABOUT: Am Men Sci 2 d3.

SIEBEL, JOHN EWALD (1845-1919). Chemistry; Physics. OCCUP: Consultant; Editor and publisher; Educator. WORKS ABOUT: Am Men Sci 2 d3; Wallace; Who Was Who Am 1.

SILLIMAN, BENJAMIN, SR. (1779-1864). Natural history; Chemistry; Geology; Mineralogy. OCCUP: Professor; Lecturer; Editor. MSS: Ind NUCMC (60-2695, 65-11, 74-1201). WORKS ABOUT: Dict Am Biog; Dict Am Med Biog; Dict Sci Biog; Elliott; Natl Ac Scis Biog Mem 1; Natl Cycl Am Biog 2; Who Was Who Am h. See also: Barr; Biog Geneal Master Ind; Ireland Ind Sci; Pelletier.

SILLIMAN, BENJAMIN, JR. (1816-1885). Chemistry; Geology. OCCUP: Professor; Editor; Consultant. MSS: Ind NUCMC (61-2572, 74-1201). WORKS ABOUT: Dict Am Biog; Dict Sci Biog; Elliott; Natl Ac Scis Biog Mem 7; Natl Cycl Am Biog 2; Who Was Who Am h. See also: Barr; Biog Geneal Master Ind; Pelletier.

SILLIMAN, BENJAMIN DOUGLAS (1805-1901). Chemistry. OCCUP: Lawyer; Public official. Scientific work: see Roy Soc Cat. MSS: Ind NUCMC. WORKS ABOUT: Appleton's; Fisher v1:315-16 (mentioned); Natl Cycl Am Biog 6; Twentieth Cent Biog Dict.

SILLIMAN, JUSTUS MITCHELL (1842-1896). Engineering. OCCUP: Professor. MSS: Ind NUCMC (?). WORKS ABOUT: Appleton's; Natl Cycl Am Biog 11; Sci 3(1896):629; Twentieth Cent Biog Dict.

SIMON, WILLIAM (1844-1916). Chemistry. OCCUP: Industrial scientist; Professor. WORKS ABOUT: Am Chem and Chem Eng; Am Men Sci 1-2 d3; Natl Cycl Am Biog 17; Sci 44(1916):165; Wallace; Who Was Who Am 1.

SIMPSON, CHARLES BAIRD (1877-1907). Entomology. OCCUP: U.S. government scientist; Government scientist (foreign). WORKS ABOUT: Am Men Sci 1 d3; Sci 25(1907):198.

SIMPSON, GEORGE B. (-1901). Paleontology. OCCUP: Museum affiliate (?). WORKS ABOUT: Sci 14(1901):660.

SIMPSON, WILLIAM KELLY (1855-1914). Laryngology. OCCUP: Surgeon; Professor. WORKS ABOUT: Dict Am Biog; Who Was Who Am 1. See also: Barr.

SINCLAIR, JAMES (1852?-1915). Zoology. OCCUP: Editor; Author. WORKS ABOUT: Sci 42(1915):826; Who Was Who Lit.

SINCLAIR, JOHN ELBRIDGE (1838-1915). Mathematics. OCCUP: Professor. WORKS ABOUT: Sci 42(1915):416; Who Was Who Am 1.

SINKLER, WHARTON (1845-1910). Medicine; Neurology. OCCUP: Physician. MSS: Ind NUCMC (?). WORKS ABOUT: Am Men Sci 1 d3; Sci 31(1910):499 (as Sinclair); Who Was Who Am 1.

SKINNER, AARON NICHOLS (1845-1918). Astronomy. OCCUP: Professor; Observatory affiliate. WORKS ABOUT: Am Men Sci 1-2 d3; Dict Am Biog; Elliott (s); Natl Cycl Am Biog 20; Who Was Who Am 1. See also: Barr; Biog Geneal Master Ind.

SKINNER, JOSEPH JOHN (1842-1919). Mathematics. OCCUP: Professor. WORKS ABOUT: Am Men Sci 1-2 d3; Wallace.

SLACK, DAVID B. (1798-1871). Biology. OCCUP: Physician. Scientific work: see Roy Soc Cat. WORKS ABOUT: Brown Hist Cat 1904; Wallace.

SLACK, JOHN HAMILTON (1834-1874). Zoology; Pisciculture. OCCUP: Physician; Pisciculturist; U.S. and state government scientist. WORKS ABOUT: Ac Nat Scis Philadelphia Proc 26(1875):140-42; Wallace.

SLADE, DANIEL DENISON (1823-1896). Osteology. OCCUP: Physician; Professor; Museum affiliate. WORKS ABOUT: Adams O; Allibone; Allibone Supp; Appleton's; Sci 3(1896):282; Wallace.

SLATER, SAMUEL (1768-1835). Technology. OCCUP: Manufacturer; Businessman. MSS: Ind NUCMC (60-1720). WORKS ABOUT: Dict Am Biog; Natl Cycl Am Biog 4, 24; Who Was Who Am h. See also: Biog Geneal Master Ind; Ireland Ind Sci; Pelletier.

SLINGERLAND, MARK VERNON (1864-1909). Entomology. OCCUP: Professor. WORKS ABOUT: Am Men Sci 1 d3; Mallis; Natl Cycl Am Biog 13; Osborn Herb p203, 346; Sci 29(1909):498; Wallace; Who Was Who Am 1.

SMALL, WILLIAM (1734-1775). Natural philosophy. OCCUP: Professor; Physician. WORKS ABOUT: Bell p75; Ganter.

SMITH, ABRAM ALEXANDER (1847-1915). Medicine. OCCUP: Physician; Professor. WORKS ABOUT: Am Men Sci 1-2 d3; Sci 42(1915):898; Who Was Who Am 1.

SMITH, ALLEN P. (-1898). Surgery. OCCUP: Surgeon. WORKS ABOUT: Sci 8(1898):129.

SMITH, ANDREW HEERMANCE (1837-1910). Medicine. OCCUP: Physician. WORKS ABOUT: Am Men Sci 1 d3; Dict Am Med Biog; Natl Cycl Am Biog 5; Wallace; Who Was Who Am 1.

SMITH, ARTHUR GEORGE (1868-1916). Mathematics; Mechanics. OCCUP: Professor. WORKS ABOUT: Am Men Sci 2 d3; Sci 44(1916):707; Who Was Who Am 1.

SMITH, AUGUSTUS WILLIAM (1802-1866). Astronomy. OCCUP: Professor; College president. WORKS ABOUT: Adams O; Allibone; Appleton's; Drake; Natl Cycl Am Biog 9; Twentieth Cent Biog Dict; Wallace.

SMITH, CLINTON DeWITT (1854-1916). Agriculture. OCCUP: Professor; State government scientist; Academic administrator. MSS: Ind NUCMC. WORKS ABOUT: Am Men Sci 2 d3; Natl Cycl Am Biog 17; Who Was Who Am 1.

SMITH, DANIEL B. (1792-1883). Chemistry. OCCUP: Pharmacist; Educator. MSS: Ind NUCMC. WORKS ABOUT: Dict Am Biog; Natl Cycl Am Biog 5; Who Was Who Am h. See also: Biog Geneal Master Ind; Pelletier.

SMITH, EDWARD DARRELL (1778-1819). Chemistry; Mineralogy. OCCUP: Physician; Planter; Professor. Scientific work: see Roy Soc Cat. WORKS ABOUT: Drake; LaBorde p96-101; Princeton Univ Gen Cat p109 (listed).

SMITH, EDWIN (1851-1912). Astronomy; Geodesy. OCCUP: U.S. government scientist. WORKS ABOUT: Am Journ Sci 185(1913):120; Am Men Sci 1-2 d3; Sci 36(1912):822; Twentieth Cent Biog Dict; Who Was Who Am 1.

SMITH, ERMINNIE ADELE PLATT (1836-1886). Ethnology; Geology. OCCUP: Independent wealth (through marriage); Lecturer; Collector. WORKS ABOUT: Dict Am Biog; Elliott; Natl Cycl Am Biog 13; Notable Am Wom; Siegel and Finley; Who Was Who Am h. See also: Biog Geneal Master Ind; Pelletier.

SMITH, EUGENE (1860-1912). Civil engineering; Zoology. OCCUP: Engineer; City government engineer. WORKS ABOUT: Am Men Sci 1-2 d3; Sci 37(1913):16; Wallace; Who Was Who Am 1.

SMITH, EUGENE ALFRED (1864-1914). Medicine; Surgery. OCCUP: Surgeon; Professor. WORKS ABOUT: Am Men Sci 1-2 d3.

SMITH, HAMILTON LANPHERE (1818-1903). Astronomy; Microscopy; Physics. OCCUP: Businessman; Professor. WORKS ABOUT: Elliott; Natl Cycl Am Biog 12; Who Was Who Am 1. See also: Barr; Biog Geneal Master Ind.

SMITH, HERBERT HUNTINGTON (1851-1919). Zoology. OCCUP: Natural history collector; Museum curator. WORKS ABOUT: Adams O; Allibone Supp; Am Journ Sci 197(1919):390; Am Lit Yearbook; Am Men Sci 2 d3; Osborn Herb p230, 359; Sci 49(1919):305, 481-83; Wallace; Who Was Who Am 1.

SMITH, JAMES (1740-1812). Chemistry. OCCUP: Physician; Professor. Am Chem and Chem Eng.

SMITH, JAMES G. (-1900). Telegraphy. OCCUP: Not determined; Inventor. WORKS ABOUT: Sci 11(1900):476.

SMITH, JOB LEWIS (1827-1897). Medicine. OCCUP: Physician; Professor. WORKS ABOUT: Dict Am Biog; Dict Am Med Biog; Who Was Who Am h. See also: Barr; Biog Geneal Master Ind.

SMITH, JOHN BERNHARD (1858-1912). Entomology. OCCUP: U.S. and state government scientist; Professor. MSS: Ind NUCMC. WORKS ABOUT: Am Men Sci 1-2 d3; Dict Am Biog; Natl Cycl Am Biog 13, 15; Who Was Who Am 1. See also: Barr; Biog Geneal Master Ind; Ireland Ind Sci; Pelletier.

SMITH, JOHN CHARLES (1850-   ). Zoology. OCCUP: Merchant. WORKS ABOUT: Am Men Sci 1-3 d4.

SMITH, JOHN HOLMES (1857-1919). Anatomy. OCCUP: Surgeon; Professor. WORKS ABOUT: Am Men Sci 1-2 d3.

SMITH, JOHN LAWRENCE (1818-1883). Chemistry; Mineralogy. OCCUP: Physician; Independent wealth (through marriage); Professor. MSS: Ind NUCMC. WORKS ABOUT: Dict Am Biog; Elliott; Natl Ac Scis Biog Mem 2; Natl Cycl Am Biog 6; Who Was Who Am h. See also: Barr; Biog Geneal Master Ind; Ireland Ind Sci; Pelletier.

SMITH, MATILDA H. (-1910). Philanthropy. OCCUP: Not determined; Benefactor (to scientific societies). WORKS ABOUT: Sci 32(1910):171, 34(1911):871.

SMITH, NATHAN (1762-1829). Medicine; Surgery. OCCUP: Physician; Professor. MSS: Ind NUCMC. WORKS ABOUT: Dict Am Biog; Dict Am Med Biog; Natl Cycl Am Biog 3; Who Was Who Am h. See also: Barr; Biog Geneal Master Ind; Ireland Ind Sci; Pelletier.

SMITH, OTIS DAVID (1831-1905). Mathematics. OCCUP: Educator; Professor. WORKS ABOUT: Am Men Sci 1-2 d3; Natl Cycl Am Biog 26; Who Was Who Am 1.

SMITH, SANDERSON (1832-1915). Conchology; Malacology. OCCUP: Engineer (?); U.S. government science associate; Collector. MSS: Ind NUCMC. WORKS ABOUT: Am Journ Sci 189(1915):685-86; Sci 41(1915):750, 787-88; Staten Island Assoc Arts Scis Proc 6(1916):141-43.

SMITH, THEODATE LOUISE (1859-1914). Psychology. OCCUP: Teacher; Academic researcher; Librarian. WORKS ABOUT [birth dates may differ]: Am Men Sci 1-2 d3; Dict Am Biog; Siegel and Finley. See also: Biog Geneal Master Ind.

SMITH, THOMAS PETERS (ca.1776-1802). Chemistry. OCCUP: Not determined. MSS: Ind NUCMC (?). WORKS ABOUT: Am Chem and Chem Eng.

SMITH, WILLIAM R. (1828?-1912). Botany. OCCUP: Museum (botanic garden) administrator. WORKS ABOUT: Sci 36(1912):79.

SMITH, WILLIAM THAYER (1839-1909). Physiology. OCCUP: Physician; Professor. WORKS ABOUT: Am Men Sci 1 d3; Natl Cycl Am Biog 9, 25; Who Was Who Am 1. See also: Biog Geneal Master Ind.

SMITHSON, JAMES LOUIS MACIE (1765-1829). Chemistry; Mineralogy; Philanthropy. OCCUP: Inherited wealth; Engaged in independent study. MSS: Ind NUCMC (72-1255). WORKS ABOUT: Dict Sci Biog; Natl Cycl Am Biog 3; Who Was Who Am h. See also: Barr; Biog Geneal Master Ind; Ireland Ind Sci; Pelletier.

SMYTH, BERNARD BRYAN (1843-1913). Biology; Glaciology. OCCUP: Professor; Curator. MSS: Ind NUCMC. WORKS ABOUT: Am Men Sci 1-2 d3.

SMYTH, WILLIAM (1797-1868). Mathematics. OCCUP: Educator; Professor. WORKS ABOUT: Dict Am Biog; Elliott; Natl Cycl Am Biog 10; Who Was Who Am h. See also: Biog Geneal Master Ind.

SNELL, EBENEZER STRONG (1801-1876). Mathematics; Physics. OCCUP: Professor. WORKS ABOUT: Elliott; Natl Cycl Am Biog 5. See also: Barr; Biog Geneal Master Ind.

SNOW, FRANCIS HUNTINGTON (1840-1908). Natural history; Entomology. OCCUP: Professor; Academic administrator. MSS: Ind NUCMC. WORKS ABOUT: Am Men Sci 1 d3; Dict Am Biog; Elliott (s); Natl Cycl Am Biog 9; Who Was Who Am 1. See also: Barr; Biog Geneal Master Ind; Ireland Ind Sci; Pelletier.

SNOW, WILLIAM APPLETON (1869-1899). Entomology. OCCUP: Professor. WORKS ABOUT: Kansas Univ Q 8(1899):189; Sci 10(1899):661-62.

SNYDER, ZACHARIA XENOPHON (1850-1915). Biology. OCCUP: Professor; College president. WORKS ABOUT: Am Men Sci 1-2 d3; Who Was Who Am 1.

SONDERICKER, JEROME (1859-1904). Mechanics. OCCUP: Professor. WORKS ABOUT: Sci 20(1904):159.

SOOYSMITH, CHARLES (1856-1916). Civil engineering. OCCUP: Engineer; Businessman. WORKS ABOUT: Dict Am Biog; Who Was Who Am 1. See also: Barr.

SOULE, CAROLINE GRAY (1855-1920). Zoology; Entomology. OCCUP: Educator. WORKS ABOUT: Am Men Sci 2 d3; Who Was Who Am 4.

SOULE, RICHARD HERMAN (1849-1908). Mechanical engineering. OCCUP: Engineer. WORKS ABOUT: Am Men Sci 1 d3.

SOULE, WILLIAM (1834-1914). Chemistry; Physics. OCCUP: Professor. WORKS ABOUT: Am Men Sci 1-2 d3.

SOUTHACK, CYPRIAN (1662-1745). Cartography. OCCUP: Shipmaster. MSS: Ind NUCMC (69-908). WORKS ABOUT: Dict Am Biog; Who Was Who Am h.

SOUTHARD, ELMER ERNEST (1876-1920). Neuropathology. OCCUP: Physician; Professor; Administrator. MSS: Ind NUCMC (62-3749). WORKS ABOUT: Am Men Sci 1-2 d3; Dict Am Biog; Dict Am Med Biog; Natl Cycl Am Biog 19; Who Was Who Am 1. See also: Biog Geneal Master Ind.

SPALDING, LYMAN (1775-1821). Chemistry; Electricity; Medicine (vaccination). OCCUP: Physician; Academic instructor (medicine). MSS: <u>Ind NUCMC</u> (71-329). WORKS ABOUT: <u>Dict Am Biog</u>; <u>Dict Am Med Biog</u>; <u>Natl Cycl Am Biog</u> 2; <u>Who Was Who Am</u> h. See also: <u>Biog Geneal Master Ind</u>; Ireland <u>Ind Sci</u>.

SPALDING, VOLNEY MORGAN (1849-1918). Botany. OCCUP: Professor; Research institution associate. MSS: <u>Ind NUCMC</u>. WORKS ABOUT: <u>Am Men Sci</u> 1-2 d3; <u>Dict Am Biog</u>; Elliott (s); <u>Natl Cycl Am Biog</u> 10; <u>Who Was Who Am</u> 1. See also: Barr; <u>Biog Geneal Master Ind</u>.

SPANGLER, HENRY WILSON (1858-1912). Mechanical engineering. OCCUP: Naval engineer; Professor. WORKS ABOUT: <u>Am Men Sci</u> 1-2 d3; <u>Dict Am Biog</u>; <u>Who Was Who Am</u> 1. See also: Barr; <u>Biog Geneal Master Ind</u>.

SPANUTIUS, FREDERICK WILLIAM (1868?-1915). Chemistry. OCCUP: Academic instructor; Industrialist. WORKS ABOUT: <u>Sci</u> 42(1915):50.

SPENCE, DAVID WENDELL (-1917). Civil engineering. OCCUP: Professor; Academic administrator. WORKS ABOUT: <u>Sci</u> 46(1917):111.

SPENCER, CHARLES A. (1813-1881). Optics. OCCUP: Instrument maker. WORKS ABOUT: <u>Am Journ Sci</u> 55(1848):237-40, 63(1852):31-32; <u>Natl Cycl Am Biog</u> 13; <u>Three Am</u> p15-33.

SPERRY, FRANCIS LOUIS (-1906). Mineralogy. OCCUP: Mining engineer. WORKS ABOUT: <u>Sci</u> 23(1906):891.

SPEYERS, CLARENCE LIVINGSTON (1863-1912). Chemistry. OCCUP: Professor; Research institution associate. WORKS ABOUT: Adams O; <u>Am Men Sci</u> 1-2 d3.

SPILSBURY, EDMUND GYBBON (1845-1920). Metallurgy. OCCUP: Engineer; Industrial and mining manager. WORKS ABOUT: <u>Dict Am Biog</u>; <u>Natl Cycl Am Biog</u> 13; <u>Who Was Who Am</u> 1.

SPITZKA, EDWARD CHARLES (1852-1914). Anatomy; Neurology. OCCUP: Physician. WORKS ABOUT: <u>Am Men Sci</u> 1-2 d3; <u>Dict Am Biog</u>; <u>Dict Am Med Biog</u>; <u>Natl Cycl Am Biog</u> 19; <u>Who Was Who Am</u> 1. See also: <u>Biog Geneal Master Ind</u>.

SPRAGUE, CHARLES JAMES (1823-1903). Botany. OCCUP: Banker; Author. WORKS ABOUT: Allibone; Appleton's; Duyckinck; <u>Who Was Who Am</u> 1.

SQUIBB, EDWARD ROBINSON (1819-1900). Chemistry; Pharmacology. OCCUP: Naval physician; Manufacturer. WORKS ABOUT: <u>Dict Am Biog</u>; <u>Dict Am Med Biog</u>; Elliott; <u>Natl Cycl Am Biog</u> 19; <u>Who Was Who Am</u> h. See also: Barr; <u>Biog Geneal Master Ind</u>; Ireland <u>Ind Sci</u>; Pelletier.

SQUIER, EPHRAIM GEORGE (1821-1888). Archeology. OCCUP:
Editor and publisher; Public official (diplomatic service);
Researcher. MSS: Ind NUCMC (60-2970, 61-2202, 61-2949, 64-
1580, 75-1730, 75-1998, 83-1102). WORKS ABOUT: Dict Am
Biog; Natl Cycl Am Biog 4; Who Was Who Am h. See also: Biog
Geneal Master Ind.

STABLER, RICHARD H. (ca.1820-1878). Pharmacy. OCCUP:
Pharmacist; Professor. WORKS ABOUT: Am Pharm Assoc Proc
26(1878):865; Pennyslvania Cat Med (listed).

STACY, ORVILLE BRIGGS (1832?-1912). Mathematics. OCCUP:
Professor. WORKS ABOUT: Sci 36(1912):79.

STALKER, MILLIKEN (1842?-1909). Zoology; Veterinary
science. OCCUP: Academic instructor (professor?). WORKS
ABOUT: Sci 30(1909):237.

STALLO, JOHANN BERNHARD (1823-1900). Philosophy of
science. OCCUP: Professor; Lawyer; Public official. MSS:
Ind NUCMC (?). WORKS ABOUT: Dict Am Biog; Dict Sci Biog;
Elliott; Natl Cycl Am Biog 11; Who Was Who Am 1. See also:
Barr; Biog Geneal Master Ind.

STAMPER, ALVA WALKER (1871-1918). Mathematics. OCCUP:
Professor. WORKS ABOUT: Am Men Sci 2 d3.

STANLEY, ANTHONY DUMOND (1810-1853). Mathematics. OCCUP:
Professor. WORKS ABOUT: Adams O; Allibone; Am Journ Sci
65(1853):464; Appleton's; Drake; Wallace.

STANLEY, WILLIAM (1858-1916). Electrical engineering.
OCCUP: Industrial engineer; Manufacturer; Inventor. WORKS
ABOUT: Am Men Sci 1-2 d3; Dict Am Biog; Natl Cycl Am Biog
24; Who Was Who Am 1. See also: Barr; Ireland Ind Sci.

STARKEY [or STIRK], GEORGE (1628-1665). Chemistry; Alchemy;
Medicine. OCCUP: Physician; Researcher. WORKS ABOUT: Dict
Sci Biog. See also: Pelletier.

STARKWEATHER, GEORGE PRATT (1873-1901). Mechanics. OCCUP:
Professor. WORKS ABOUT: Sci 13(1901):516.

STEARNS, HERMAN DeCLERCQ (1865-1907). Physics. OCCUP:
Professor. WORKS ABOUT: Am Men Sci 1 d3; Sci 26(1907):647.

STEARNS, ROBERT EDWARDS CARTER (1827-1909). Natural
history; Zoology. OCCUP: U.S. government scientist; Museum
curator. MSS: Ind NUCMC (72-1257). WORKS ABOUT: Am Men Sci
1 d3; Dict Am Biog; Elliott (s); Who Was Who Am 1. See
also: Barr.

STEARNS, SILAS (1859-1888). Ichthyology. OCCUP:
Businessman; U.S. government scientist. MSS: Ind NUCMC.
WORKS ABOUT: Am Journ Sci 136(1888):303-04.

STEEL, JOHN HONEYWOOD (1780-1838). Geology; Study of mineral waters. OCCUP: Physician; Public official. WORKS ABOUT: Anderson p137, 149, 164, 176, 331, 444-45 (listed); Sylvester p201-02 (mentioned); Wallace.

STEELE, JOEL DORMAN (1836-1886). Science textbook writing. OCCUP: Educator; Writer. MSS: Ind NUCMC. WORKS ABOUT: Dict Am Biog; Elliott; Natl Cycl Am Biog 3; Who Was Who Am h. See also: Biog Geneal Master Ind; Pelletier.

STEELE, WARREN MERRILL (1874-1910). Philosophy. OCCUP: Professor. WORKS ABOUT: Am Men Sci 1-2 d3.

STEINER, LEWIS HENRY (1827-1892). Chemistry. OCCUP: Physician; Professor; Librarian. MSS: Ind NUCMC (66-1494, 67-1698). WORKS ABOUT: Dict Am Biog; Natl Cycl Am Biog 11; Who Was Who Am h. See also: Biog Geneal Master Ind; Pelletier.

STEIRINGER, LUTHER (1845?-1903). Electricity. OCCUP: Not determined. WORKS ABOUT: Sci 18(1903):158.

STERN, LOUIS C. (-1918). Civil engineering; Sanitation. OCCUP: City and state government engineer; Academic instructor. WORKS ABOUT: Sci 48(1918):644.

STERNBERG, GEORGE MILLER (1838-1915). Bacteriology; Epidemiology; Medicine. OCCUP: Army surgeon; Administrator. MSS: Ind NUCMC (66-1482, 69-2075). WORKS ABOUT: Am Men Sci 1-2 d3; Dict Am Biog; Dict Am Med Biog; Natl Cycl Am Biog 4; Who Was Who Am 1. See also: Barr; Biog Geneal Master Ind; Ireland Ind Sci; Pelletier.

STETEFELDT, CARL AUGUST (1838-1896). Metallurgy. OCCUP: Consultant; Inventor. WORKS ABOUT: Dict Am Biog; Who Was Who Am h. See also: Biog Geneal Master Ind.

STEVENS, EDWARD (ca.1755-1834). Medicine; Physiology. OCCUP: Physician; Professor. WORKS ABOUT: Dict Sci Biog.

STEVENS, ISAAC INGALLS (1818-1862). Exploration. OCCUP: Army officer; Engineer; Public official. MSS: Ind NUCMC (61-1348, 70-1388). WORKS ABOUT: Dict Am Biog; Natl Cycl Am Biog 12; Who Was Who Am h. See also: Barr; Biog Geneal Master Ind.

STEVENS, JOHN (1749-1838). Invention. OCCUP: Engineer; Inventor; Businessman. MSS: Ind NUCMC. WORKS ABOUT: Dict Am Biog; Natl Cycl Am Biog 11; Who Was Who Am h. See also: Barr; Biog Geneal Master Ind; Ireland Ind Sci; Pelletier.

STEVENS, JOHN FLOYD (-1918). Electrical engineering. OCCUP: Professor; Industrial engineer. WORKS ABOUT: Sci 48(1918):443.

STEVENS, NETTIE MARIA (1861-1912). Zoology; Genetics.
OCCUP: Academic researcher; Teacher. WORKS ABOUT: Am Men
Sci 1-2 d3; Notable Am Wom; Ogilvie; Siegel and Finley. See
also: Barr; Biog Geneal Master Ind.

STEVENS, ROBERT LIVINGSTON (1787-1856). Engineering;
Invention. OCCUP: Engineer; Businessman; Inventor. WORKS
ABOUT: Dict Am Biog; Natl Cycl Am Biog 11; Who Was Who Am
h. See also: Barr; Biog Geneal Master Ind; Ireland Ind Sci;
Pelletier.

STEVENSON, JAMES (1840-1888). Ethnology; Exploration.
OCCUP: Explorer; U.S. government scientist. MSS: Ind
NUCMC (?). WORKS ABOUT: Dict Am Biog; Natl Cycl Am Biog 12;
Who Was Who Am h. See also: Barr; Biog Geneal Master Ind.

STEVENSON, MATILDA COXE EVANS (1849-1915). Anthropology;
Archeology; Ethnology. OCCUP: Explorer; Collector; U.S.
government scientist. MSS: Ind NUCMC (?). WORKS ABOUT: Am
Men Sci 1-2 d3; Dict Am Biog; Natl Cycl Am Biog 20; Notable
Am Wom; Siegel and Finley; Who Was Who Am 1. See also:
Barr; Biog Geneal Master Ind; Pelletier.

STEWART, DAVID (1813-1899). Chemistry; Pharmacy. OCCUP:
Pharmacist; Professor; Physician. WORKS ABOUT: Am Chem and
Chem Eng.

STEWART, DAVID DENISON (1858-1905). Medicine. OCCUP:
Physician; Professor. WORKS ABOUT: Am Men Sci 1 d3;
Wallace; Who Was Who Am 1.

STEWART, HUGH ANGUS (1882-1917). Pathology. OCCUP:
Physician; Professor. WORKS ABOUT: Am Men Sci 2 d3.

STEWART, VERN BONHAM (1888-1918). Plant pathology. OCCUP:
Professor; U.S. government scientist. WORKS ABOUT: Sci
48(1918):616.

STILES, EZRA (1727-1795). Astronomy; Chemistry;
Electricity. OCCUP: Clergyman; Professor (ecclesiastical
history); College president. MSS: Ind NUCMC (80-140). WORKS
ABOUT: Dict Am Biog; Natl Cycl Am Biog 1; Who Was Who Am
h. See also: Biog Geneal Master Ind.

STILES, WILLIAM AUGUSTUS (1837-1897). Botany. OCCUP:
Editor; Public official. WORKS ABOUT: Gard and Forest
10(1897):399-400; Sci 6(1897):591.

STILLE, ALFRED (1813-1900). Medicine. OCCUP: Physician;
Professor. MSS: Ind NUCMC. WORKS ABOUT: Dict Am Biog; Dict
Am Med Biog; Natl Cycl Am Biog 9; Who Was Who Am 1. See
also: Barr; Biog Geneal Master Ind; Ireland Ind Sci;
Pelletier.

STILLMAN, THOMAS BLISS (1852-1915). Chemistry; Chemical engineering. OCCUP: Professor; Consultant. WORKS ABOUT: Am Men Sci 1-2 d3; Dict Am Biog; Natl Cycl Am Biog 16; Who Was Who Am 1. See also: Barr; Biog Geneal Master Ind; Pelletier.

STIMPSON, WILLIAM (1832-1872). Natural history; Marine zoology; Conchology. OCCUP: Natural history explorer/collector; Curator; Administrator. MSS: Ind NUCMC. WORKS ABOUT: Dict Am Biog; Dict Sci Biog; Elliott; Natl Ac Scis Biog Mem 8; Who Was Who Am h. See also: Barr; Biog Geneal Master Ind.

STIMSON, LEWIS ATTERBURY (1844-1917). Medicine. OCCUP: Surgeon; Professor. WORKS ABOUT: Am Men Sci 2 d3; Dict Am Biog; Who Was Who Am 1. See also: Barr; Biog Geneal Master Ind; Ireland Ind Sci.

STINSON, MARGARET E. (-1912). Chemistry. OCCUP: Academic affiliate. WORKS ABOUT: Sci 35(1912):689.

STOCKWELL, JOHN NELSON (1832-1920). Astronomy; Mathematics. OCCUP: Professor; Engaged in independent research. WORKS ABOUT: Am Men Sci 1-2 d3; Dict Am Biog; Elliott (s); Natl Cycl Am Biog 9; Who Was Who Am 1. See also: Barr; Biog Geneal Master Ind.

STODDARD, JOHN FAIR (1825-1873). Mathematics. OCCUP: Educator; Writer. WORKS ABOUT: Dict Am Biog; Elliott; Who Was Who Am h. See also: Biog Geneal Master Ind.

STODDARD, JOHN TAPPAN (1852-1919). Chemistry. OCCUP: Professor. WORKS ABOUT: Am Men Sci 1-2 d3; Dict Am Biog; Elliott (s); Who Was Who Am 1. See also: Barr; Biog Geneal Master Ind; Pelletier.

STODDARD, ORANGE NASH (1812-1892). General science. OCCUP: Professor. Scientific work: see Roy Soc Cat. WORKS ABOUT: Tobey and Thompson p197-98.

STODDER, CHARLES (ca.1808-1884). Microscopy. OCCUP: Instrument maker. WORKS ABOUT: Am Mo Micr Journ 5(1884):55-56; Boston Evening Transcript (January 15, 1884):8.

STONE, GEORGE HAPGOOD (1841-1917). Mining geology. OCCUP: Professor; U.S. government scientist; Mining expert. WORKS ABOUT: Am Journ Sci 194(1917):86; Am Men Sci 1-2; Wallace; Who Was Who Am 1.

STONE, LIVINGSTON (1835-1912). Pisciculture. OCCUP: Clergyman; U.S. government scientist; Pisciculturist. WORKS ABOUT: Allibone Supp; Wallace; Who Was Who Am 1.

STONE, ROYCE (1836?-1905). Engineering. OCCUP: Army officer; Engineer. WORKS ABOUT: Sci 22(1905):221.

STORER, DAVID HUMPHREYS (1804-1891). Natural history; Zoology; Ichthyology. OCCUP: Physician; Professor (medicine); State government scientist. MSS: Ind NUCMC. WORKS ABOUT: Dict Am Biog; Dict Am Med Biog; Elliott; Natl Cycl Am Biog 11; Who Was Who Am h. See also: Biog Geneal Master Ind; Pelletier.

STORER, FRANCIS HUMPHREYS (1832-1914). Chemistry; Agricultural chemistry. OCCUP: Professor; Academic administrator. MSS: Ind NUCMC. WORKS ABOUT: Am Men Sci 1-2 d3; Dict Am Biog; Elliott (s); Natl Cycl Am Biog 11; Who Was Who Am 1. See also: Barr; Biog Geneal Master Ind; Pelletier.

STRATFORD, WILLIAM (1844-1908). Biology. OCCUP: Professor. WORKS ABOUT: Sci 27(1908):234-35.

STRATTON, JAMES THOMPSON (1830-1903). Geography. OCCUP: Surveyor; Public official. MSS: Ind NUCMC (75-375). WORKS ABOUT: Sci 17(1903):797.

STRAUGHN, M. N. (-1919). Chemistry. OCCUP: U.S. government scientist. WORKS ABOUT: Sci 49(1919):213.

STRECKER, (FERDINAND HEINRICH) HERMAN (1836-1901). Entomology. OCCUP: Businessman; Artist; Collector. MSS: Ind NUCMC (66-86, 66-764). WORKS ABOUT: Elliott; Natl Cycl Am Biog 10; Who Was Who Am 1. See also: Barr; Biog Geneal Master Ind; Pelletier.

STREERUWITZ, WILLIAM H. RITTER VON (1833-   ). Civil engineering; Geology; Mining engineering. OCCUP: Engineer; Professor; State government scientist. WORKS ABOUT: Who Was Who Am 4.

STRINGHAM, WASHINGTON IRVING (1847-1909). Mathematics. OCCUP: Professor. WORKS ABOUT: Am Men Sci 1 d3; Dict Am Biog; Elliott (s). See also: Barr; Biog Geneal Master Ind; Ireland Ind Sci.

STRONG, EDWIN ATSON (1834-1920). Physics. OCCUP: Educator. WORKS ABOUT: Who Was Who Am 4.

STRONG, MOSES (-1877). Geology. OCCUP: State government scientist. WORKS ABOUT: Am Journ Sci 114(1877):336.

STRONG, MOSES McCURE (1810-1894). Surveying. OCCUP: Lawyer; Surveyor; Public official. MSS: Ind NUCMC (62-2886). WORKS ABOUT: Dict Am Biog; Who Was Who Am h. See also: Biog Geneal Master Ind.

STRONG, THEODORE (1790-1869). Mathematics. OCCUP: Professor; Academic administrator. WORKS ABOUT: Dict Am Biog; Elliott; Natl Ac Scis Biog Mem 2; Natl Cycl Am Biog 9; Who Was Who Am h. See also: Barr; Biog Geneal Master Ind; Ireland Ind Sci; Pelletier.

STUART, AMBROSE PASCAL SEVILON (1820-1899). Chemistry.
OCCUP: Educator; Professor. WORKS ABOUT: Appleton's; <u>Sci</u>
10(1899):423; <u>Who Was Who Am</u> 1.

STUBENRAUCH, ARNOLD VALENTINE (1871-1917). Pomology. OCCUP:
Professor; U.S. government scientist. WORKS ABOUT: <u>Sci</u>
45(1917):235-36; <u>Who Was Who Am</u> 1.

STUNTZ, STEPHEN CONRAD (1875-1918). Bibliography; Botany.
OCCUP: Librarian; U.S. government scientist. MSS: <u>Ind</u>
<u>NUCMC</u> (?). WORKS ABOUT: <u>Am Men Sci</u> 2 d3; Wallace; <u>Who Was</u>
<u>Who Am</u> 1.

STURTEVANT, EDWARD LEWIS (1842-1898). Agricultural science;
Botany. OCCUP: Farmer; State government scientist;
Researcher. MSS: <u>Ind NUCMC</u> (70-1130, 72-1577). WORKS ABOUT:
<u>Dict Am Biog</u>; Elliott; <u>Who Was Who Am</u> h. See also: <u>Biog</u>
<u>Geneal Master Ind</u>.

SUCKLEY, GEORGE (1830-1869). Natural history. OCCUP:
Physician; Army surgeon. MSS: <u>Ind NUCMC</u> (61-2604, 78-900).
WORKS ABOUT: Allibone; Appleton's; Kelly and Burrage.

SULLIVAN, JOHN LANGDON (1777-1865). Engineering. OCCUP:
Businessman; Civil engineer; Physician. MSS: <u>Ind NUCMC</u>.
WORKS ABOUT: Allibone; Appleton's; Drake; Sullivan p3-12;
Yale <u>Obit Rec</u> (1870):380.

SULLIVAN, LILLIE (-1903). Art; Entomology. OCCUP:
Artist; U.S. government scientist. WORKS ABOUT: <u>Sci</u>
18(1903):61.

SULLIVANT, WILLIAM STARLING (1803-1873). Botany; Bryology.
OCCUP: Businessman; Researcher. MSS: <u>Ind NUCMC</u>. WORKS
ABOUT: <u>Dict Am Biog</u>; Elliott; Natl Ac Scis <u>Biog Mem</u> 1; <u>Natl</u>
<u>Cycl Am Biog</u> 8; <u>Who Was Who Am</u> h. See also: Barr; <u>Biog</u>
<u>Geneal Master Ind</u>; Ireland <u>Ind Sci</u>.

SUMMERS, THOMAS OSMOND (ca.1852-1899). Anatomy. OCCUP:
Professor; Physician. MSS: <u>Ind NUCMC</u> (?). WORKS ABOUT: Am
Med Assoc <u>Journ</u> 32(1899):1460; <u>Sci</u> 9(1899):917.

SUMNER, THOMAS HUBBARD (1807-1851). Navigation. OCCUP:
Shipmaster. WORKS ABOUT: Struik.

SUTHERLAND, CHARLES (1830?-1895). Surgery. OCCUP: Army
surgeon; Administrator. WORKS ABOUT: <u>Natl Cycl Am Biog</u> 4;
<u>Sci</u> 1(1895):585; <u>Twentieth Cent Biog Dict</u>.

SUTTON, WALTER STANBOROUGH (1877-1916). Biology; Medicine.
OCCUP: Surgeon; Professor. MSS: <u>Ind NUCMC</u> (80-423). WORKS
ABOUT: <u>Dict Sci Biog</u>. See also: Barr.

SUTTON, WILLIAM JOHN (1859-1915). Geology. OCCUP: Academic
instructor; Industrial scientist. WORKS ABOUT: <u>Am Men Sci</u>
1-2 d3.

SWALLOW, GEORGE CLINTON (1817-1899). Geology. OCCUP: Professor; State government scientist; Editor. MSS: Ind NUCMC (?). WORKS ABOUT: Dict Am Biog; Elliott; Who Was Who Am 1. See also: Barr; Ireland Ind Sci.

SWAN, CHARLES H. (-1899). Civil and sanitary engineering. OCCUP: Engineer. WORKS ABOUT: Sci 9(1899):629.

SWAN, JOSHUA AUGUSTUS (1823-1871). Biology. OCCUP: Clergyman; Administrator. WORKS ABOUT: Am Journ Sci 103(1872):77; Wallace.

SWEET, ELNATHAN (1837-1903). Civil engineering. OCCUP: Engineer; State government engineer. WORKS ABOUT: Appleton's; Sci 17(1903):238; Who Was Who Am 1.

SWEET, JOHN EDSON (1832-1916). Mechanics. OCCUP: Inventor; Academic instructor; Manufacturer. WORKS ABOUT: Am Men Sci 1-2 d3; Dict Am Biog; Natl Cycl Am Biog 13; Who Was Who Am 1. See also: Barr; Biog Geneal Master Ind; Ireland Ind Sci.

SWEETLAND, ARTHUR E. (-1903). Meteorology. OCCUP: Observatory affiliate. WORKS ABOUT: Sci 17(1903):799.

SWIFT, LEWIS (1820-1913). Astronomy. OCCUP: Observatory director. MSS: Ind NUCMC. WORKS ABOUT: Am Men Sci 1-2 d3; Dict Am Biog; Elliott (s); Natl Cycl Am Biog 4; Who Was Who Am 1. See also: Barr; Biog Geneal Master Ind.

SWIFT, ROBERT (1799-1872). Conchology; Ornithology. OCCUP: Merchant; Collector. WORKS ABOUT: Am Journ Sci 104(1872):160; Appleton's.

SWINBURNE, RALPH (ca.1805-1895). Engineering. OCCUP: Engineer. WORKS ABOUT: Sci 1(1895):697.

SYLVESTER, JAMES JOSEPH (1814-1897). Mathematics. OCCUP: Professor; Actuary; Editor. MSS: Ind NUCMC (60-2152). WORKS ABOUT: Dict Am Biog; Dict Sci Biog; Elliott; Who Was Who Am h. See also: Barr; Biog Geneal Master Ind; Ireland Ind Sci; Pelletier.

T

TAGLIABUE, GIUSEPPE (1812-1878). Thermometer manufacturing. OCCUP: Instrument maker; Inventor. WORKS ABOUT: Dict Am Biog; Who Was Who Am h. See also: Biog Geneal Master Ind.

TANNER, HENRY SCHENCK (1786-1858). Cartography; Geography. OCCUP: Map-maker. MSS: Ind NUCMC (?). WORKS ABOUT: Dict Am Biog; Who Was Who Am h. See also: Biog Geneal Master Ind.

TANNER, ZERA LUTHER (1835-1906). Ichthyology. OCCUP: Naval officer. WORKS ABOUT: Natl Cycl Am Biog 8; Sci 41(1915):749; Who Was Who Am 1.

TARR, RALPH STOCKMAN (1864-1912). Geography; Geology. OCCUP: Professor. MSS: Ind NUCMC (62-3904, 66-1046). WORKS ABOUT: Am Men Sci 1-2 d3; Dict Am Biog; Natl Cycl Am Biog 10; Who Was Who Am 1. See also: Barr; Biog Geneal Master Ind.

TASSIN, WIRT de VIVIER (1868-1915). Chemistry; Metallurgy. OCCUP: Consultant; Museum curator; Businessman. WORKS ABOUT: Am Journ Sci 190(1915):670; Am Men Sci 1-2 d3; Natl Cycl Am Biog 16; Sci 42(1915):688; Who Was Who Am 1.

TATHAM, WILLIAM (1752-1819). Geography. OCCUP: Lawyer; Public official; Civil engineer. MSS: Ind NUCMC. WORKS ABOUT: Dict Am Biog; Who Was Who Am h. See also: Biog Geneal Master Ind.

TATUM, SLEDGE (-1916). Geology; Geography. OCCUP: U.S. government scientist. WORKS ABOUT: Sci 43(1916):166.

TAYLOR, CHARLOTTE DeBERNIER SCARBROUGH (1806-1861). Entomology. OCCUP: Independent wealth; Engaged in independent study. WORKS ABOUT: Dict Am Biog; Elliott; Natl Cycl Am Biog 2; Siegel and Finley; Who Was Who Am h.

TAYLOR, EDWARD RANDOLPH (1844-1917). Chemistry. OCCUP: Professor; Industrial scientist and officer. WORKS ABOUT: Am Men Sci 2 d3; Sci 46(1917):58-59; Who Was Who Am 1.

TAYLOR, FREDERICK WINSLOW (1856-1915). Mechanical engineering; Efficiency engineering. OCCUP: Industrial engineer and manager; Consultant; Inventor. MSS: Ind NUCMC (61-1973). WORKS ABOUT: Am Men Sci 2 d3; Dict Am Biog; Dict Sci Biog; Natl Cycl Am Biog 14, 23; Who Was Who Am 1. See also: Barr; Biog Geneal Master Ind; Pelletier.

TAYLOR, HERBERT DOUGLAS (1888?-1918). Medicine. OCCUP: Research institution scientist. WORKS ABOUT: Sci 48(1918):392, 411.

TAYLOR, RICHARD COWLING (1789-1851). Geology. OCCUP: Surveyor; Mining expert. WORKS ABOUT: Dict Am Biog; Elliott; Natl Cycl Am Biog 9; Who Was Who Am h. See also: Barr; Biog Geneal Master Ind.

TAYLOR, ROSE M. (-1918). Botany. OCCUP: Academic instructor. WORKS ABOUT: Sci 48(1918):616.

TAYLOR, THOMAS (1820-1910). Chemistry; Botany; Microscopy. OCCUP: Inventor; U.S. government scientist; Physician. WORKS ABOUT: Twentieth Cent Biog Dict; Wallace; Whos Who Am (1908-1909).

TAYLOR, THOMAS MAYNARD (1874-1907). Chemistry. OCCUP: Academic instructor. WORKS ABOUT: Am Men Sci 1 d3.

TAYLOR WILLIAM BOWER (1821-1895). Astronomy; Mathematics; Physics. OCCUP: U.S. government scientist; Librarian; Editor. WORKS ABOUT: Appleton's; Phil Soc Washington Bull 13(1900):418-26; Smithsonian Rep (1896):645-56.

TAYLOR, WILLIAM HENRY (1835-1917). Chemistry. OCCUP: Physician; Professor; State government scientist. WORKS ABOUT: Natl Cycl Am Biog 18; Wallace.

TAYLOR, WILLIAM JOHNSON (ca.1833-1864). Mineralogy. OCCUP: Professor. WORKS ABOUT: Am Journ Sci 87(1864):447; Smith E p230-31.

TEGMEYER, JOHN H. (1821?-1901). Engineering. OCCUP: Civil engineer. WORKS ABOUT: Sci 14(1901):77.

TEMPLE, JAMES CLARENCE (1882-1916). Bacteriology. OCCUP: State government scientist. WORKS ABOUT: Am Men Sci 2 d3.

TEMPLE, ROBERT (1831?-1901). Engineering. OCCUP: Engineer. WORKS ABOUT: Sci 15(1902):37.

TENNEY, SAMUEL (1748-1816). Physics. OCCUP: Physician; Legislator. MSS: Ind NUCMC. WORKS ABOUT: Allibone; Appleton's; Biog Dir Am Congress; Drake; Lanman; Natl Cycl Am Biog 5; Who Was Who Am h.

TENNEY, SANBORN (1827-1877). Natural history; Geology. OCCUP: Professor. WORKS ABOUT: Adams O; Allibone; Allibone Supp; Am Journ Sci 114(1877):168; Appleton's; Wallace.

TESCHEMACHER, JAMES ENGLEBERT (1790-1853). Botany; Geology; Mineralogy. OCCUP: Businessman. WORKS ABOUT: Elliott. See also: Barr; Biog Geneal Master Ind.

THAYER, CHARLES PAINE (1843-1910). Medicine; Anatomy. OCCUP: Physician; Professor; Public official. WORKS ABOUT: Sci 31(1910):294; Who Was Who Am 1.

THOBURN, WILBUR WILSON (-1899). Biomechanics. OCCUP: Professor. WORKS ABOUT: Sci 9(1899):301.

THOMAS, BENJAMIN FRANKLIN (1850-1911). Physics. OCCUP: Professor. WORKS ABOUT: Am Men Sci 1-2 d3; Elliott (s); Sci 34(1911):74, 35(1912):647; Who Was Who Am 1.

THOMAS, CYRUS (1825-1910). Entomology; Ethnology. OCCUP: Lawyer; U.S. government scientist. MSS: Ind NUCMC (?). WORKS ABOUT: Am Men Sci 1 d3; Dict Am Biog; Elliott (s); Natl Cycl Am Biog 13; Who Was Who Am 1. See also: Barr; Biog Geneal Master Ind; Ireland Ind Sci.

THOMAS, DAVID (1776-1859). Natural history. OCCUP: Civil engineer. WORKS ABOUT: Adams O; Appleton's; Bartonia 9(October 1926):42; Drake; Wallace.

THOMAS, EDWARD (-1832). Astronomy; Optical instrumentation. OCCUP: Not determined; Engineer. WORKS ABOUT: Am Journ Sci 19(1831):57, 22(1832):380.

THOMAS, MASON BLANCHARD (1866-1912). Botany. OCCUP: Professor; Academic administrator. MSS: Ind NUCMC. WORKS ABOUT: Am Men Sci 1-2 d3; Sci 35(1912):414; Who Was Who Am 1.

THOMAS, ROBERT P. (1821-1864). Materia medica. OCCUP: Physician; Professor. WORKS ABOUT: Allibone; Am Journ Pharm 36(1864):277-80; Natl Cycl Am Biog 5; Philadelphia College Pharm First Cent.

THOMAS, WILLIAM D. (-1901). Psychology. OCCUP: Professor. WORKS ABOUT: Sci 13(1901):877.

THOME, JOHN MACON (1843-1908). Astronomy. OCCUP: Observatory director. WORKS ABOUT: Sci 28(1908):752.

THOMPSON, ALMON HARRIS (1839-1906). Geography; Topographical engineering. OCCUP: U.S. government scientist. WORKS ABOUT: Elliott; Who Was Who Am 1.

THOMPSON, ALTON HOWARD (1849-1914). Dental anatomy; Anthropology. OCCUP: Dentist; Professor. WORKS ABOUT: Am Men Sci 1-2 d3; Wallace.

THOMPSON, BENJAMIN (COUNT RUMFORD) (1753-1814). Physics. OCCUP: Public official (foreign). MSS: Ind NUCMC (73-189). WORKS ABOUT: Dict Am Biog; Dict Sci Biog; Elliott; Natl Cycl Am Biog 5; Who Was Who Am h. See also: Barr; Biog Geneal Master Ind; Ireland Ind Sci; Pelletier.

THOMPSON, (JAMES) MAURICE (1844-1901). Engineering; Geology. OCCUP: Lawyer; Author; State government scientist. MSS: Ind NUCMC (62-934, 79-754). WORKS ABOUT: Dict Am Biog; Natl Cycl Am Biog 11; Who Was Who Am 1. See also: Barr; Biog Geneal Master Ind; Ireland Ind Sci.

THOMPSON, SAMUEL RANKIN (1833-1896). Agriculture; Physics. OCCUP: Educator; Professor. MSS: Ind NUCMC. WORKS ABOUT: Dict Am Biog; Natl Cycl Am Biog 7:517; Who Was Who Am h. See also: Biog Geneal Master Ind.

THOMPSON, WILLIAM ABDIEL (1762-1847). Geology. OCCUP: Lawyer; Judge. Scientific work: see Roy Soc Cat. WORKS ABOUT: Dexter Biog Sketches 4(1907):238-40; Quinlan p515-22.

THOMPSON, ZADOCK (1796-1856). Natural history; Mathematics. OCCUP: Professor; State government scientist; Author. MSS: Ind NUCMC (62-1752, 81-325). WORKS ABOUT: Dict Am Biog; Elliott; Natl Cycl Am Biog 6; Who Was Who Am h. See also: Barr; Biog Geneal Master Ind; Ireland Ind Sci.

THOMSON, WILLIAM (1833-1907). Ophthalmology. OCCUP: Physician; Professor. WORKS ABOUT: Am Men Sci 1 d3; Appleton's; Sci 26(1907):263; Who Was Who Am 1.

THOMSON, WILLIAM HANNA (1833-1918). Medicine. OCCUP: Physician; Professor. WORKS ABOUT: Adams O; Allibone Supp; Am Men Sci 1-2 d3; Natl Cycl Am Biog 23; Wallace; Who Was Who Am 1.

THRASHER, WILLIAM M. (-1900). Mathematics. OCCUP: Professor. WORKS ABOUT: Sci 11(1900):676.

THURBER, EUGENE CARLETON (ca.1865-1896). Ornithology. OCCUP: Not determined; Natural history collector. WORKS ABOUT: Auk 13(1896):349.

THURBER, GEORGE (1821-1890). Botany; Horticulture. OCCUP: Professor; U.S. government scientist; Editor. MSS: Ind NUCMC. WORKS ABOUT: Dict Am Biog; Elliott; Who Was Who Am h. See also: Biog Geneal Master Ind.

THURSTON, ROBERT HENRY (1839-1903). Mechanical engineering; Engineering education; Testing of materials. OCCUP: Naval engineer; Professor; Academic administrator. MSS: Ind NUCMC (62-4323). WORKS ABOUT: Dict Am Biog; Dict Sci Biog; Natl Cycl Am Biog 4; Who Was Who Am 1. See also: Barr; Biog Geneal Master Ind.

TIDD, MARSHALL M. (1827-1895). Engineering. OCCUP: Civil engineer. WORKS ABOUT: Am Soc Civil Eng <u>Trans</u> 37(1897):568-70; <u>Sci</u> 2(1895):270.

TIFFANY, LOUIS McLANE (1844-1916). Medicine. OCCUP: Surgeon; Professor. WORKS ABOUT: <u>Dict Am Biog</u>; <u>Natl Cycl Am Biog</u> 12, 24; <u>Who Was Who Am</u> 1. See also: Barr.

TIGHT, WILLIAM GEORGE (1865-1910). Geology. OCCUP: Professor; College president. WORKS ABOUT: <u>Am Men Sci</u> 1 d3; <u>Sci</u> 31(1910):141; <u>Who Was Who Am</u> 1.

TILGHMAN, RICHARD ALBERT (1824-1899). Chemistry. OCCUP: Industrial scientist; Researcher. WORKS ABOUT: <u>Dict Am Biog</u>; Elliott; <u>Who Was Who Am</u> h.

TILLMAN, SAMUEL DYER (1815-1875). Chemistry. OCCUP: Lawyer; Professor. WORKS ABOUT: Adams O; <u>Am Journ Sci</u> 110(1875):402; Appleton's; Wallace.

TINGLE, JOHN BISHOP (1866-1918). Chemistry. OCCUP: Professor; Editor. WORKS ABOUT: <u>Am Men Sci</u> 1-2 d3; <u>Macmillan Dict Canadian Biog</u>; <u>Sci</u> 48(1918):188; <u>Who Was Who Am</u> 3.

TOBEY, EDWARD NELSON (1871-1915). Bacteriology. OCCUP: Academic instructor; City government scientist. WORKS ABOUT: <u>Am Men Sci</u> 2 d3; <u>Sci</u> 42(1915):306, 416.

TODD, HENRY DAVIS (1838-1907). Astronomy; Mathematics. OCCUP: Naval officer; Professor. WORKS ABOUT: <u>Am Journ Sci</u> 173(1907):324; <u>Am Men Sci</u> 1 d3; <u>Natl Cycl Am Biog</u> 13, 19; <u>Sci</u> 25(1907):438-39; <u>Twentieth Cent Biog Dict</u>; <u>Who Was Who Am</u> 1.

TOLLES, ROBERT BRUCE (ca.1825-1883). Optics. OCCUP: Instrument maker. WORKS ABOUT: <u>Natl Cycl Am Biog</u> 13; <u>Sci</u> 1st ser 2(1883):726, 35(1912):444-46; <u>Three Am</u> p35-51.

TOOKER, WILLIAM WALLACE (1848-1917). Ethnology; Indian language. OCCUP: Pharmacist; Businessman. MSS: <u>Ind NUCMC</u> (62-1707). WORKS ABOUT: <u>Am Men Sci</u> 1-2 d3; <u>Sci</u> 46(1917):135; Wallace.

TORREY, BRADFORD (1843-1912). Ornithology. OCCUP: Clerk (?); Editor; Author. WORKS ABOUT: <u>Dict Am Biog</u>; Elliott; <u>Natl Cycl Am Biog</u> 10; <u>Who Was Who Am</u> 1. See also: Barr; <u>Biog Geneal Master Ind</u>.

TORREY, HENRY AUGUSTUS (1871-1910). Chemistry. OCCUP: Professor. MSS: <u>Ind NUCMC</u>. WORKS ABOUT: <u>Am Men Sci</u> 1 d3; <u>Sci</u> 31(1910):499, 32(1910):50-51.

TORREY, JOHN (1796-1873). Botany; Chemistry. OCCUP: Physician; Professor; U.S. government scientist. MSS: <u>Ind NUCMC</u> (66-87, 71-2042). WORKS ABOUT: <u>Dict Am Biog</u>; <u>Dict Sci Biog</u>; Elliott; Natl Ac Scis <u>Biog Mem</u> 1; <u>Natl Cycl Am Biog</u> 6; <u>Who Was Who Am</u> h. See also: Barr; <u>Biog Geneal Master Ind</u>; Ireland <u>Ind Sci</u>; Pelletier.

TOTTEN, JOSEPH GILBERT (1788-1864). Military engineering; Conchology. OCCUP: Army officer; Engineer. MSS: <u>Ind NUCMC</u>. WORKS ABOUT: <u>Dict Am Biog</u>; Natl Ac Scis <u>Biog Mem</u> 1; <u>Natl Cycl Am Biog</u> 4; <u>Who Was Who Am</u> h. See also: <u>Biog Geneal Master Ind</u>.

TOWNSEND, FITZHUGH (1872?-1906). Electrical engineering. OCCUP: Academic instructor. WORKS ABOUT: <u>Sci</u> 24(1906):830.

TOWNSEND, JOHN KIRK (1809-1851). Ornithology. OCCUP: Natural history explorer/ collector; Curator. MSS: <u>Ind NUCMC</u>. WORKS ABOUT: <u>Dict Am Biog</u>; Elliott; <u>Who Was Who Am</u> h. See also: <u>Biog Geneal Master Ind</u>; Ireland <u>Ind Sci</u>; Pelletier.

TRACY, CLARISSA TUCKER (1818-1905). Botany. OCCUP: Teacher; Educator. MSS: <u>Ind NUCMC</u>. WORKS ABOUT: Elliott; Siegel and Finley; <u>Who Was Who Am</u> 1.

TRACY, SAMUEL MILLS (1847-1920). Agriculture; Botany. OCCUP: Professor; U.S. and state government scientist. WORKS ABOUT: <u>Am Men Sci</u> 1-2 d3; <u>Natl Cycl Am Biog</u> 18; Wallace; <u>Who Was Who Am</u> 1.

TRASK, JOHN BOARDMAN (1824-1879). Geology. OCCUP: Physician; Editor; State government scientist. WORKS ABOUT: Elliott. See also: Barr.

TRAUTWINE, JOHN CRESSON (1810-1883). Engineering; Mineralogy. OCCUP: Engineer; Architect. MSS: <u>Ind NUCMC</u>. WORKS ABOUT: <u>Dict Am Biog</u>; Elliott; <u>Natl Cycl Am Biog</u> 5; <u>Who Was Who Am</u> h. See also: <u>Biog Geneal Master Ind</u>.

TRAVIS, ISAAC N. (-1897). Taxidermy. OCCUP: Taxidermist; Museum affiliate. WORKS ABOUT: <u>Sci</u> 6(1897):405.

TREADWELL, DANIEL (1791-1872). Invention; Physics; Technical education. OCCUP: Businessman; Inventor; Professor. MSS: <u>Ind NUCMC</u> (65-1285). WORKS ABOUT: <u>Dict Am Biog</u>; Elliott; <u>Natl Cycl Am Biog</u> 10; <u>Who Was Who Am</u> h. See also: Barr; <u>Biog Geneal Master Ind</u>.

TREADWELL, FREDERICK PEARSON (1857-1918). Chemistry. OCCUP: Professor; Author. WORKS ABOUT: <u>Sci</u> 48(1918):244.

TRIMBLE, HENRY (1853-1898). Chemistry. OCCUP: Pharmacist; Professor. WORKS ABOUT: Adams O; <u>Am Journ Pharm</u> 70(1898):537-43; Franklin Inst <u>Journ</u> 146(1898):307-12; <u>Natl Cycl Am Biog</u> 5; Soc Chem Indus <u>Journ</u> 17(1898):904; Wallace.

TRIPLER, CHARLES EASTMAN (1849-1906). Physics. OCCUP: Businessman; Researcher; Inventor. WORKS ABOUT: Lincoln Lib Social Studies; Natl Cycl Am Biog 11; New York Times (June 23, 1906):7:4; Who Was Who Am 4.

TROOST, GERARD (1776-1850). Natural history; Geology; Mineralogy; Paleontology. OCCUP: Pharmacist; Professor; Natural history explorer/ collector. MSS: Ind NUCMC. WORKS ABOUT: Dict Am Biog; Dict Sci Biog; Elliott; Natl Cycl Am Biog 7:349; Who Was Who Am h. See also: Barr; Biog Geneal Master Ind; Ireland Ind Sci; Pelletier.

TROUVELOT, ETIENNE LEOPOLD (1827-1895). Astronomy; Natural history; Entomology. OCCUP: Artist; Observatory affiliate. WORKS ABOUT: Dict Sci Biog. See also: Barr.

TROWBRIDGE, CHARLES CHRISTOPHER (1870-1918). Physics. OCCUP: Professor. WORKS ABOUT: Am Journ Sci 196(1918):550; Am Men Sci 1-2 d3; Natl Cycl Am Biog 18; Sci 47(1918):587; Who Was Who Am 1.

TROWBRIDGE, WILLIAM PETIT (1828-1892). Engineering; Geophysics; Instrumentation. OCCUP: Professor; U.S. government scientist; Industrial officer. MSS: Ind NUCMC. WORKS ABOUT: Dict Am Biog; Elliott; Natl Ac Scis Biog Mem 3; Natl Cycl Am Biog 4; Who Was Who Am h. See also: Barr; Biog Geneal Master Ind.

TRUDEAU, EDWARD LIVINGSTON (1848-1915). Medicine. OCCUP: Physician; Researcher. MSS: Ind NUCMC. WORKS ABOUT: Am Men Sci 1-2 d3; Dict Am Biog; Dict Am Med Biog; Natl Cycl Am Biog 13; Who Was Who Am 1. See also: Biog Geneal Master Ind; Ireland Ind Sci.

TRUE, FREDERICK WILLIAM (1858-1914). Zoology. OCCUP: Museum curator; Librarian. MSS: Ind NUCMC (78-901). WORKS ABOUT: Am Men Sci 1-2 d3; Dict Am Biog; Elliott (s); Natl Cycl Am Biog 19; Who Was Who Am 1. See also: Barr.

TRUMAN, JAMES (1826-1914). Surgery. OCCUP: Dentist; Professor; Academic administrator. WORKS ABOUT: Am Men Sci 2 d3; Natl Cycl Am Biog 25; Who Was Who Am 1.

TRUMBULL, GURDON (1841-1903). Ornithology. OCCUP: Artist. WORKS ABOUT: Elliott; Who Was Who Am 1. See also: Barr; Biog Geneal Master Ind.

TRUMBULL, JAMES HAMMOND (1821-1897). Indian philology; History. OCCUP: Public official; Librarian; Author. MSS: Ind NUCMC (62-2162, 67-823). WORKS ABOUT: Dict Am Biog; Natl Ac Scis Biog Mem 7; Natl Cycl Am Biog 9; Who Was Who Am h. See also: Barr; Biog Geneal Master Ind.

TRYON, GEORGE WASHINGTON (1838-1888). Conchology. OCCUP: Businessman; Editor; Engaged in independent study. MSS: Ind NUCMC (66-359). WORKS ABOUT: Dict Am Biog; Elliott; Who Was Who Am h. See also: Biog Geneal Master Ind.

TUCKERMAN, EDWARD (1817-1886). Botany. OCCUP: Professor. MSS: Ind NUCMC (60-2207, 62-3178). WORKS ABOUT: Dict Am Biog; Elliott; Natl Ac Scis Biog Mem 3; Natl Cycl Am Biog 5; Who Was Who Am h. See also: Barr; Biog Geneal Master Ind.

TUFTS, FRANK LEO (1871-1909). Physics. OCCUP: Professor. WORKS ABOUT: Am Men Sci 1 d3; Sci 29(1909):654.

TULLY, WILLIAM (1785-1859). Materia medica; Medicine. OCCUP: Physician; Professor. WORKS ABOUT: Dict Am Biog; Who Was Who Am h. See also: Biog Geneal Master Ind.

TUOMEY, MICHAEL (1805-1857). Geology. OCCUP: State government scientist; Professor. MSS: Ind NUCMC. WORKS ABOUT: Elliott; Natl Cycl Am Biog 13; Who Was Who Am h. See also: Barr; Biog Geneal Master Ind; Ireland Ind Sci.

TUTTLE, CHARLES WESLEY (1829-1881). Astronomy. OCCUP: Observatory affiliate; Lawyer. MSS: Ind NUCMC (83-1609). WORKS ABOUT: Dict Am Biog; Elliott; Who Was Who Am h. See also: Biog Geneal Master Ind.

TUTTLE, DAVID KITCHELL (1835-1915). Chemistry; Metallurgy. OCCUP: Professor; Industrial scientist; U.S. government scientist. WORKS ABOUT: Am Men Sci 1 d3; Who Was Who Am 1.

TUTTLE, HORACE PARNELL (1839-1893). Astronomy. OCCUP: Naval officer; Observatory affiliate. WORKS ABOUT: Allibone; Appleton's.

TUTTLE, JAMES PERCIVAL (1857-1913). Surgery. OCCUP: Physician; Professor. WORKS ABOUT: Kelly and Burrage; Sci 37(1913):250-51; Wallace.

TWINING, ALEXANDER CATLIN (1801-1884). Astronomy; Engineering; Invention. OCCUP: Engineer; Professor; Inventor. MSS: Ind NUCMC (76-640). WORKS ABOUT: Dict Am Biog; Elliott; Natl Cycl Am Biog 19; Who Was Who Am h. See also: Biog Geneal Master Ind.

TYSON, JAMES (1841-1919). Medicine. OCCUP: Physician; Professor. WORKS ABOUT: Am Men Sci 1-2 d3; Dict Am Biog; Dict Am Med Biog; Natl Cycl Am Biog 9; Who Was Who Am 1. See also: Biog Geneal Master Ind.

TYSON, PHILIP THOMAS (1799-1877). Chemistry; Geology. OCCUP: Businessman (?); State government scientist. WORKS ABOUT: Elliott; Natl Cycl Am Biog 13. See also: Biog Geneal Master Ind; Ireland Ind Sci.

U

UHLER, PHILIP REESE (1835-1913). Entomology; Geology.
OCCUP: Librarian. WORKS ABOUT: <u>Am Men Sci</u> 1-2 d3; <u>Dict Am
Biog</u>; Elliott (s); <u>Natl Cycl Am Biog</u> 8; <u>Who Was Who Am</u> 1.
See also: Barr; <u>Biog Geneal Master Ind</u>; Ireland <u>Ind Sci</u>;
Pelletier.

ULKE, HENRY (1821-1910). Entomology. OCCUP: Artist. WORKS
ABOUT: Appleton's <u>Supp</u> 1901; Groce and Wallace; Mallis; <u>Pop
Sci Mo</u> 59(1909):12; <u>Who Was Who Am</u> 1.

UNDERWOOD, LUCIEN MARCUS (1853-1907). Botany. OCCUP:
Professor. MSS: <u>Ind NUCMC</u>. WORKS ABOUT: <u>Am Men Sci</u> 1 d3;
<u>Dict Am Biog</u>; Elliott (s); <u>Natl Cycl Am Biog</u> 12; <u>Who Was
Who Am</u> 1. See also: Barr; <u>Biog Geneal Master Ind</u>.

UP DE GRAFF, THAD. STEVENS (1839-1885). Microscopy. OCCUP:
Physician. WORKS ABOUT: Am Soc Micr <u>Proc</u> (1885):216-22.

UPHAM, EDWIN PORTER (1845-1918). Archeology. OCCUP: Museum
affiliate. WORKS ABOUT: <u>Am Men Sci</u> 2 d3; U S Natl Mus <u>Rep</u>
(1919):49.

UPTON, WINSLOW (1853-1914). Astronomy; Meteorology. OCCUP:
Professor; Observatory director. MSS: <u>Ind NUCMC</u>. WORKS
ABOUT: <u>Am Men Sci</u> 1-2 d3; <u>Dict Am Biog</u>; <u>Natl Cycl Am Biog</u>
12; <u>Who Was Who Am</u> 1. See also: Barr; <u>Biog Geneal Master
Ind</u>.

V

VAN AMRINGE, JOHN HOWARD (1835-1915). Mathematics. OCCUP: Professor; Academic administrator. MSS: Ind NUCMC (66-1611). WORKS ABOUT: Dict Am Biog; Natl Cycl Am Biog 13, 29; Who Was Who Am 1. See also: Barr; Biog Geneal Master Ind; Ireland Ind Sci.

VAN BRUNT, CORNELIUS (1827?-1903). Botany. OCCUP: Not determined. WORKS ABOUT: Sci 18(1903):509.

VAN DEPOELE, CHARLES JOSEPH (1846-1892). Electricity; Invention. OCCUP: Manufacturer; Inventor. MSS: Ind NUCMC (80-1980). WORKS ABOUT: Dict Am Biog; Natl Cycl Am Biog 13; Who Was Who Am h. See also: Biog Geneal Master Ind; Pelletier.

VANDERBILT, GEORGE WASHINGTON (1862-1914). Forestry. OCCUP: Businessman; Benefactor; Landowner. MSS: Ind NUCMC. WORKS ABOUT: Dict Am Biog; Natl Cycl Am Biog 6; Who Was Who Am 1. See also: Biog Geneal Master Ind.

VAN GIESON, IRA THOMPSON (1866-1913). Neurology; Pathology. OCCUP: Physician; Administrator; Academic instructor. WORKS ABOUT: Am Men Sci 1; Bulloch p369; Kelly and Burrage; Sci 37(1913):515.

VAN HISE, CHARLES RICHARD (1857-1918). Geology. OCCUP: Professor; College president; U.S. government scientist. MSS: Ind NUCMC (62-2894). WORKS ABOUT: Am Men Sci 1 d3; Dict Am Biog; Dict Sci Biog; Elliott (s); Natl Ac Scis Biog Mem [Mem 17]; Natl Cycl Am Biog 10, 19; Who Was Who Am 1. See also: Biog Geneal Master Ind; Pelletier.

VAN NOSTRAND, HENRY D. (-1896). Conchology. OCCUP: Businessman. WORKS ABOUT: Sci 4(1896):568.

VAN RENSSELAER, JEREMIAH (1793-1871). Medicine; Natural history; Geology. OCCUP: Physician. WORKS ABOUT: Elliott; Natl Cycl Am Biog 7:525. See also: Pelletier (as Rensselaer).

VAN RENSSELAER, STEPHEN (1764-1839). Philanthropy; Education; Founder of Rensselaer Polytechnic Institute. OCCUP: Inherited wealth; Legislator; Benefactor. MSS: <u>Ind NUCMC</u>. WORKS ABOUT: <u>Dict Am Biog</u>; <u>Natl Cycl Am Biog</u> 2; <u>Who Was Who Am</u> h. See also: Barr; <u>Biog Geneal Master Ind</u>.

VANUXEM, LARDNER (1792-1848). Geology. OCCUP: Professor; State government scientist; Independent wealth. MSS: <u>Ind NUCMC</u>. WORKS ABOUT: <u>Dict Am Biog</u>; <u>Dict Sci Biog</u>; Elliott; <u>Natl Cycl Am Biog</u> 8; <u>Who Was Who Am</u> h. See also: Barr; <u>Biog Geneal Master Ind</u>; Ireland <u>Ind Sci</u>; Pelletier.

VAN VLECK, JOHN MONROE (1833-1912). Astronomy; Mathematics. OCCUP: Professor; Academic administrator. WORKS ABOUT: <u>Am Journ Sci</u> 185(1913):120; <u>Am Men Sci</u> 2 d3; Appleton's; <u>Sci</u> 36(1912):669, 43(1916):924; <u>Who Was Who Am</u> 1.

VASEY, GEORGE (1822-1893). Botany. OCCUP: Physician; Museum curator; U.S. government scientist. MSS: <u>Ind NUCMC</u>. WORKS ABOUT: <u>Dict Am Biog</u>; Elliott; <u>Who Was Who Am</u> h. See also: Barr; <u>Biog Geneal Master Ind</u>.

VASEY, HARVEY ELMER (1890-1918). Botany. OCCUP: Professor. WORKS ABOUT: <u>Sci</u> 48(1918):616.

VAUGHAN, BENJAMIN (1751-1835). Agriculture; General science. OCCUP: Merchant; Public official; Author. MSS: <u>Ind NUCMC</u> (61-1135). WORKS ABOUT: <u>Dict Am Biog</u>; <u>Who Was Who Am</u> h. See also: Barr; <u>Biog Geneal Master Ind</u>; Ireland <u>Ind Sci</u>.

VAUGHAN, DANIEL (ca.1818-1879). Astronomy; Chemistry; Mathematics; Physiology. OCCUP: Professor. MSS: <u>Ind NUCMC</u>. WORKS ABOUT: <u>Dict Am Biog</u>; Elliott; <u>Natl Cycl Am Biog</u> 13; <u>Who Was Who Am</u> h. See also: Barr; <u>Biog Geneal Master Ind</u>.

VAUGHAN, JOHN (1756-1841). General science. OCCUP: Merchant. MSS: <u>Ind NUCMC</u> (68-1837). WORKS ABOUT: Bell p75-76.

VAUGHAN, JOHN (1775-1807). Chemistry. OCCUP: Physician. WORKS ABOUT: Adams O; Allibone; <u>Am Chem and Chem Eng</u>; Appleton's.

VAUGHAN, VICTOR CLARENCE, JR. (1879-1919). Pathology; Medicine. OCCUP: Physician; Administrator; Professor. WORKS ABOUT: <u>Am Men Sci</u> 2 d3; <u>Sci</u> 50(1919):17.

VAUGHN, WILLIAM JAMES (1834-1912). Astronomy; Mathematics. OCCUP: Professor; College president; Librarian. WORKS ABOUT: <u>Am Men Sci</u> 1-2 d3; <u>Sci</u> 36(1912):901; <u>Who Was Who Am</u> 1.

VAUX, WILLIAM SANSOM (1811-1882). Archeology; Mineralogy. OCCUP: Not determined; Collector; Benefactor. WORKS ABOUT: <u>Am Journ Sci</u> 123(1882):498; Appleton's.

VEEDER, MAJOR ALBERT (1848-1915). Sanitary science. OCCUP: Physician; Public official. WORKS ABOUT: Am Men Sci 1-2 d3; Who Was Who Am 1.

VENABLE, CHARLES SCOTT (1827-1900). Mathematics. OCCUP: Professor. MSS: Ind NUCMC (60-707, 71-456). WORKS ABOUT: Dict Am Biog; Elliott; Natl Cycl Am Biog 10; Who Was Who Am 1. See also: Barr; Biog Geneal Master Ind.

VERRILL, CLARENCE SIDNEY (-1918). Mining. OCCUP: Mining engineer. WORKS ABOUT: Sci 48(1918):544.

VESTAL, GEORGE (1857?-1898). Agriculture; Horticulture. OCCUP: Professor. WORKS ABOUT: Sci 8(1898):668.

VINAL, STUART C. (1895?-1918). Entomology. OCCUP: State government scientist. WORKS ABOUT: Sci 48(1918):443.

VOLK, ERNEST (1845-1919). Archeology; Botany. OCCUP: Explorer; Collector; Museum associate. WORKS ABOUT: Am Men Sci 1-2 d3; Sci 50(1919):451-53.

VOLNEY, CARL WALTER (1838-   ). Chemistry. OCCUP: Manufacturer. WORKS ABOUT: Am Men Sci 1.

VON SCHWEINITZ, LEWIS DAVID. See SCHWEINITZ, LEWIS DAVID VON.

VOORHEES, EDWARD BURNETT (1856-1911). Agriculture; Agricultural chemistry. OCCUP: State government scientist; Professor. WORKS ABOUT: Am Men Sci 1-4 d5; Dict Am Biog; Natl Cycl Am Biog 13; Who Was Who Am 1. See also: Biog Geneal Master Ind.

W

WACHSMUTH, CHARLES (1829-1896). Paleontology. OCCUP: Businessman; Researcher; Collector. MSS: Ind NUCMC. WORKS ABOUT: Dict Am Biog; Elliott; Natl Cycl Am Biog 7:159; Who Was Who Am h. See also: Barr; Biog Geneal Master Ind; Ireland Ind Sci.

WADSWORTH, OLIVER FAIRFIELD (1838-1911). Ophthalmology. OCCUP: Surgeon; Professor. MSS: Ind NUCMC (?). WORKS ABOUT: Am Men Sci 1-2 d3; Natl Cycl Am Biog 20; Who Was Who Am 1.

WAGNER, CLINTON (1837-1914). Medicine; Laryngology. OCCUP: Physician; Professor. WORKS ABOUT: Dict Am Biog; Natl Cycl Am Biog 1. See also: Barr; Biog Geneal Master Ind.

WAGNER, WILLIAM (1796-1885). Natural history; Paleontology; Science benefactor. OCCUP: Businessman; Collector; Lecturer. MSS: Ind NUCMC (66-1503). WORKS ABOUT: Dict Am Biog; Elliott; Natl Cycl Am Biog 6; Who Was Who Am h. See also: Biog Geneal Master Ind.

WAHL, WILLIAM HENRY (1848-1909). Electrochemistry; Metallurgy; Science journalism. OCCUP: Administrator; Editor; Teacher. WORKS ABOUT: Am Men Sci 1 d3; Dict Am Biog; Natl Cycl Am Biog 21; Who Was Who Am 1. See also: Barr; Biog Geneal Master Ind; Pelletier.

WAILES, BENJAMIN LEONARD COVINGTON (1797-1862). Natural history. OCCUP: Planter; Public official. MSS: Ind NUCMC (60-1548, 63-132). WORKS ABOUT: Dict Am Biog; Elliott; Who Was Who Am h. See also: Biog Geneal Master Ind.

WAIT, LUCIEN AUGUSTUS (1846-1913). Mathematics. OCCUP: Professor. WORKS ABOUT: Am Men Sci 1 d3; Who Was Who Am 1. See also: Barr.

WAKE, CHARLES STANILAND (1835-1910). Ethnology; Philosophy. OCCUP: Museum affiliate; Administrator; Editor. WORKS ABOUT: Adams O; Allibone Supp; Am Men Sci 1 d3; Sci 32(1910):14; Who Was Who Am 1.

WALCOTT, HELENA B. (MRS. CHARLES D.) (-1911). Geology. OCCUP: Not determined (personal scientific assistant). WORKS ABOUT: Sci 34(1911):109-10.

WALDO, FRANK (1857-1920). Engineering; Meteorology. OCCUP: U.S. government scientist; Editor; Farmer. WORKS ABOUT: Adams O; Am Men Sci 1 d3; Wallace; Who Was Who Am 1.

WALKER, FRANCIS AMASA (1840-1897). Mathematics; Statistics; Economics. OCCUP: Public official; Professor (political economy and history); College president. MSS: Ind NUCMC. WORKS ABOUT: Dict Am Biog; Natl Ac Scis Biog Mem 5; Natl Cycl Am Biog 5; Who Was Who Am h. See also: Barr; Biog Geneal Master Ind; Pelletier.

WALKER, SEARS COOK (1805-1853). Astronomy; Mathematics. OCCUP: Teacher; Actuary; U.S. government scientist. MSS: Ind NUCMC (?). WORKS ABOUT: Dict Am Biog; Elliott; Natl Cycl Am Biog 8; Who Was Who Am h. See also: Barr; Biog Geneal Master Ind; Ireland Ind Sci.

WALKER, WILLIAM JOHNSON (1790-1865). Surgery. OCCUP: Physician; Businessman; Benefactor. WORKS ABOUT: Dict Am Biog; Who Was Who Am h. See also: Barr; Biog Geneal Master Ind.

WALLER, ELWYN (1846-1919). Chemistry. OCCUP: Professor; Consulant (analytical chemistry). WORKS ABOUT: Am Men Sci 1-2 d3; Appleton's; Elliott (s); Natl Cycl Am Biog 13; Twentieth Cent Biog Dict; Who Was Who Am 1.

WALLING, HENRY FRANCIS (1825-1888). Map-making; Topography. OCCUP: Civil engineer; U.S. government scientist. WORKS ABOUT: Appleton's; Groce and Wallace; Phil Soc Washington Bull 11(1892):492-96.

WALPOLE, FREDERICK A. (1861-1904). Botany. OCCUP: Artist; U.S. government scientist. WORKS ABOUT: Sci 19(1904):838.

WALSH, BENJAMIN DANN (1808-1869). Entomology. OCCUP: Farmer; Businessman; State government scientist. MSS: Ind NUCMC (74-983). WORKS ABOUT: Dict Am Biog; Elliott; Who Was Who Am h. See also: Barr; Biog Geneal Master Ind; Ireland Ind Sci; Pelletier.

WALTER, THOMAS (ca.1740-1789). Botany. OCCUP: Planter. WORKS ABOUT: Dict Am Biog; Elliott; Who Was Who Am h. See also: Biog Geneal Master Ind.

WARD, HENRY AUGUSTUS (1834-1906). Natural history. OCCUP: Natural history collector; Professor; Businessman. MSS: Ind NUCMC (61-1278). WORKS ABOUT: Am Men Sci 1 d3; Dict Am Biog; Elliott (s); Natl Cycl Am Biog 3, 28; Who Was Who Am 1. See also: Barr; Biog Geneal Master Ind.

WARD, J. F. (1831?-1902). Engineering. OCCUP: Engineer. WORKS ABOUT: Sci 15(1902):158.

WARD, LESTER FRANK (1841-1913). Botany; Geology; Sociology. OCCUP: U.S. government employee and scientist; Professor (sociology). MSS: Ind NUCMC (62-4525). WORKS ABOUT: Am Men Sci 1-2 d3; Dict Am Biog; Elliott (s); Natl Cycl Am Biog 13; Who Was Who Am h (add), 1, 4 (add). See also: Barr; Biog Geneal Master Ind.

WARD, RICHARD HALSTED (1837-1917). Biology; Microscopy. OCCUP: Physician; Professor. WORKS ABOUT: Am Men Sci 1-2 d3; Dict Am Biog; Natl Cycl Am Biog 13; Who Was Who Am 1. See also: Barr; Biog Geneal Master Ind.

WARD, SAMUEL BALDWIN (1842-1915). Medicine. OCCUP: Physician; Professor. WORKS ABOUT: Am Men Sci 1-2 d3; Natl Cycl Am Biog 1; Sci 41(1915):859; Who Was Who Am 1.

WARD, W. H. (-1897). Horticulture. OCCUP: Not determined. WORKS ABOUT: Sci 5(1897):103.

WARDER, JOHN ASTON (1812-1883). Forestry; Horticulture. OCCUP: Physician; Farmer; Public official. WORKS ABOUT: Dict Am Biog; Elliott; Natl Cycl Am Biog 4; Who Was Who Am h. See also: Biog Geneal Master Ind.

WARDER, ROBERT BOWNE (1848-1905). Chemistry. OCCUP: Professor; State government scientist. WORKS ABOUT: Elliott; Natl Cycl Am Biog 13; Who Was Who Am 1. See also: Biog Geneal Master Ind.

WARNER, JAMES D. (1827-1911). Mathematics. OCCUP: Mechanical engineer. WORKS ABOUT: Am Men Sci 2 d3.

WARNER, WILLIAM RICHARD (1836-1901). Chemistry. OCCUP: Pharmacist; Businessman. WORKS ABOUT: Biog Geneal Master Ind (gives death date); Natl Cycl Am Biog 2.

WARREN, CYRUS MOORS (1824-1891). Chemistry. OCCUP: Manufacturer; Professor. WORKS ABOUT: Dict Am Biog; Elliott; Natl Cycl Am Biog 10; Who Was Who Am h. See also: Barr; Biog Geneal Master Ind.

WARREN, GOUVERNEUR KEMBLE (1830-1882). Geology; Topographical engineering. OCCUP: Army officer; Engineer. MSS: Ind NUCMC (79-1026). WORKS ABOUT: Dict Am Biog; Elliott; Natl Ac Scis Biog Mem 2; Natl Cycl Am Biog 4; Who Was Who Am h. See also: Barr; Biog Geneal Master Ind.

WARREN, JOHN (1753-1815). Medicine. OCCUP: Surgeon; Professor. MSS: Ind NUCMC. WORKS ABOUT: Dict Am Biog; Dict Am Med Biog; Natl Cycl Am Biog 10; Who Was Who Am h. See also: Biog Geneal Master Ind; Ireland Ind Sci; Pelletier.

WARREN, JOHN COLLINS (1778-1856). Anatomy; Paleontology. OCCUP: Physician; Professor. MSS: Ind NUCMC (62-3750, 65-1290). WORKS ABOUT: Dict Am Biog; Dict Am Med Biog; Elliott; Natl Cycl Am Biog 6; Who Was Who Am h. See also: Barr; Biog Geneal Master Ind; Ireland Ind Sci; Pelletier.

WARREN, JONATHAN MASON (1811-1867). Anatomy; Medicine. OCCUP: Physician. MSS: Ind NUCMC. WORKS ABOUT: Adams O; Allibone; Appleton's; Arnold; Drake; Leonardo p445-46; Wallace.

WARREN, JOSEPH WEATHERHEAD (1849-1916). Physiology. OCCUP: Physician; Professor; Public official. WORKS ABOUT: Am Men Sci 1-2 d3; Elliott (s); Who Was Who Am 1.

WARREN, LEBEN (1836-1905). Mathematics. OCCUP: Professor. WORKS ABOUT: Sci 21(1905):677.

WARREN, SAMUEL EDWARD (1831-1909). Geometry. OCCUP: Professor; Teacher; Author. WORKS ABOUT: Adams O; Allibone; Allibone Supp; Am Men Sci 1 d3; Appleton's; Biog Dict Am Educ; Natl Cycl Am Biog 4; Twentieth Cent Biog Dict; Wallace; Who Was Who Am 1.

WARRING, CHARLES BARTLETT (1825-1907). Physics; Science and religion. OCCUP: Teacher; Educator. WORKS ABOUT: Adams O; Allibone Supp; Am Men Sci 1 d3; Sci 26(1907):61; Wallace; Who Was Who Am 1.

WATERHOUSE, BENJAMIN (1754-1846). Medicine; Natural history. OCCUP: Physician; Professor. MSS: Ind NUCMC (62-3751). WORKS ABOUT: Dict Am Biog; Dict Am Med Biog; Elliott; Natl Cycl Am Biog 9; Who Was Who Am h. See also: Biog Geneal Master Ind; Ireland Ind Sci; Pelletier.

WATKINS, JOHN ELFRETH (1852-1903). Engineering. OCCUP: Engineer; Administrator; Museum curator. WORKS ABOUT: Dict Am Biog; Who Was Who Am 1. See also: Barr; Biog Geneal Master Ind.

WATKINS, OSCAR S. (1883-1919). Chemistry. OCCUP: State government scientist. WORKS ABOUT: Am Men Sci 2 d3.

WATSON, BENJAMIN MARSTON (1848-1918). Horticulture. OCCUP: Academic instructor. WORKS ABOUT: Am Men Sci 1-2 d3; Harvard Quinquennial Cat.

WATSON, JAMES CRAIG (1838-1880). Astronomy. OCCUP: Professor; Observatory director. MSS: Ind NUCMC (65-627). WORKS ABOUT: Dict Am Biog; Elliott; Natl Ac Scis Biog Mem 3; Natl Cycl Am Biog 7:70; Who Was Who Am h. See also: Barr; Biog Geneal Master Ind; Ireland Ind Sci.

WATSON, SERENO (1826-1892). Botany. OCCUP: Editor; U.S. government scientist; Curator. MSS: Ind NUCMC. WORKS ABOUT: Dict Am Biog; Dict Sci Biog; Elliott; Natl Ac Scis Biog Mem 5; Natl Cycl Am Biog 6; Who Was Who Am h. See also: Barr; Biog Geneal Master Ind.

WATSON, WILLIAM (1834-1915). Engineering; Mathematics. OCCUP: Professor; Author. WORKS ABOUT: Am Men Sci 1-2 d3; Dict Am Biog; Natl Cycl Am Biog 12; Who Was Who Am 1. See also: Barr; Biog Geneal Master Ind.

WATT, GEORGE (1820-1893). Chemistry. OCCUP: Dentist; Professor; Editor. WORKS ABOUT: Am Chem and Chem Eng.

WATTS, WILLIAM LORD (1850-   ). Mining engineering. OCCUP: Engineer; State government engineer. WORKS ABOUT: Who Was Who Am 4 (see also: Whos Who Am [1920-1921]).

WEAVER, WILLIAM DIXON (1857-1919). Electrical engineering. OCCUP: Naval engineer; Editor. MSS: Ind NUCMC. WORKS ABOUT: Dict Am Biog; Who Was Who Am 1.

WEBB, JOHN BURKITT (1841-1912). Engineering; Physics. OCCUP: Professor; Inventor; Consultant. MSS: Ind NUCMC. WORKS ABOUT: Am Men Sci 1-2 d3; Dict Am Biog; Who Was Who Am 4.

WEBBER, CHARLES WILKINS (1819-1856). Natural history. OCCUP: Author; Explorer. WORKS ABOUT: Dict Am Biog; Natl Cycl Am Biog 4; Who Was Who Am h. See also: Biog Geneal Master Ind.

WEBER, HENRY ADAM (1845-1912). Chemistry. OCCUP: Professor; State government scientist. WORKS ABOUT: Am Men Sci 1-2 d3; Dict Am Biog; Natl Cycl Am Biog 19; Who Was Who Am 4. See also: Barr; Biog Geneal Master Ind; Pelletier.

WEBSTER, FRANCIS MARION (1849-1916). Entomology. OCCUP: U.S. and state government scientist. MSS: Ind NUCMC (?). WORKS ABOUT: Am Men Sci 1-2 d3; Elliott (s); Mallis; Natl Cycl Am Biog 13; Sci 43(1916):64, 162-64; Who Was Who Am 1.

WEBSTER, FREDERIC SMITH (1849-   ). Natural history; Zoology. OCCUP: Museum affiliate; Businessman; Administrator. WORKS ABOUT: Pop Sci Mo 59(1901):12; Who Was Who Am 4.

WEBSTER, HARRISON EDWIN (1841-1906). Biology; Geology. OCCUP: Professor; College president. WORKS ABOUT: Appleton's; Natl Cycl Am Biog 7:172; Sci 23(1906):989.

WEBSTER, JOHN WHITE (1793-1850). Chemistry. OCCUP: Professor. MSS: Ind NUCMC (65-1291). WORKS ABOUT: Dict Am Biog; Elliott; Who Was Who Am h. See also: Biog Geneal Master Ind; Pelletier.

WEBSTER, NATHAN BURNHAM (1821-1900). Education; Physical science. OCCUP: Educator; Civil engineer. WORKS ABOUT: Adams O; Allibone Supp; Appleton's; Sci 13(1901):78; Twentieth Cent Biog Dict; Wallace; Who Was Who Am 1.

WEBSTER, NOAH (1758-1843). Meteorology; Lexicography. OCCUP: Lawyer; Author; Editor. MSS: Ind NUCMC (72-223, 78-53, 83-543). WORKS ABOUT: Dict Am Biog; Dict Am Med Biog; Natl Cycl Am Biog 2; Who Was Who Am h. See also: Biog Geneal Master Ind.

WEEDON, WILLIAM STONE (1877-1912). Chemistry. OCCUP: Academic instructor; Industrial scientist. WORKS ABOUT: <u>Am Men Sci</u> 1-2 d3; <u>Sci</u> 36(1912):273.

WEEKS, JOSEPH DAME (1840-1896). Statistics; Technical journalism; Metallurgy. OCCUP: Editor; Researcher. WORKS ABOUT: <u>Dict Am Biog</u>; <u>Natl Cycl Am Biog</u> 13; <u>Who Was Who Am</u> h. See also: Barr; <u>Biog Geneal Master Ind</u>.

WEIGHTMAN, WILLIAM (1813-1904). Chemistry. OCCUP: Industrial scientist; Manufacturer. WORKS ABOUT: <u>Dict Am Biog</u>; Elliott; <u>Natl Cycl Am Biog</u> 14.

WEIL, RICHARD (1876-1917). Pathology. OCCUP: Physician; Professor. WORKS ABOUT: <u>Am Men Sci</u> 2 d3; <u>Dict Am Biog</u>. See also: Barr; <u>Biog Geneal Master Ind</u>; Ireland <u>Ind Sci</u>.

WEINSTEIN, JOSEPH (1862?-1917). Chemistry. OCCUP: Academic instructor. WORKS ABOUT: <u>Sci</u> 46(1917):14.

WEIR, JAMES, JR. (1856-1906). Animal intelligence; Medicine. OCCUP: Physician; Author. WORKS ABOUT: Adams O; Knight; Wallace; <u>Who Was Who Am</u> 1.

WELD, LAENAS GIFFORD (1862-1919). Astronomy; Mathematics. OCCUP: Professor; Academic administrator; State government scientist. WORKS ABOUT: <u>Am Men Sci</u> 1-2 d3; <u>Natl Cycl Am Biog</u> 18; Wallace; <u>Who Was Who Am</u> 1.

WELLINGTON, ARTHUR MELLEN (1847-1895). Civil engineering. OCCUP: Engineer; Editor. WORKS ABOUT: <u>Dict Am Biog</u>; <u>Natl Cycl Am Biog</u> 11; <u>Who Was Who Am</u> h. See also: Barr; <u>Biog Geneal Master Ind</u>.

WELLMAN, SAMUEL THOMAS (1847-1919). Engineering. OCCUP: Industrial engineer; Businessman; Inventor. MSS: <u>Ind NUCMC</u> (68-908). WORKS ABOUT: <u>Dict Am Biog</u>; <u>Natl Cycl Am Biog</u> 13; <u>Who Was Who Am</u> 1. See also: Barr.

WELLS, DAVID AMES (1828-1898). Chemistry; Geology; Science editing and writing. OCCUP: Publisher; Author; Public official. MSS: <u>Ind NUCMC</u>. WORKS ABOUT: <u>Dict Am Biog</u>; Elliott; <u>Natl Cycl Am Biog</u> 10; <u>Who Was Who Am</u> h. See also: Barr; <u>Biog Geneal Master Ind</u>.

WELLS, ELIAB HORATIO (1836-   ). Geology; Mathematical physics. OCCUP: Professor; Civil engineer. WORKS ABOUT: <u>Am Men Sci</u> 1-2 d3.

WELLS, F. J. (-1904). Agricultural physics. OCCUP: Professor. WORKS ABOUT: <u>Sci</u> 19(1904):743.

WELLS, HORACE (1815-1848). Anesthesia; Dentistry. OCCUP: Dentist. MSS: <u>Ind NUCMC</u> (60-1002). WORKS ABOUT: <u>Dict Am Biog</u>; <u>Dict Am Med Biog</u>; <u>Natl Cycl Am Biog</u> 6; <u>Who Was Who Am</u> h. See also: Barr; <u>Biog Geneal Master Ind</u>; Ireland <u>Ind Sci</u>; Pelletier.

WELLS, SAMUEL ROBERTS (1820-1875). Phrenology. OCCUP: Editor and publisher. MSS: Ind NUCMC. WORKS ABOUT: Dict Am Biog; Elliott; Who Was Who Am h. See also: Biog Geneal Master Ind.

WELLS, WEBSTER (1851-1916). Mathematics. OCCUP: Professor. WORKS ABOUT: Adams O; Am Men Sci 1-2 d3; Biog Dict Am Educ; Natl Cycl Am Biog 17; Wallace; Who Was Who Am 1.

WELLS, WILLIAM CHARLES (1757-1817). Medicine; Meteorology; Physics; Physiology. OCCUP: Physician. WORKS ABOUT: Dict Am Biog; Dict Sci Biog; Elliott; Natl Cycl Am Biog 12; Who Was Who Am h. See also: Biog Geneal Master Ind; Ireland Ind Sci; Pelletier.

WENDELL, OLIVER CLINTON (1845-1912). Astronomy. OCCUP: Engineer; Observatory affiliate; Professor. WORKS ABOUT: Am Men Sci 2 d3; Sci 36(1912):914; Who Was Who Am 1.

WENTWORTH, GEORGE ALBERT (1835-1906). Mathematics; Mathematics textbook writing; Physics. OCCUP: Teacher; Author. WORKS ABOUT: Am Men Sci 1 d3; Dict Am Biog; Natl Cycl Am Biog 10; Who Was Who Am 1. See also: Biog Geneal Master Ind.

WENZELL, WILLIAM THEODORE (1829-1913). Chemistry. OCCUP: Professor. WORKS ABOUT: Am Men Sci 1-2 d3; Sci 38(1913):623.

WERNLE, C. HENRY (1831?-1902). Instrumentation. OCCUP: Instrument maker; U.S. government employee. WORKS ABOUT: Sci 15(1902):878.

WESBROOK, FRANK FAIRCHILD (1868-1918). Pathology. OCCUP: Professor; State government scientist; College president. WORKS ABOUT: Am Men Sci 2 d3; Dict Am Biog; Natl Cycl Am Biog 14; Who Was Who Am 1. See also: Barr; Biog Geneal Master Ind; Ireland Ind Sci.

WEST, BENJAMIN (1730-1813). Almanac making; Astronomy. OCCUP: Businessman; Professor; Public official. WORKS ABOUT: Dict Am Biog; Elliott; Natl Cycl Am Biog 8; Who Was Who Am h. See also: Biog Geneal Master Ind.

WEST, E. P. (1820-1892). Archeology; Geology. OCCUP: Lawyer; Judge; Academic affiliate. WORKS ABOUT: Kansas Ac Sci Trans 13(1893):68-69.

WEST, THOMAS DYSON (1851-1915). Mechanical engineering. OCCUP: Industrial manager; Industrialist. WORKS ABOUT: Allibone Supp; Am Men Sci 2 d3; Wallace; Who Was Who Am 1.

WESTINGHOUSE, GEORGE (1846-1914). Engineering; Invention. OCCUP: Inventor; Manufacturer. MSS: Ind NUCMC. WORKS ABOUT: Am Men Sci 2 d3; Dict Am Biog; Natl Cycl Am Biog 11, 15; Who Was Who Am 1. See also: Barr; Biog Geneal Master Ind; Ireland Ind Sci; Pelletier.

WESTON, EDMUND BROWNELL (1850-1916). Civil engineering. OCCUP: Engineer; City government engineer; Industrial officer. WORKS ABOUT: Am Men Sci 2 d3; Wallace; Who Was Who Am 1.

WETHERILL, CHARLES MAYER (1825-1871). Chemistry. OCCUP: Consultant (analytical chemistry); U.S. government scientist; Professor. WORKS ABOUT: Dict Am Biog; Elliott; Natl Cycl Am Biog 13; Who Was Who Am h. See also: Barr; Biog Geneal Master Ind; Ireland Ind Sci; Pelletier.

WETHERILL, SAMUEL (1821-1890). Chemical engineering. OCCUP: Industrial scientist; Inventor; Manufacturer. WORKS ABOUT: Dict Am Biog; Natl Cycl Am Biog 7:506; Who Was Who Am h. See also: Biog Geneal Master Ind; Pelletier.

WHARTON, JOSEPH (1826-1909). Chemistry. OCCUP: Industrial manager; Manufacturer; Benefactor. MSS: Ind NUCMC (71-427, 82-1755). WORKS ABOUT: Dict Am Biog; Natl Cycl Am Biog 13; Who Was Who Am 1. See also: Barr; Biog Geneal Master Ind.

WHEATLEY, CHARLES MOORE (1822-1882). Geology; Mineralogy; Conchology. OCCUP: Mining manager; Industrialist; Collector. MSS: Ind NUCMC (76-955). WORKS ABOUT: Adams O; Allibone; Am Journ Sci 123(1882):498; Appleton's; Wallace.

WHEELER, AMOS DEAN (1803-1876). Mathematics. OCCUP: Educator; Clergyman. Scientific work: see Roy Soc Cat. WORKS ABOUT: Durfee; Williams Gen Cat.

WHEELER, CHARLES FAY (1842-1910). Botany. OCCUP: Businessman; Professor; U.S. government scientist. WORKS ABOUT: Am Men Sci 1 d3; Sci 31(1910):410, 32(1910):72-75.

WHEELER, CHARLES GILBERT (1836-1912). Chemistry; Geology; Mining geology. OCCUP: Consultant; Professor. WORKS ABOUT: Adams O; Allibone Supp; Am Journ Sci 183(1912):296; Am Men Sci 2 d3; Sci 35(1912):214; Wallace; Who Was Who Am 1.

WHEELER, CLAUDE L. (-1916). Medicine. OCCUP: Physician (?); Editor. WORKS ABOUT: Sci 45(1917):14.

WHEELER, EBENEZER SMITH (1839-1913). Engineering. OCCUP: Engineer; U.S. government scientist. MSS: Ind NUCMC (81-1950). WORKS ABOUT: Am Men Sci 1-3 d4; Who Was Who Am 1.

WHEELER, ERNEST S. (1869?-1909). Tropical medicine. OCCUP: Physician (?); Academic instructor. WORKS ABOUT: Sci 29(1909):332.

WHEELER, GEORGE MONTAGUE (1842-1905). Topographical engineering. OCCUP: Army officer; Engineer. MSS: Ind NUCMC. WORKS ABOUT: Dict Am Biog; Elliott; Who Was Who Am 4. See also: Biog Geneal Master Ind.

WHEELER, HENRY LORD (1867-1914). Chemistry. OCCUP: Professor. WORKS ABOUT: Am Men Sci 1-2 d3; Who Was Who Am 1.

WHELPLEY, JAMES DAVENPORT (1817-1872). Physical science. OCCUP: Physician; Editor; Businessman. Scientific work: see Roy Soc Cat. WORKS ABOUT: Am Ac Arts Scis Proc 8(1868-1873):482-84; Appleton's.

WHIPPLE, AMIEL WEEKS (1816-1863). Topographical engineering. OCCUP: Army officer; Engineer. MSS: Ind NUCMC. WORKS ABOUT: Dict Am Biog; Elliott; Natl Cycl Am Biog 10; Who Was Who Am h. See also: Biog Geneal Master Ind.

WHIPPLE, SQUIRE (1804-1888). Civil engineering. OCCUP: Engineer; Instrument maker. MSS: Ind NUCMC. WORKS ABOUT: Dict Am Biog; Elliott; Natl Cycl Am Biog 9; Who Was Who Am h. See also: Biog Geneal Master Ind.

WHISTLER, GEORGE WASHINGTON (1800-1849). Civil engineering. OCCUP: Army officer; Engineer. MSS: Ind NUCMC. WORKS ABOUT: Dict Am Biog; Natl Cycl Am Biog 1, 9; Who Was Who Am h. See also: Barr; Biog Geneal Master Ind.

WHITE, CHARLES ABIATHAR (1826-1910). Natural history; Geology; Paleontology. OCCUP: U.S. and state government scientist; Professor; Curator. MSS: Ind NUCMC (62-4165). WORKS ABOUT: Am Men Sci 1 d3; Dict Am Biog; Elliott (s); Natl Ac Scis Biog Mem 7; Natl Cycl Am Biog 6; Who Was Who Am 1. See also: Barr; Biog Geneal Master Ind; Ireland Ind Sci.

WHITE, CHARLES JOYCE (1839-1917). Mathematics. OCCUP: Professor; Academic administrator. MSS: Ind NUCMC. WORKS ABOUT: Adams O; Allibone; Am Men Sci 1-2 d3; Sci 45(1917):184; Wallace; Who Was Who Am 1.

WHITE, JAMES CLARKE (1833-1916). Dermatology. OCCUP: Physician; Professor. WORKS ABOUT: Am Men Sci 1-2 d3; Dict Am Biog; Dict Am Med Biog; Natl Cycl Am Biog 19; Who Was Who Am 1. See also: Barr; Biog Geneal Master Ind.

WHITE, JAMES WILLIAM (1850-1916). Surgery. OCCUP: Surgeon; Professor. WORKS ABOUT: Am Men Sci 1-2 d3; Dict Am Biog; Natl Cycl Am Biog 17; Who Was Who Am 1. See also: Barr; Biog Geneal Master Ind; Ireland Ind Sci.

WHITE, JOHN (fl. 1585-1593). Natural history; Art; Cartography. OCCUP: Artist; Governor. WORKS ABOUT: Dict Am Biog; Who Was Who Am h. See also: Biog Geneal Master Ind.

WHITE, LAURA BRADSTREET (-1919). Chemistry. OCCUP: Teacher. WORKS ABOUT: Sci 49(1919):234.

WHITE, MOSES CLARKE (1819-1900). Microscopy. OCCUP: Physician; Missionary; Professor. WORKS ABOUT: Atkinson; Chamberlain v3:500; Sci 12(1900):693.

WHITE, THEODORE GREELY (1872?-1901). Geology; Botany. OCCUP: Not determined; Student; Editor. WORKS ABOUT: Sci 14(1901):77.

WHITEHEAD, CABELL (1863-1908). Chemistry; Metallurgy. OCCUP: U.S. government scientist; Banker; Mining affiliate. WORKS ABOUT: <u>Am Men Sci</u> 1 d3; <u>Who Was Who Am</u> 1.

WHITEHEAD, RICHARD HENRY (1865-1916). Anatomy. OCCUP: Professor; Academic administrator. WORKS ABOUT: <u>Am Men Sci</u> 1-2 d3; <u>Sci</u> 43(1916):236; Wallace; <u>Who Was Who Am</u> 1.

WHITEHEAD, WILLIAM RIDDICK (1831-1902). Medicine. OCCUP: Physician; Professor. WORKS ABOUT: <u>Dict Am Med Biog</u>; <u>Natl Cycl Am Biog</u> 10; <u>Sci</u> 16(1902):678.

WHITFIELD, ROBERT PARR (1828-1910). Geology; Paleontology; Stratigraphy. OCCUP: Curator. MSS: <u>Ind NUCMC</u>. WORKS ABOUT: <u>Am Men Sci</u> 1 d3; <u>Dict Am Biog</u>; <u>Dict Sci Biog</u>; Elliott (s); <u>Natl Cycl Am Biog</u> 5; <u>Who Was Who Am</u> 1. See also: Barr; <u>Biog Geneal Master Ind</u>.

WHITING, HAROLD (1855-1895). Physics. OCCUP: Professor. WORKS ABOUT: Am Ac Arts Scis <u>Proc</u> 31(1896):356-58; <u>Sci</u> 1(1895):667; Wallace.

WHITING, HENRY L. (1821-1897). Topography. OCCUP: U.S. government scientist; Public official. WORKS ABOUT: <u>Sci</u> 5(1897):300-02.

WHITLOCK, GEORGE CLINTON (1808-1864). Mathematics. OCCUP: Professor; Clergyman. Scientific work: see Roy Soc <u>Cat</u>. WORKS ABOUT: Adams O; Allibone; Wallace.

WHITMAN, CHARLES OTIS (1842-1910). Biology; Zoology. OCCUP: Professor; Research institution administrator. MSS: <u>Ind NUCMC</u>. WORKS ABOUT: <u>Am Men Sci</u> 1-2 d3; <u>Dict Am Biog</u>; <u>Dict Sci Biog</u>; Elliott (s); Natl Ac Scis <u>Biog Mem</u> 7; <u>Natl Cycl Am Biog</u> 11; <u>Who Was Who Am</u> 1. See also: Barr; <u>Biog Geneal Master Ind</u>.

WHITMAN, FRANK PERKINS (1853-1919). Physics. OCCUP: Professor. WORKS ABOUT: <u>Am Journ Sci</u> 199(1920):226; <u>Am Men Sci</u> 1-2 d3; <u>Who Was Who Am</u> 1.

WHITNEY, ELI (1765-1825). Invention. OCCUP: Manufacturer; Inventor. MSS: <u>Ind NUCMC</u> (61-3473, 83-384). WORKS ABOUT: <u>Dict Am Biog</u>; <u>Natl Cycl Am Biog</u> 4; <u>Who Was Who Am</u> h. See also: Barr; <u>Biog Geneal Master Ind</u>; Ireland <u>Ind Sci</u>; Pelletier.

WHITNEY, JOSIAH DWIGHT (1819-1896). Chemistry; Geology. OCCUP: Consultant (mining); Professor; State government scientist. MSS: <u>Ind NUCMC</u>. WORKS ABOUT: <u>Dict Am Biog</u>; <u>Dict Sci Biog</u>; Elliott; <u>Natl Cycl Am Biog</u> 9; <u>Who Was Who Am</u> h. See also: Barr; <u>Biog Geneal Master Ind</u>; Pelletier.

WHITTLESEY, CHARLES (1808-1886). Geology. OCCUP: Army officer; Mining engineer; U.S. and state government scientist. MSS: <u>Ind NUCMC</u> (75-1803). WORKS ABOUT: Elliott. See also: Barr; <u>Biog Geneal Master Ind</u>.

WIECHMANN, FERDINAND GERHARD (1858-1919). Chemistry. OCCUP: Academic instructor; Industrial scientist; Consultant. WORKS ABOUT: <u>Am Men Sci</u> 1-2 d3; <u>Dict Am Biog</u>; <u>Natl Cycl Am Biog</u> 14; <u>Who Was Who Am</u> 1. See also: Barr; <u>Biog Geneal Master Ind</u>; Pelletier.

WIENER, JOSEPH (1828-1904). Pathology. OCCUP: Physician (?); Professor. WORKS ABOUT: <u>Sci</u> 20(1904):255.

WIGHTMAN, WILLIAM (-1909). Public health; Sanitation. OCCUP: Surgeon; Public official. WORKS ABOUT: <u>Sci</u> 29(1909):893, 30(1909):516-17.

WILBERT, MARTIN INVENTIUS (1855-1916). Chemistry; Pharmacology. OCCUP: Not determined; Laboratory affiliate. WORKS ABOUT: <u>Sci</u> 44(1916):850.

WILKES, CHARLES (1798-1877). Astronomy; Exploration; Geophysics. OCCUP: Naval officer; Explorer. MSS: <u>Ind NUCMC</u> (62-4650). WORKS ABOUT: <u>Dict Am Biog</u>; Elliott; <u>Natl Cycl Am Biog</u> 2; <u>Who Was Who Am</u> h. See also: Barr; <u>Biog Geneal Master Ind</u>; Ireland <u>Ind Sci</u>; Pelletier.

WILKINSON, A. WILLIAMS (1832-1908). Chemistry. OCCUP: Not determined; Invention. WORKS ABOUT: <u>Sci</u> 27(1908):237.

WILLARD, DE FOREST (1846-1910). Medicine. OCCUP: Surgeon; Professor. WORKS ABOUT: <u>Am Men Sci</u> 2 d3; <u>Dict Am Biog</u>; <u>Natl Cycl Am Biog</u> 12, 15; <u>Who Was Who Am</u> 1. See also: Barr; <u>Biog Geneal Master Ind</u>; Ireland <u>Ind Sci</u>.

WILLARD, JOSEPH (1738-1804). Astronomy; Mathematics. OCCUP: Clergyman; College president. MSS: <u>Ind NUCMC</u>. WORKS ABOUT: <u>Dict Am Biog</u>; <u>Natl Cycl Am Biog</u> 6; <u>Who Was Who Am</u> h. See also: <u>Biog Geneal Master Ind</u>.

WILLEY, HENRY (1824-1907). Botany. OCCUP: Lawyer; Journalist. WORKS ABOUT: Adams O; Allibone <u>Supp</u>; <u>Am Men Sci</u> 1 d3; Appleton's; <u>Natl Cycl Am Biog</u> 10; Wallace; <u>Who Was Who Am</u> 1.

WILLIAMS, GEORGE HUNTINGTON (1856-1894). Mineralogy; Petrology. OCCUP: Professor. WORKS ABOUT: <u>Dict Am Biog</u>; Elliott; <u>Natl Cycl Am Biog</u> 42; <u>Who Was Who Am</u> h. See also: Barr; <u>Biog Geneal Master Ind</u>.

WILLIAMS, HENRY SHALER (1847-1918). Geology; Paleontology; Stratigraphy. OCCUP: Professor. MSS: <u>Ind NUCMC</u> (80-2214). WORKS ABOUT: <u>Am Men Sci</u> 1-2 d3; <u>Dict Am Biog</u>; <u>Dict Sci Biog</u>; Elliott (s); <u>Natl Cycl Am Biog</u> 21; <u>Who Was Who Am</u> 1. See also: Barr; <u>Biog Geneal Master Ind</u>.

WILLIAMS, HENRY WILLARD (1821-1895). Ophthalmology. OCCUP: Physician; Professor. MSS: <u>Ind NUCMC</u>. WORKS ABOUT: <u>Dict Am Biog</u>; <u>Dict Am Med Biog</u>; <u>Natl Cycl Am Biog</u> 3; <u>Who Was Who Am</u> h. See also: Barr; <u>Biog Geneal Master Ind</u>.

WILLIAMS, JOHN FRANCIS (1862?-1891). Mineralogy. OCCUP: Curator; State government scientist; Professor. WORKS ABOUT: <u>Am Journ Sci</u> 142(1891):524; <u>Sci</u> 1st ser 18(1891):300.

WILLIAMS, JONATHAN (1750-1815). Natural philosophy. OCCUP: Merchant; Army officer; Engineer. MSS: <u>Ind NUCMC</u> (83-1296). WORKS ABOUT: <u>Dict Am Biog</u>; <u>Natl Cycl Am Biog</u> 3; <u>Who Was Who Am</u> h. See also: <u>Biog Geneal Master Ind</u>.

WILLIAMS, LEONARD WORCESTER (1875-1912). Comparative anatomy. OCCUP: Professor. WORKS ABOUT: <u>Am Men Sci</u> 1-2 d3; <u>Sci</u> 36(1912):431.

WILLIAMS, LEWIS WHITE (1807?-1876). Mineralogy. OCCUP: Not determined; Collector. WORKS ABOUT: <u>Am Journ Sci</u> 106(1873):398.

WILLIAMS, RUFUS PHILLIPS (1851-1911). Chemistry. OCCUP: Teacher. WORKS ABOUT: Adams O; Allibone <u>Supp</u>; <u>Am Men Sci</u> 1-2 d3; Wallace; <u>Who Was Who Am</u> 1.

WILLIAMS, SAMUEL (1743-1817). Astronomy; Mathematics. OCCUP: Clergyman; Professor; Editor. MSS: <u>Ind NUCMC</u> (65-1296, 81-333). WORKS ABOUT: Adams O; Allibone; Appleton's; Drake; Duyckinck; <u>Natl Cycl Am Biog</u> 1; Wallace.

WILLIAMS, SAMUEL GARDNER (1827-1900). Geology. OCCUP: Educator; Professor (geology, pedagogy). MSS: <u>Ind NUCMC</u> (?). WORKS ABOUT: <u>Natl Cycl Am Biog</u> 8; Wallace.

WILLIAMS, STEPHEN WEST (1790-1855). Botany; Medicine. OCCUP: Physician; Academic instructor (medicine); Author. WORKS ABOUT: <u>Dict Am Biog</u>; <u>Dict Am Med Biog</u>; <u>Natl Cycl Am Biog</u> 1; <u>Who Was Who Am</u> h. See also: <u>Biog Geneal Master Ind</u>.

WILLIAMSON, HUGH (1735-1819). Astronomy; Climatology; Medicine. OCCUP: Physician; Businessman; Public official. MSS: <u>Ind NUCMC</u> (?). WORKS ABOUT: <u>Dict Am Biog</u>; Elliott; <u>Natl Cycl Am Biog</u> 2; <u>Who Was Who Am</u> h. See also: <u>Biog Geneal Master Ind</u>; Ireland <u>Ind Sci</u>.

WILLISTON, SAMUEL WENDELL (1851-1918). Entomology; Medicine; Paleontology. OCCUP: Professor. MSS: <u>Ind NUCMC</u> (80-425). WORKS ABOUT [birth dates may differ]: <u>Am Men Sci</u> 1-2 d3; <u>Dict Am Biog</u>; <u>Dict Sci Biog</u>; Elliott (s); Natl Ac Scis <u>Biog Mem</u> [<u>Mem</u> 17]; <u>Natl Cycl Am Biog</u> 30; <u>Who Was Who Am</u> 1. See also: Barr; <u>Biog Geneal Master Ind</u>; Ireland <u>Ind Sci</u>; Pelletier.

WILLSON, ROBERT NEWTON (1873-1916). Pathology. OCCUP: Physician; Academic instructor (medicine). WORKS ABOUT: <u>Am Men Sci</u> 2 d3; Wallace; <u>Who Was Who Am</u> 1.

WILLSON, THOMAS LEOPOLD (1860-1915). Chemistry. OCCUP: Industrialist; Inventor. WORKS ABOUT: <u>Am Chem and Chem Eng</u>; <u>Sci</u> 21(1905):882.

WILLYOUNG, ELMER GRANT (1865-1919). Electrical engineering; Physics. OCCUP: Industrial scientist and officer; Consultant. WORKS ABOUT: Am Men Sci 1-2 d3.

WILSON, ALEXANDER (1766-1813). Ornithology. OCCUP: Skilled laborer; Teacher; Editor. MSS: Ind NUCMC (66-93). WORKS ABOUT: Dict Am Biog; Dict Sci Biog; Elliott; Natl Cycl Am Biog 7:440; Who Was Who Am h. See also: Barr; Biog Geneal Master Ind; Ireland Ind Sci; Pelletier.

WILSON, EZRA HERBERT (1858-1906). Bacteriology. OCCUP: City government scientist; Laboratory administrator. WORKS ABOUT: Am Men Sci 1 d3.

WILSON, HERBERT MICHAEL (1860-1920). Engineering. OCCUP: Engineer; U.S. government engineer; Administrator. WORKS ABOUT: Adams O; Am Lit Yearbook; Am Men Sci 1-3 d4; Wallace; Who Was Who Am 1.

WILSON, JAMES (1836-1920). Agriculture. OCCUP: Farmer; Public official; Professor. MSS: Ind NUCMC. WORKS ABOUT [birth dates may differ]: Am Men Sci 2 d3; Dict Am Biog; Natl Cycl Am Biog 11, 14; Who Was Who Am 1. See also: Barr; Biog Geneal Master Ind; Ireland Ind Sci.

WILSON, JOSEPH MILLER (1838-1902). Civil engineering. OCCUP: Engineer; Architect. MSS: Ind NUCMC (?). WORKS ABOUT: Dict Am Biog; Natl Cycl Am Biog 7:492; Who Was Who Am 1. See also: Barr; Biog Geneal Master Ind.

WILSON, THOMAS (1832-1902). Anthropology. OCCUP: Lawyer; Curator; Professor. WORKS ABOUT: Natl Cycl Am Biog 11; Ohio Authors; Pop Sci Mo 41(1892):300-04; Wallace; Who Was Who Am 1.

WILSON, THOMAS BELLERBY (1807-1865). Zoology. OCCUP: Physician; Collector. MSS: Ind NUCMC (66-94). WORKS ABOUT: Am Journ Pharm 37(1865):320; Am Journ Sci 89(1865):373; Appleton's; Drake; Ent Soc Philadelphia Proc 5(1865); Natl Cycl Am Biog 13.

WILSON, THOMAS M. (-1908). Medicine; Pathology. OCCUP: Academic instructor; Student. WORKS ABOUT: Sci 29(1909):23.

WINCHELL, ALEXANDER (1824-1891). Education; Geology. OCCUP: Educator; Professor; State government scientist. MSS: Ind NUCMC (65-654). WORKS ABOUT: Dict Am Biog; Dict Sci Biog; Elliott; Natl Cycl Am Biog 6, 16; Who Was Who Am h. See also: Barr; Biog Geneal Master Ind; Ireland Ind Sci.

WINCHELL, NEWTON HORACE (1839-1914). Archeology; Geology. OCCUP: State government scientist; Professor; Editor. MSS: Ind NUCMC (70-1647). WORKS ABOUT: Am Men Sci 1-2 d3; Dict Am Biog; Dict Sci Biog; Elliott (s); Natl Cycl Am Biog 31; Who Was Who Am 1. See also: Barr; Biog Geneal Master Ind; Ireland Ind Sci.

WINDLE, FRANCIS (1845?-1917). Botany; Horticulture; Ornithology. OCCUP: Lawyer; Teacher; State government scientist. WORKS ABOUT: Sci 46(1917):451.

WING, AUGUSTUS (1809-1876). Geology. OCCUP: Clergyman; Teacher (?). WORKS ABOUT: Am Journ Sci 111(1876):334, 113(1877):332-47, 405-19, 114(1877):36-37.

WING, CHARLES HALLET (1836-1915). Chemistry. OCCUP: Professor. WORKS ABOUT: Sci 42(1915):416; Who Was Who Am 1.

WINLOCK, ANNA (1857-1904). Astronomy. OCCUP: Observatory affiliate. WORKS ABOUT: Ogilvie; Sci 19(1904):157.

WINLOCK, JOSEPH (1826-1875). Astronomy; Mathematics. OCCUP: Professor; U.S. government scientist; Observatory director. MSS: Ind NUCMC. WORKS ABOUT: Dict Am Biog; Dict Sci Biog; Elliott; Natl Ac Scis Biog Mem 1; Natl Cycl Am Biog 9; Who Was Who Am h. See also: Barr; Biog Geneal Master Ind.

WINLOCK, WILLIAM CRAWFORD (1859-1896). Astronomy. OCCUP: Observatory affiliate; Professor; Administrator. WORKS ABOUT: Elliott; Natl Cycl Am Biog 9. See also: Barr.

WINSLOW, CHARLES FREDERICK (1811-1877). General science; Cosmography. OCCUP: Physician; Public official (diplomatic service); Author. WORKS ABOUT: Adams O; Allibone; Am Journ Sci 114(1877):168; Appleton's; Drake; Wallace.

WINSTON, CHARLES HENRY (1831-   ). Physics. OCCUP: Educator; Professor. WORKS ABOUT: Who Was Who Am 4.

WINTER, WILLIAM PHILLIPS (1866-1919). Chemistry. OCCUP: Professor. WORKS ABOUT: Am Men Sci 2 d3.

WINTERHALTER, ALBERT GUSTAVUS (1856-1920). Astronomy; Engineering. OCCUP: Naval officer. WORKS ABOUT: Who Was Who Am 1.

WINTHROP, JOHN, JR. (1605/06-1676). Natural philosophy; Medicine. OCCUP: Public official; Governor; Businessman. MSS: Ind NUCMC. WORKS ABOUT: Dict Am Biog; Dict Sci Biog; Elliott; Natl Cycl Am Biog 10; Who Was Who Am h. See also: Biog Geneal Master Ind; Ireland Ind Sci; Pelletier.

WINTHROP, JOHN (1714-1779). Astronomy; Mathematics; Physics. OCCUP: Professor. MSS: Ind NUCMC. WORKS ABOUT: Dict Am Biog; Dict Sci Biog; Elliott; Natl Cycl Am Biog 7:165; Who Was Who Am h. See also: Barr; Biog Geneal Master Ind; Ireland Ind Sci; Pelletier.

WISLIZENUS, FREDERICK ADOLPHUS (1810-1889). Meteorology; Natural history. OCCUP: Physician; Natural history explorer. MSS: Ind NUCMC (?). WORKS ABOUT: Dict Am Biog; Elliott; Who Was Who Am h. See also: Barr; Biog Geneal Master Ind; Ireland Ind Sci.

WISTAR, CASPAR (1761-1818). Anatomy. OCCUP: Physician; Professor. MSS: Ind NUCMC (61-944). WORKS ABOUT: Dict Am Biog; Dict Am Med Biog; Dict Sci Biog; Elliott; Natl Cycl Am Biog 1; Who Was Who Am h. See also: Biog Geneal Master Ind; Ireland Ind Sci; Pelletier.

WISTAR, ISAAC JONES (1827-1905). Anatomy; Penology; Philanthropy. OCCUP: Lawyer; Businessman; Benefactor. MSS: Ind NUCMC. WORKS ABOUT: Appleton's; Natl Cycl Am Biog 12; Sci 22(1905):415; Twentieth Cent Biog Dict; Wallace; Who Was Who Am 1.

WITHINGTON, CHARLES FRANCIS (1852-1917). Medicine. OCCUP: Physician; Academic instructor. WORKS ABOUT: Am Men Sci 1-2 d3; Who Was Who Am 1.

WITTHAUS, RUDOLPH AUGUST (1846-1915). Chemistry; Toxicology. OCCUP: Professor. WORKS ABOUT: Am Men Sci 1-2 d3; Dict Am Biog; Dict Am Med Biog; Elliott (s); Natl Cycl Am Biog 11; Who Was Who Am 1. See also: Barr; Biog Geneal Master Ind; Pelletier.

WOELFEL, ALBERT (1871-1920). Physiology. OCCUP: Physician; Academic instructor; Administrator. WORKS ABOUT: Am Men Sci 2 d3; Natl Cycl Am Biog 19.

WOLCOTT, TOWNSEND (1857-1910). Electrical engineering. OCCUP: Engineer; U.S. government engineer. WORKS ABOUT: Am Men Sci 1 d3; Wallace.

WOLF, THEODORE R. (-1909). Chemistry. OCCUP: Professor. WORKS ABOUT: Sci 30(1909):111.

WOLFE, HARRY KIRKE (1858-1918). Psychology. OCCUP: Professor; Educator. MSS: Ind NUCMC (76-1707). WORKS ABOUT: Am Men Sci 1-2 d3; Dict Am Biog; Natl Cycl Am Biog 18; Who Was Who Am 1. See also: Barr; Biog Geneal Master Ind.

WOLFE, JAMES JACOB (1875-1920). Biology; Cytology. OCCUP: Professor. WORKS ABOUT: Am Men Sci 2 d3; Who Was Who Am 1.

WOLLE, FRANCIS (1817-1893). Botany; Invention. OCCUP: Teacher; Educator; Clergyman. MSS: Ind NUCMC (?). WORKS ABOUT: Adams O; Allibone Supp; Am Micr Soc Proc 15(1893):245-46; Am Mo Micr Journ 14(1893):181-82; Appleton's; Bot Gaz 18(1893):109-10; Natl Cycl Am Biog 1; Notarisia 9(1894):8; Soc Belge Micr Bull 20(1893):71-72; Torrey Bot Club Bull 20(1893):211-12; Wallace.

WOOD, ALPHONSO (1810-1881). Botany. OCCUP: Teacher; Educator; Civil engineer. MSS: Ind NUCMC. WORKS ABOUT: Adams O; Allibone Supp; Am Journ Sci 123(1882):333, 127(1884):242; Appleton's; Bot Zentralblatt 7(1881):223-24; Drake; Natl Cycl Am Biog 14; Torrey Bot Club Bull 8(1881):12, 53-56; Wallace.

WOOD, BARNABAS (1819-1875). Chemistry (metals). OCCUP: Dentist; Editor. Scientific work: see Roy Soc <u>Cat</u>. WORKS ABOUT: <u>Albany Med Ann</u> 3(1882):379-80.

WOOD, DeVOLSON (1832-1897). Engineering; Mathematics. OCCUP: Professor; Inventor. MSS: <u>Ind NUCMC</u>. WORKS ABOUT: Adams O; Allibone <u>Supp</u>; Appleton's; <u>Biog Dict Am Educ</u>; <u>Natl Cycl Am Biog</u> 13; Preston; Rensselaer <u>Biog Rec</u>; <u>Sci</u> 6(1897):28, 204-06; Wallace; <u>Who Was Who Am</u> h.

WOOD, EDWARD STICKNEY (1846-1905). Chemistry; Medicine. OCCUP: Professor; Hospital scientist; Public official. WORKS ABOUT: <u>Dict Am Biog</u>; Elliott; <u>Who Was Who Am</u> 1. See also: Barr; Pelletier.

WOOD, HORATIO CHARLES (1841-1920). Medicine; Pharmacology; Therapeutics. OCCUP: Physician; Professor. WORKS ABOUT: <u>Am Men Sci</u> 1-2 d3; <u>Dict Am Biog</u>; <u>Dict Am Med Biog</u>; <u>Dict Sci Biog</u>; Natl Ac Scis <u>Biog Mem</u> 33; <u>Natl Cycl Am Biog</u> 13; <u>Who Was Who Am</u> 1. See also: <u>Biog Geneal Master Ind</u>; Ireland <u>Ind Sci</u>.

WOOD, HUDSON A. (1841-1903). Mathematics. OCCUP: Teacher. WORKS ABOUT: <u>Sci</u> 18(1903):477; <u>Who Was Who Am</u> 4.

WOOD, JOHN (ca.1775-1822). Map making. OCCUP: Author; Cartographer. MSS: <u>Ind NUCMC</u> (?). WORKS ABOUT: <u>Dict Am Biog</u>; <u>Who Was Who Am</u> h. See also: <u>Biog Geneal Master Ind</u>.

WOOD, JOHN CLAIRE (1871-1916). Zoology. OCCUP: Surveyor; Civil engineer. WORKS ABOUT: <u>Auk</u> ns 33(1916):459-60; <u>Sci</u> 44(1919):564.

WOOD, WILLIAM (ca.1822-1885). Ornithology. OCCUP: Not determined; Physician (?). Scientific work: see Roy Soc <u>Cat</u>. WORKS ABOUT: <u>Auk</u> 2(1885):391.

WOODBRIDGE, LUTHER DANA (1850-1899). Anatomy; Physiology. OCCUP: Physician (?); Professor. WORKS ABOUT: <u>Sci</u> 10(1899):702.

WOODBURY, CHARLES JEPTHA HILL (1851-1916). Civil engineering. OCCUP: Industrial engineer; Consultant. WORKS ABOUT: <u>Am Men Sci</u> 1-2 d3; <u>Dict Am Biog</u>; <u>Natl Cycl Am Biog</u> 12; <u>Who Was Who Am</u> 1. See also: Barr; <u>Biog Geneal Master Ind</u>.

WOODBURY, DANIEL PHINEAS (1812-1864). Engineering. OCCUP: Army officer; Engineer. MSS: <u>Ind NUCMC</u>. WORKS ABOUT: <u>Dict Am Biog</u>; <u>Natl Cycl Am Biog</u> 1; <u>Who Was Who Am</u> h. See also: <u>Biog Geneal Master Ind</u>.

WOODHOUSE, JAMES (1770-1809). Chemistry. OCCUP: Professor. MSS: <u>Ind NUCMC</u>. WORKS ABOUT: <u>Dict Am Biog</u>; Elliott; <u>Who Was Who Am</u> h. See also: Barr; <u>Biog Geneal Master Ind</u>; Ireland <u>Ind Sci</u>; Pelletier.

WOODHOUSE, SAMUEL WASHINGTON (1821-1904). Natural history; Ornithology. OCCUP: Physician; Army surgeon; Natural history explorer/ collector. MSS: Ind NUCMC. WORKS ABOUT: Elliott; Who Was Who Am Hist. See also: Barr.

WOODMAN, DURAND (1859-1907). Chemistry. OCCUP: Industrial employee; Consultant. WORKS ABOUT: Am Men Sci 1 d3; Who Was Who Am 1.

WOODMAN, HENRY J. (-1903). Biology. OCCUP: Not determined; Natural history collector. WORKS ABOUT: Sci 17(1903):916.

WOODROW, JAMES (1828-1907). Natural science; Science and religion. OCCUP: Clergyman; Professor; College president. MSS: Ind NUCMC (70-1899, 77-356). WORKS ABOUT: Dict Am Biog; Natl Cycl Am Biog 11; Who Was Who Am 1. See also: Barr; Biog Geneal Master Ind.

WOODRUFF, CHARLES EDWARD (1860-1915). Anthropology. OCCUP: Army physician. WORKS ABOUT: Am Men Sci 2 d3; Dict Am Biog; Who Was Who Am 1. See also: Barr; Biog Geneal Master Ind.

WOODS, GRANVILLE T. (1856-1910). Invention. OCCUP: Laborer; Inventor; Manufacturer. WORKS ABOUT: Adams R; Carwell; Haber; Klein; Young.

WOODWARD, AMOS E. (-1891). Geology. OCCUP: State government scientist. WORKS ABOUT: Sci 1st ser 18(1891):199.

WOODWARD, ANTHONY (-1915). Geology. OCCUP: Librarian. WORKS ABOUT: Sci 41(1915):240.

WOODWARD, CALVIN MILTON (1837-1914). Mathematics. OCCUP: Professor; Academic administrator. WORKS ABOUT: Am Men Sci 1-2 d3; Dict Am Biog; Natl Cycl Am Biog 9; Who Was Who Am 1. See also: Barr; Biog Geneal Master Ind.

WOODWARD, JOSEPH JANVIER (1833-1884). Medicine; Pathology. OCCUP: Physician; Army officer. MSS: Ind NUCMC. WORKS ABOUT: Dict Am Biog; Dict Am Med Biog; Natl Ac Scis Biog Mem 2; Natl Cycl Am Biog 11; Who Was Who Am h. See also: Biog Geneal Master Ind; Ireland Ind Sci.

WOODWARD, JULIUS HAYDEN (1858-1916). Ophthalmology. OCCUP: Physician; Professor. WORKS ABOUT: Sci 44(1916):53; Who Was Who Am 1.

WOODWORTH, WILLIAM McMICHAEL (1864-1912). Zoology. OCCUP: Academic instructor; Museum curator; Editor. MSS: Ind NUCMC. WORKS ABOUT: Am Journ Sci 184(1912):228; Am Men Sci 1-2 d3.

WOOLF, ALBERT EDWARD (1846-1920). Chemistry; Electrical engineering. OCCUP: Not determined; Inventor. WORKS ABOUT: Am Men Sci 1-2 d3; Who Was Who Am 1.

WOOLF, SOLOMON (1841-1911). Geometry. OCCUP: Professor.
WORKS ABOUT: Appleton's; <u>Sci</u> 33(1911):889.

WORCESTER, JOSEPH EMERSON (1784-1865). Geography;
Lexicography. OCCUP: Teacher; Author; Editor. WORKS ABOUT:
<u>Dict Am Biog</u>; <u>Natl Cycl Am Biog</u> 6; <u>Who Was Who Am</u> h. See
also: <u>Biog Geneal Master Ind</u>.

WORMLEY, THEODORE GEORGE (1826-1897). Chemistry;
Toxicology. OCCUP: Physician; Professor; State government
scientist. MSS: <u>Ind NUCMC</u> (68-1496). WORKS ABOUT: <u>Dict Am
Biog</u>; <u>Natl Cycl Am Biog</u> 13; <u>Who Was Who Am</u> h. See also:
Barr; <u>Biog Geneal Master Ind</u>; Pelletier.

WORTHEN, AMOS HENRY (1813-1888). Geology. OCCUP: Merchant;
State government scientist; Curator. MSS: <u>Ind NUCMC</u>. WORKS
ABOUT: <u>Dict Am Biog</u>; Elliott; Natl Ac Scis <u>Biog Mem</u> 3; <u>Natl
Cycl Am Biog</u> 6; <u>Who Was Who Am</u> h. See also: Barr; <u>Biog
Geneal Master Ind</u>; Ireland <u>Ind Sci</u>.

WORTHEN, GEORGE CARLTON (1871-1919). Botany. OCCUP:
Academic affiliate. WORKS ABOUT: <u>Am Journ Sci</u>
197(1919):454; <u>Sci</u> 49(1919):398.

WRIGHT, ALBERT ALLEN (1846-1905). Geology; Natural history.
OCCUP: Professor; State government scientist. MSS: <u>Ind
NUCMC</u>. WORKS ABOUT: Elliott; <u>Who Was Who Am</u> 1. See also:
Barr; <u>Biog Geneal Master Ind</u>.

WRIGHT, ARTHUR WILLIAMS (1836-1915). Physics. OCCUP:
Professor; Laboratory administrator. MSS: <u>Ind NUCMC</u> (74-
1209). WORKS ABOUT: <u>Am Journ Sci</u> 191(1916):152, 361-66; <u>Am
Men Sci</u> 1-2 d3; Appleton's; Elliott (s); Natl Ac Scis <u>Biog
Mem</u> 15; <u>Natl Cycl Am Biog</u> 13; Preston; <u>Sci</u> 42(1915):932,
43(1916):270-72, 650; <u>Who Was Who Am</u> 1.

WRIGHT, CARROLL DAVIDSON (1840-1909). Statistics; Social
economics. OCCUP: Lawyer; Public official; College
president. MSS: <u>Ind NUCMC</u> (70-1500). WORKS ABOUT: <u>Am Men
Sci</u> 1 d3; <u>Dict Am Biog</u>; <u>Natl Cycl Am Biog</u> 19; <u>Who Was Who
Am</u> 1. See also: Barr; <u>Biog Geneal Master Ind</u>.

WRIGHT, CHARLES (1811-1885). Botany. OCCUP: Teacher;
Natural history explorer/ collector; U.S. government
scientist MSS: <u>Ind NUCMC</u>. WORKS ABOUT: <u>Dict Am Biog</u>;
Elliott; <u>Who Was Who Am</u> h. See also: Barr; Ireland <u>Ind Sci</u>;
Pelletier.

WRIGHT, CHAUNCEY (1830-1875). Mathematics; Philosophy of
science. OCCUP: U.S. government scientist; Editor; Academic
instructor. MSS: <u>Ind NUCMC</u>. WORKS ABOUT: <u>Dict Am Biog</u>;
Elliott; <u>Natl Cycl Am Biog</u> 1; <u>Who Was Who Am</u> h. See also:
<u>Biog Geneal Master Ind</u>; Pelletier.

WRIGHT, ELIZUR (1762-1845). Mathematics. OCCUP: Farmer; Teacher; Public official. MSS: Ind NUCMC. Scientific work: see Roy Soc Cat. WORKS ABOUT: Allibone; Dexter Biog Sketches 4(1907):210-12; Wright [biography of his son; see its index for page references].

WRIGHT, ELIZUR, JR. (1804-1885). Actuarial science. OCCUP: Journalist; Public official; Actuary. MSS: Ind NUCMC (60-1487, 66-1467, 75-1828). WORKS ABOUT: Dict Am Biog; Natl Cycl Am Biog 2; Who Was Who Am h. See also: Biog Geneal Master Ind.

WRIGHT, HAMILTON KEMP (1867-1917). Medical science. OCCUP: Government scientist; Researcher. WORKS ABOUT: Dict Am Biog; Natl Cycl Am Biog 22; Who Was Who Am 1. See also: Biog Geneal Master Ind.

WRIGHT, JOSEPH EDMUND (1878-1910). Mathematics. OCCUP: Professor. WORKS ABOUT: Natl Union Cat; Sci 31(1910):342.

WRIGHT, JULIA McNAIR (1840-1903). Nature writing, especially for young readers. OCCUP: Author. Scientific work: see Roy Soc Cat. WORKS ABOUT: Adams O; Allibone; Allibone Supp; Am Wom; Appleton's; Biog Dict Synopsis Books; Burke and Howe; Ireland Ind Wom; Wallace; Who Was Who Am 1.

WRIGHT, WALTER CHANNING (1846-1917). Actuarial science; Mathematics. OCCUP: Actuary. MSS: Ind NUCMC. WORKS ABOUT: Am Men Sci 1-2 d3.

WRIGHT, WILBUR (1867-1912). Aeronautics. OCCUP: Manufacturer; Inventor. MSS: Ind NUCMC (60-588, 61-1702). WORKS ABOUT: Dict Am Biog; Dict Sci Biog; Natl Cycl Am Biog 14; Who Was Who Am 1. See also: Barr; Biog Geneal Master Ind; Ireland Ind Sci; Pelletier.

WRIGHT, WILLIAM GREENWOOD (1829?-1912). Entomology. OCCUP: Businessman; Natural history collector. WORKS ABOUT: Am Journ Sci 185(1913):336; Essig p802-04; Sci 37(1913):142-43.

WRIGHT, WILLIAM JANES (1831-1906). Mathematics; Philosophy. OCCUP: Clergyman; Professor. WORKS ABOUT: Adams O; Am Men Sci 1 d3; Appleton's; Appleton's Supp 1901; Who Was Who Am 1.

WURTTEMBERG, FRIEDRICH PAUL WILHELM, DUKE OF. See PAUL WILHELM, DUKE OF WURTTEMBERG.

WURTZ, HENRY (1828-1910). Chemistry. OCCUP: Consultant; Professor; Editor. WORKS ABOUT: Dict Am Biog; Elliott; Natl Cycl Am Biog 7:519. See also: Barr; Biog Geneal Master Ind.

WYATT, FRANCIS (1855?-1916). Bacteriology; Chemistry. OCCUP: Not determined. WORKS ABOUT: Sci 43(1916):347; Wallace.

WYLIE, THEOPHILUS ADAM (1810-1895). Chemistry; Physics. OCCUP: Professor (physical sciences, languages); Clergyman. WORKS ABOUT: Adams O; Appleton's; <u>Sci</u> 1(1895):723.

WYMAN, ARTHUR D. (-1904). Chemistry. OCCUP: Academic affiliate. WORKS ABOUT: <u>Sci</u> 20(1904):476.

WYMAN, JEFFRIES (1814-1874). Anatomy; Ethnology; Physiology. OCCUP: Physician; Professor; Curator. MSS: <u>Ind NUCMC</u>. WORKS ABOUT: <u>Dict Am Biog</u>; <u>Dict Am Med Biog</u>; <u>Dict Sci Biog</u>; Elliott; Natl Ac Scis <u>Biog Mem</u> 2; <u>Natl Cycl Am Biog</u> 2; <u>Who Was Who Am</u> h. See also: Barr; <u>Biog Geneal Master Ind</u>; Ireland <u>Ind Sci</u>; Pelletier.

WYMAN, MORRILL (1812-1903). Medicine. OCCUP: Physician. MSS: <u>Ind NUCMC</u>. WORKS ABOUT: <u>Dict Am Biog</u>; <u>Dict Am Med Biog</u>; <u>Natl Cycl Am Biog</u> 28; <u>Who Was Who Am</u> h. See also: Barr; <u>Biog Geneal Master Ind</u>.

WYMAN, WALTER (1848-1911). Public health. OCCUP: Physician; Public official; Administrator. WORKS ABOUT: <u>Am Men Sci</u> 1 d3; Appleton's <u>Supp</u> 1901; <u>Natl Cycl Am Biog</u> 12; <u>Sci</u> 34(1911):755; Wallace; <u>Who Was Who Am</u> 1.

X

XANTUS, JANOS (1825-1894). Ornithology. OCCUP: Natural history explorer/ collector; Museum affiliate. MSS: <u>Ind</u> <u>NUCMC</u> (78-905). WORKS ABOUT: <u>Dict Am Biog</u>; Elliott; <u>Who Was</u> <u>Who Am</u> h. See also: <u>Biog Geneal Master Ind</u>; Ireland <u>Ind</u> <u>Sci</u>; Pelletier.

Y

YANDELL, LUNSFORD PITTS (1805-1878). Paleontology. OCCUP: Physician; Professor; Editor. MSS: Ind NUCMC. WORKS ABOUT: Dict Am Biog; Dict Am Med Biog; Elliott; Natl Cycl Am Biog 4; Who Was Who Am h. See also: Biog Geneal Master Ind.

YARNALL, MORDECAI (1816-1879). Astronomy. OCCUP: Professor; Observatory affiliate. WORKS ABOUT: Elliott. See also: Biog Geneal Master Ind.

YATES, LORENZO GORDIN (1837-1909). Natural history; Botany; Zoology. OCCUP: State government scientist; Teacher (?); Museum associate. MSS: Ind NUCMC (?). WORKS ABOUT: Adams O; Am Men Sci 1 d3; Wallace; Who Was Who Am 1.

YEATES, WILLIAM SMITH (1856-1908). Geology; Mineralogy. OCCUP: Museum curator; Professor; State government scientist. WORKS ABOUT: Elliott; Natl Cycl Am Biog 13; Who Was Who Am 1.

YENDELL, PAUL SEBASTIAN (1844-1918). Astrophysics. OCCUP: Draftsman; City and state government employee. WORKS ABOUT: Am Men Sci 1-2 d3; Sci 47(1918):168 (as Yandel).

YOUMANS, EDWARD LIVINGSTON (1821-1887). Popularization of science; Science journalism. OCCUP: Lecturer; Editor. MSS: Ind NUCMC. WORKS ABOUT: Dict Am Biog; Elliott; Natl Cycl Am Biog 2; Who Was Who Am h. See also: Barr; Biog Geneal Master Ind; Pelletier.

YOUMANS, WILLIAM JAY (1838-1901). Popularization of science; Science journalism. OCCUP: Physician; Editor. WORKS ABOUT: Dict Am Biog; Elliott; Natl Cycl Am Biog 2; Who Was Who Am h. See also: Barr; Biog Geneal Master Ind.

YOUNG, AARON (1819-1898). Botany. OCCUP: Physician; State government scientist; Public official (diplomatic service). MSS: Ind NUCMC (76-959). WORKS ABOUT: Dict Am Biog; Elliott; Who Was Who Am h.

YOUNG, AUGUSTUS (1784-1857). Geology. OCCUP: Lawyer; Legislator; State government scientist. WORKS ABOUT [birth dates may differ]: Adams O; Allibone; Appleton's; <u>Biog Dir Am Congress</u>; Lanman; <u>Natl Cycl Am Biog</u> 3; Wallace; <u>Who Was Who Am</u> h.

YOUNG, CHARLES AUGUSTUS (1834-1908). Astronomy. OCCUP: Professor. MSS: <u>Ind NUCMC</u> (76-1846). WORKS ABOUT: <u>Am Men Sci</u> 1 d3; <u>Dict Am Biog</u>; <u>Dict Sci Biog</u>; Elliott (s); Natl Ac Scis <u>Biog Mem</u> 7; <u>Natl Cycl Am Biog</u> 6; <u>Who Was Who Am</u> 1. See also: Barr; <u>Biog Geneal Master Ind</u>; Ireland <u>Ind Sci</u>.

YOUNG, DAVID (1781-1852). Almanac making; Astronomy. OCCUP: Teacher; Author. WORKS ABOUT: <u>Dict Am Biog</u>; <u>Who Was Who Am</u> h. See also: <u>Biog Geneal Master Ind</u>.

YOUNG, HERBERT A. (ca.1857-1894). Botany. OCCUP: Civil engineer. WORKS ABOUT: Torrey Bot Club <u>Bull</u> 22(1895):51.

YOUNG, JOHN RICHARDSON (1782-1804). Physiology. OCCUP: Physician. WORKS ABOUT: <u>Dict Am Biog</u>; <u>Dict Sci Biog</u>; <u>Who Was Who Am</u> h.

YOUNG, MARY SOPHIE (1872-1919). Botany. OCCUP: Academic instructor; Curator. MSS: <u>Ind NUCMC</u> (71-487). WORKS ABOUT: <u>Sci</u> 49(1919):377.

Z

ZABRISKIE, JEREMIAH LOTT (1835?-1910). Microscopy. OCCUP: Clergyman (?). WORKS ABOUT: Sci 31(1910):615.

ZABRISKIE, JOHN BARREA (1805-1850). Medicine; Physics. OCCUP: Physician; Public official. Scientific work: see Roy Soc Cat. WORKS ABOUT: Natl Am Soc Am Families v2:442-44; Stiles v2:888.

ZALINSKI, EDMUND LOUIS GRAY (1849-1909). Military science. OCCUP: Army officer; Inventor; Professor. MSS: Ind NUCMC (?). WORKS ABOUT: Am Men Sci 1 d3; Appleton's; Natl Cycl Am Biog 7:248; Preston; Sci 29(1909):498; Twentieth Cent Biog Dict; Who Was Who Am 1.

ZENTMAYER, JOSEPH (1826-1888). Manufacturing of scientific instruments. OCCUP: Instrument maker; Inventor. WORKS ABOUT: Dict Am Biog; Natl Cycl Am Biog 13; Who Was Who Am h.

ZOLLICKOFFER, WILLIAM (1793-1853). Medical botany. OCCUP: Physician (?); Academic instructor. Scientific work: see Roy Soc Cat. WORKS ABOUT: Cordell Med Ann p92, 93, 637, 685, 687, 703 (mentioned); Kelly and Burrage.

Index of Names by Scientific Field

In the listings that follow, the names (with year of birth) are arranged under the scientific field(s) in which the individuals were active. The listings are based on the information in the Biographical Index entries, although a certain degree of consolidation or systemization has been imposed. The field titles are arranged alphabetically. A person's name may be under one or several fields. For persons in certain composite fields, the names are listed under both (e.g., agricultural chemists are under agriculture and chemistry). There may be repetition of particular names under both a general <u>and</u> a related specific field, if they are so designated in the Biographical Index. Conversely, a name may appear under only a general <u>or</u> a specific related field title (e.g., under entomology but not under zoology, or laryngology but not under medicine). To cover any subject area completely, both general and specific field headings should be consulted.

ACTUARIAL SCIENCE
McCay, Charles F. (1810)
McClintock, Emory (1840)
Meech, Levi W. (1821)
Messenger, Hiram J. (1855)
Wright, Elizur, Jr. (1804)
Wright, Walter C. (1846)

AERONAUTICS
Durant, Charles F. (1805)
Jeffries, John (1744/45)
Lowe, Thaddeus S. C. (1832)
Montgomery, John J. (1858)
Selfridge, Thomas E. (1882)
Wright, Wilbur (1867)

AGRICULTURE
Adlum, John (1759)
Alvord, Henry E. (1844)
Atwater, Wilbur O. (1844)
Bordley, John B. (1727)
Brewer, William H. (1828)
Browne, Daniel J. (1804)
Buel, Jesse (1778)
Coe, Howard S. (1888)
Collier, Peter (1835)
Couper, James H. (1794)
Dodge, Jacob R. (1823)
Eliot, Jared (1685)
Gaylord, Willis (1792)
Geddes, George (1809)
Gold, Theodore S. (1818)
Goodrich, Chauncey E. (1801)
Hilgard, Eugene W. (1833)
Holmes, Ezekiel (1801)
Hopkins, Cyril G. (1866)
Hoskins, Thomas H. (1828?)
Jefferson, Thomas (1743)
Johnson, Samuel W. (1830)
Johnson, Willis G. (1866)
Jones, William L. (1827)

Judd, Orange (1822)
King, Franklin H. (1848)
Klippart, John H. (1823)
Mapes, Charles V. (1836)
Mapes, James J. (1806)
Martin, George A. (1831?)
Miles, Manly (1826)
Morrow, G. E. (1840?)
Myers, Jesse J. (1876?)
Nettleton, Edwin S. (1831)
Norton, John P. (1822)
Pettit, James H. (1876)
Pinckney, Elizabeth L.
  (1722?)
Puryear, Bennet (1826)
Ravenel, St. Julien (1819)
Reid, James L. (1844)
Ruffin, Edmund (1794)
Schweitzer, Johann .P. (1840)
Shaw, Thomas (1843)
Smith, Clinton D. (1854)
Storer, Francis H. (1832)
Sturtevant, Edward L. (1842)
Thompson, Samuel R. (1833)
Tracy, Samuel M. (1847)
Vaughan, Benjamin (1751)
Vestal, George (1857?)
Voorhees, Edward B. (1856)
Wells, F. J. (   )
Wilson, James (1836)

AGRONOMY
Hopkins, Cyril G. (1866)

ALCHEMY
Brewster, Jonathan (1593)
Bulkeley, Gershom (1636)
Danforth, Samuel (1696)
Hitchcock, Ethan A. (1798)

ALMANAC MAKING
Ames, Nathaniel (1708)
Daboll, Nathan (1750)
Leavitt, Dudley (1772)
Leeds, Daniel (1652)
Peirce, William (ca.1590)
West, Benjamin (1730)
Young, David (1781)

ANAESTHESIOLOGY
Morton, William T. G. (1819)
Wells, Horace (1815)

ANATOMY
Allen, Harrison (1841)
Baker, Frank (1841)
Bard, John (1716)
Beach, Henry H. A. (1843)

Brockway, Fred J. (1860)
Cabell, James L. (1813)
Capshaw, Walter L. (   )
Carmalt, Churchill (1866)
Chovet, Abraham (1704)
Davidge, John B. (1768)
Dorsey, John S. (1783)
Draper, John C. (1835)
Dwight, Thomas (1843)
Elmer, Jonathan (1745)
Forbes, William S. (1831)
Ford, Corydon L. (1813)
Gage, Susanna P. (1857)
Gerrish, Frederic H. (1845)
Gibson, James A. (1867?)
Goddard, Paul B. (1811)
Godman, John D. (1794)
Harlan, Richard (1796)
Hoeve, Heikobus J. H. (1882)
Holt, Jacob F. (   )
Horner, William E. (1793)
Hoskins, Thomas H. (1828?)
Howard, William (1793)
Jackson, John B. S. (1806)
Jayne, Horace F. (1859)
Johnston, Christopher (1822)
Longstreth, Morris (1846)
McBurney, Charles (1845)
McClellan, George (1796)
McClellan, George (1849)
McConnell, J. C. (   )
Mall, Franklin P. (1862)
March, Alden (1795)
Mathews, William P. (1868)
Mears, James E. (1838)
Minot, Charles S. (1852)
Pancoast, Joseph (1805)
Pancoast, Seth (1823)
Pancoast, William H. (1835)
Pattison, Granville S.
  (1791/2)
Prentiss, Charles W. (1874)
Ramsay, Alexander (1754?)
Robinson, Frederick B.
  (1855)
Sanford, Leonard J. (1833)
Sheldon, Ralph E. (1883)
Smith, John H. (1857)
Spitzka, Edward C. (1852)
Summers, Thomas O. (ca.1852)
Thayer, Charles P. (1843)
Thompson, Alton H. (1849)
Warren, John C. (1778)
Warren, Jonathan M. (1811)
Whitehead, Richard H. (1865)
Williams, Leonard W. (1875)
Wistar, Caspar (1761)
Wistar, Isaac J. (1827)

ASTROPHYSICS. See ASTRONOMY

BACTERIOLOGY

BIOCHEMISTRY

Buckhout, William A. (1846)
Buckley, Samuel B. (1809)
Burrill, Thomas J. (1839)
Byrd, William (1674)
Carey, John (1797?)
Carpenter, William M. (1811)
Carruth, James H. (1807)
Carson, Joseph (1808)
Catesby, Mark (1683)
Chapman, Alvan W. (1809)
Chickering John W. (1831)
Clapp, Asahel (1792)
Clark, Henry J. (1826)
Clark, William S. (1826)
Clarke, Cora H. (1851)
Clayton, John (1694)
Clinton, George W. (1807)
Colden, Cadwallader (1688)
Colden, Jane (1724)
Collins, Frank S. (1848)
Collins, Zaccheus (1764)
Conrad, Solomon W. (1779)
Cooper, James G. (1830)
Cox, Jacob D. (1828)
Craig, Moses (   )
Crawe, Ithamar B. (1794)
Creevey, Caroline A. S.
   (1843)
Croom, Hardy B. (1798)
Crozier, Arthur A. (1856)
Cummings, Clara E. (1855)
Curtis, Moses A. (1808)
Curtiss, Allen H. (1845)
Cutler, Manasseh (1742)
Darby, John (1804)
Darlington, William (1782)
Davenport, George E. (1833)
Davis, Charles A. (1861)
Day, David F. (1829)
Dewey, Chester (1784)
Dodge, Charles K. (1844?)
Dodge, Raynal (1844)
Doubleday, Neltje B. D.
   (1865)
Drowne, Solomon (1753)
Dudley, William R. (1849)
Dufour, John J. (ca.1763)
Dunn, Louise B. (   )
Durand, Elias (1794)
Durant, Charles F. (1805)
Durkee, Silas (1798)
Eads, Darwin D. (   )
Eaton, Alvah A. (1865)
Eaton, Amos (1776)
Eaton, Daniel C. (1834)
Eberle, John (1787)
Edson, Arthur W. (   )
Elliott, Stephen (1771)

Ellis, Arvilla (   )
Ellis, Job B. (1829)
Emerson, George B. (1797)
Engelmann, George (1809)
Everhart, Benjamin M. (1818)
Farlow, William G. (1844)
Faxon, Charles E. (1846)
Feay, William T. (1803?)
Fendler, Augustus (1813)
Fontaine, William M. (1835)
Forbes, Francis B. (1840?)
Fosdick, Nellie (   )
Frost, Charles C. (1806)
Fuller, Andrew S. (1828).
Garden, Alexander (1730)
Gattinger, Augustin (1825)
Gibbons, William P. (1812)
Gilbert, Benjamin D. (1835)
Glatfelter, Noah M. (1837)
Gow, James E. (1877)
Gray, Asa (1810)
Green, Jacob (1790)
Greene, Benjamin D. (1793)
Greene, Edward L. (1843)
Greenleaf, Richard C. (1809)
Gregory, Emily L. (1841)
Griffith, Robert E. (1798)
Hall, Elihu (1822)
Hallowell, Susan M. (1835)
Halsted, Byron D. (1852)
Hicks, John F. (   )
Hilgard, Theodore C. (1828)
Hill, Ellsworth J. (1833)
Holton, Isaac F. (1812)
Holton, Nina (   )
Hoopes, Joshua (1788?)
Hoopes, Josiah (1832)
Horsfield, Thomas (1773)
Hosack, David (1769)
Hough, Franklin B. (1822)
Howe, Elliot C. (1828)
Howell, Thomas J. (1842)
Humphrey, James E. (1861)
Hunt, James G. (ca.1826)
Ives, Eli (1778)
James, Edwin (1797)
James, Joseph F. (1857)
James, Thomas P. (1803)
Jefferson, Thomas (1743)
Jesup, Henry G. (1826)
Johnson, Lorenzo N. (   )
Jones, Herbert L. (1866?)
Jones, William R. (1883)
Joor, Joseph F. (1848)
Kellerman, William A. (1850)
Kellogg, Albert (1813)
Knieskern, Peter D. (1798)
Krauter, Louis (   )

Walpole, Frederick A. (1861)
Walter, Thomas (ca.1740)
Ward, Lester F. (1841)
Watson, Sereon (1826)
Wheeler, Charles F. (1842)
White, Theodore G. (1872?)
Willey, Henry (1824)
Williams, Stephen W. (1790)
Windle, Francis (1845?)
Wolle, Francis (1817)
Wood, Alphonso (1810)
Worthen, George C. (1871)
Wright, Charles (1811)
Yates, Lorenzo G. (1837)
Young, Aaron (1819)
Young, Herbert A. (ca.1857)
Young, Mary S. (1872)
Zollickoffer, William (1793)

BRYOLOGY
Austin, Coe F. (1831)
Sullivant, William S. (1803)

CARTOGRAPHY; MAP MAKING
Baker, Marcus (1849)
Bien, Julius (1826)
Bonner, John (ca.1643)
Evans, Lewis (1700)
Farmer, John (1798)
Herrman, Augustine (1605)
Hoen, August (1817)
Holme, Thomas (1624)
Jefferson, Thomas (1743)
Kino, Eusebio F. (ca.1645)
Lindenkohl, Adolph (1833)
Mitchell, John (1711)
Ogden, Herbert G. (1846)
Pound, Thomas (ca.1650)
Romans, Bernard (ca.1720)
Southack, Cyprian (1662)
Tanner, Henry S. (1786)
Walling, Henry F. (1825)
White, John (fl.1585-93)
Wood, John (ca.1775)

CHEMISTRY. Also see ALCHEMY;
ENGINEERING CHEMICAL
Acker, Charles E. (1868)
Aikin, William E. A. (1807)
Allen, Jonathan A. (1787)
Allen, Oscar D. (1836)
Alpers, William C. (1851)
Amend, Bernhard G. (1821?)
Amon, Frank (   )
Andrews, William H. (   )
Antisell, Thomas (1817)
Ashley, Harrison E. (1876)
Atwater, Wilbur O. (1844)

Atwood, Luther (1826)
Atwood, William (1830)
Austen, Peter T. (1852)
Babcock, James F. (1844)
Bache, Franklin (1792)
Bacon, John, Jr. (1817)
Bailey, Jacob W. (1811)
Baird, Julian W. (1859)
Baker, Philip S. (   )
Bancroft, Edward (1744)
Barker, George F. (1835)
Barnard, Edith E. (1880)
Batchelder, Loren H. (1846)
Beck, Lewis C. (1798)
Bell, Charles J. (1854)
Bernadou, John B. (1858)
Bigelow, Artemas (1818)
Black, Greene V. (1836)
Blake, James (1815)
Blake, John H. (1808)
Bodley, Rachel L. (1831)
Bolton, Henry C. (1843)
Booth, Edward (   )
Booth, James C. (1810)
Bowen, George T. (1803)
Bower, Henry (1833)
Boye, Martin H. (1812)
Breed, Daniel (ca.1825?)
Brewer, William H. (1828)
Bridges, Robert (1806)
Brinley, Charles A. (1847)
Brown, William G. (1853)
Brumback, Arthur M. (1869)
Brumby, Richard T. (1804)
Bryan, Albert H. (1874)
Bulkeley, Gershom (1636)
Bullock, Charles (1826)
Burnett, Joseph (1820)
Burton, Beverly S. (   )
Cabot, Samuel (1850)
Caldwell, George C. (1834)
Campbell, John L. (1818)
Carbutt, John (1832)
Carpenter, George W. (1802)
Carr, Ezra S. (1819)
Carroll, Charles G. (1875)
Carter, Oscar C. S. (1857)
Casamajor, Paul (1831)
Castner, Hamilton Y. (1859)
Catlin, Charles A. (1849)
Chandler, William H. (1841)
Chauvenet, Regis (1842)
Chester, Albert H. (1843)
Chilton, George (ca.1767)
Chilton, James R. (1809?)
Clark, William S. (1826)
Clemson, Thomas G. (1807)

Weightman, William (1813)
Weinstein, Joseph (1862?)
Wells, David A. (1828)
Wenzell, William T. (1829)
Wetherill, Charles M. (1825)
Wetherill, Samuel (1821)
Wharton, Joseph (1826)
Wheeler, Charles G. (1836)
Wheeler, Henry L. (1867)
White, Laura B. (   )
Whitehead, Cabell (1863)
Whitney, Josiah D. (1819)
Wiechmann, Ferdinand G.
  (1858)
Wilbert, Martin I. (1855)
Wilkinson, A. Williams
  (1832)
Williams, Rufus P. (1851)
Willson, Thomas L. (1860)
Wing, Charles H. (1836)
Winter, William P. (1866)
Witthaus, Rudolph A. (1846)
Wolf, Theodore R. (   )
Wood, Barnabas (1819)
Wood, Edward S. (1846)
Woodhouse, James (1770)
Woodman, Durand (1859)
Woolf, Albert E. (1846)
Wormley, Theodore G. (1826)
Wurtz, Henry (1828)
Wyatt, Francis (1855?)
Wylie, Theophilus A. (1810)
Wyman, Arthur D. (   )

CLIMATOLOGY. See METEOROLOGY

CONCHOLOGY; MALACOLOGY
Adams, Charles B. (1814)
Baldwin, David D. (1831)
Banister, John (1650)
Barnes, Daniel H. (1785)
Binney, Amos (1803)
Binney, William G. (1833)
Bland, Thomas (1809)
Bryan, Elizabeth L. (1874)
Campbell, John H. (1847)
Conrad, Timothy A. (1803)
Couthouy, Joseph P. (1808)
Garrett, Andrew (1823)
Gould, Augustus A. (1805)
Gratacap, Louis P. (1851)
Griffith, Robert E. (1798)
Harford, W. G. W. (1825)
Hartman, William D. (1817)
Hemphill, Henry (1830)
Jewett, Ezekiel (1791)
Law, Annie E. (1840?)
Lea, Henry C. (1825)

Lea, Isaac (1792)
Lichtenhaler, G. W. (   )
Newcomb, Wesley (1808)
Prime, Temple (1832)
Ravenel, Edmund (1797)
Roper, Edward W. (1858)
Sampson, Francis A. (1842)
Sanborn, Francis G. (1838)
Say, Thomas (1787)
Smith, Sanderson (1832)
Stimpson, William (1832)
Swift, Robert (1799)
Totten, Joseph G. (1788)
Tryon, George W. (1838)
Van Nostrand, Henry D.
  (   )
Wheatley, Charles M. (1822)

CONSERVATION; ENVIRONMENT
Langford, Nathaniel P.
  (1832)
Marsh, George P. (1801)
Roosevelt, Robert B. (1829)
Roosevelt, Theodore (1858)

COSMOGRAPHY. See ASTRONOMY

CRANIOLOGY
Meigs, James A. (1829)
Morton, Samuel G. (1799)

CYTOLOGY
Wolfe, James J. (1875)

DENTISTRY
Black, Greene V. (1836)
Chandler, Thomas H. (1824?)
Essig, Charles J. (1827?)
Fillebrown, Thomas (1836)
Garretson, James E. (1828)
Hayden, Horace H. (1769)
Koch, Charles R. E. (1844)
Miller, Willoughby D. (1853)
Thompson, Alton H. (1849)
Wells, Horace (1815)

DERMATOLOGY
Dyer, Isadore (1865)
Gottheil, William S. (1859)
Hyde, James N. (1840)
Jackson, George T. (1852)
Keany, Francis J. (1866?)
Morrow, Prince A. (1846)
White, James C. (1833)

Mercur, James (1842)
Metcalf, William (1838)
Michie, Peter S. (1839)
Mills, Adelbert P. (1883)
Morris, Ellwood (ca.1813)
Nettleton, Edwin S. (1831)
Newton, John (1823)
Noble, Alfred (1844)
Nystrom, John W. (1824)
Oliver, Marshal (1843?)
Pike, William A. (   )
Porter, John T. (1825)
Prosser, Thomas (   )
Renwick, Edward S. (1823)
Renwick, James, Sr. (1792)
Reynolds, Edwin (1831)
Ross, Henry S. (   )
Sellers, Coleman (1827)
Silliman, Justus M. (1842)
Stevens, Robert L. (1787)
Stone, Royce (1836?)
Sullivan, John L. (1777)
Swinburne, Ralph (ca.1805)
Tegmeyer, John H. (1821?)
Temple, Robert (1831?)
Thompson, James M. (1844)
Tidd, Marshall M. (1827)
Trautwine, John C. (1810)
Trowbridge, William P.
  (1828)
Twining, Alexander C. (1801)
Waldo, Frank (1857)
Ward, J. F. (1831?)
Watkins, John E. (1852)
Watson, William (1834)
Webb, John B. (1841)
Wellman, Samuel T. (1847)
Westinghouse, George (1846)
Wheeler, Ebenezer S. (1839)
Wilson, Herbert M. (1860)
Winterhalter, Albert G.
  (1856)
Wood, DeVolson (1832)
Woodbury, Daniel P. (1812)

ENGINEERING CHEMICAL
Adams, Isaac, Jr. (1836)
Stillman, Thomas B. (1852)
Wetherill, Samuel (1821)

ENGINEERING CIVIL
Abert, Silvanus T. (1828)
Berg, Walter G. (1858)
Borden, Simeon (1798)
Briggs, Robert (1822)
Campbell, John T. (1833)
Cattell, William A. (1863)
Chittenden, Hiram M. (1858)

Coleman, Clarence (   )
Collingwood, Francis (1834)
Crandall, Charles L. (1850)
Croes, John J. R. (1834)
Davis, Joseph B. (1845)
DuBois, Augustus J. (1849)
Ellis, Theodore G. (1829)
Freeman, Thomas (   )
Fuertes, Estevan A. (1838)
Greene, Charles E. (1842)
Johnson, Edwin F. (1803)
Johnson, John B. (1850)
McMath, Robert E. (1833)
Marvin, Frank O. (1852)
Millington, John (1779)
Morison, George S. (1842)
Nichols, Othniel F. (1845)
Page, Logan W. (1870)
Pierce, Josiah, Jr. (1861)
Rafter, George W. (1851)
Randolph, Isham (1848)
Rice, George S. (1849)
Roebling, John A. (1806)
Rogers, Fairman (1833)
Romans, Bernard (ca.1720)
Smith, Eugene (1860)
Sooysmith, Charles (1856)
Spence, David W. (   )
Stern, Louis C. (   )
Streeruwitz, William H. R. v.
  (1833)
Swan, Charles H. (   )
Sweet, Elnathan (1837)
Wellington, Arthur M. (1847)
Weston, Edmund B. (1850)
Whipple, Squire (1804)
Whistler, George W. (1800)
Wilson, Joseph M. (1838)
Woodbury, Charles J. H.
  (1851)

ENGINEERING EFFICIENCY
Taylor, Frederick W. (1856)

ENGINEERING ELECTRICAL
Abbott, Arthur V. (1854)
Burton, George D. (1855)
Duncan, Louis (1862)
Foster, Horatio A. (1858)
Ganz, Albert F. (1872)
Haskins, Caryl D. (1867)
Hooper, William L. (1855)
Houston, Edwin J. (1847)
Matthews, Charles P. (1867)
Osterberg, Max (1869)
Parmly, Charles H. (1868)
Perrine, Frederic A. C.
  (1862)

Schulte, Edward D. N. (1877)
Sheldon, Samuel (1862)
Short, Sidney H. (1858)
Stanley, William (1858)
Stevens, John F. (   )
Townsend, Fitzhugh (1872?)
Weaver, William D. (1857)
Willyoung, Elmer G. (1865)
Wolcott, Townsend (1857)
Woolf, Albert E. (1846)

ENGINEERING MECHANICAL
Allen, John R. (1869)
Bull, Storm (1856)
Carpenter, Rolla C. (1852)
Christie, James (1840)
Churchill, William W. (1867)
Clarke, Benjamin F. (1831)
Frary, Hobart D. (1887)
Higgins, Milton P. (1842)
Hutton, Frederick R. (1853)
Isherwood, Benjamin F.
  (1822)
Kent, William (1851)
Kerr, Walter C. (1858)
Klein, Joseph F. (1849)
Manning, Charles H. (1844)
Melville, George W. (1841)
Merrick, John V. (1828)
Phetteplace, Thurston M.
  (1877)
Richards, Charles B. (1833)
Robinson, Stillman W. (1838)
Sellers, William (1824)
Soule, Richard H. (1849)
Spangler, Henry W. (1858)
Taylor, Frederick W. (1856)
Thurston, Robert H. (1839)
West, Thomas D. (1851)

ENGINEERING MINING
Booraem, Robert E. (1856)
Christy, Samuel B. (1853)
Clemson, Thomas G. (1807)
Comstock, Theodore B. (1849)
Douglas, James (1837)
Glenn, William (1840)
Hague, James D. (1836)
Holmes, Joseph A. (1859)
Hulbert, Edwin J. (1829)
Irving, Roland D. (1847)
Kirchhoff, Charles W. H.
  (1853)
Lyman, Benjamin S. (1835)
Maynard, George W. (1839)
Newton, Henry (1845)
Phillips, William B. (1857)
Rockwell, Alfred P. (1834)

Sheafer, Peter W. (1819)
Streeruwitz, William H. R. v.
  (1833)
Watts, William L. (1850)

ENGINEERING TOPOGRAPHICAL
Abert, John J. (1788)
Chapman, Robert H. (1868)
Emory, William H. (1811)
Gardiner, James T. (1842)
Graham, James D. (1799)
Long, Stephen H. (1784)
Thompson, Almon H. (1839)
Warren, Gouverneur K. (1830)
Wheeler, George M. (1842)
Whipple, Amiel W. (1816)

ENTOMOLOGY; ARACHNOLOGY
Abbot, John (1751)
Ashmead, William H. (1855)
Bailey, James S. (1830)
Banister, John (1650)
Barnard, William S. (1849)
Behr, Hans H. (1818)
Benton, Frank (1852)
Blanchard, Frederick (1843)
Bland, James H. B. (ca.1832)
Bolter, Andrew (1820)
Burgess, Edward (1848)
Chambers, Vactor T. (1830)
Clemens, James B. (1829/30)
Cook, Albert J. (1842)
Coquillet, Daniel W. (1856)
Craw, Albert (1850)
Crotch, George R. (1842)
Dakin, John A. (1852)
Durkee, Silas (1798)
Edwards, Henry (1830)
Edwards, William H. (1822)
Fitch, Asa (1809)
Glover, Townend (1813)
Grossbeck, John A. (1883)
Grote, Augustus R. (1841)
Hagen, Hermann A. (1817)
Hamilton, John (1827)
Harkness, William H. (1821?)
Harris, George H. (   )
Harris, Thaddeus W. (1795)
Hart, Charles A. (1859)
Harvey, Francis L. (1850)
Heidemann, Otto (1842)
Hentz, Nicholas M. (1797)
Herrick, Edward C. (1811)
Hooker, Charles W. (1883?)
Horn, George H. (1840)
Hoy, Philo R. (1816)
Hubbard, Henry G. (1850)
Hulst, George D. (1846)

ENVIRONMENT. See
CONSERVATION

ETHNOGRAPHY. See
ANTHROPOLOGY

ETHNOLOGY. See ANTHROPOLOGY

EXPLORATION

FISHERIES. See PISCICULTURE

FORESTRY

Morey, Samuel (1762)
Morse, Samuel F. B. (1791)
Perkins, Jacob (1766)
Pope, Franklin L. (1840)
Riddell, John L. (1807)
Rumsey, James (1743)
Short, Sidney H. (1858)
Stevens, John (1749)
Stevens, Robert L. (1787)
Treadwell, Daniel (1791)
Twining, Alexander C. (1801)
Van Depoele, Charles J.
  (1846)
Westinghouse, George (1846)
Whitney, Eli (1765)
Wolle, Francis (1817)
Woods, Granville T. (1856)

LARYNGOLOGY
Elsberg, Louis (1836)
Jarvis, William C. (1855)
Simpson, William K. (1855)
Wagner, Clinton (1837)

LOGIC (SYMBOLIC)
Mitchell, Oscar H. (ca.1852)

MALACOLOGY. See CONCHOLOGY

MAP MAKING. See CARTOGRPHY

MATHEMATICS. Also see
ACTUARIAL SCIENCE
Adams, Daniel (1773)
Adrain, Robert (1775)
Alexander, James (1691)
Alexander, John H. (1812)
Alexander, William (1726)
Alvord, Benjamin (1813)
Ames, Nathaniel (1708)
Baker, Marcus (1849)
Banneker, Benjamin (1731)
Barnard, Frederick A. P.
  (1809)
Barnard, John (1681)
Barnard, John G. (1815)
Bartlett, William H. C.
(1804)
Bartlett, William P. G.
  (1837)
Bass, Edgar W. (1843)
Bayma, Joseph (1816)
Becker, George F. (1847)
Beebe, William (1851)
Benner, Henry (   )
Black, Charles W. M. (   )
Bocher, Maxime (1867)
Bonnycastle, Charles (1792)

Bowditch, Jonathan I. (1806)
Bowditch, Nathaniel (1773)
Bowser, Edward A. (1837)
Broun, William L. (1827)
Brown, Benjamin G. (1837)
Buck, Samuel J. (1835)
Caldwell, Joseph (1773)
Caswell, Alexis (1799)
Chandler, Charles H. (1840)
Chauvenet, William (1820)
Coakley, George W. (1814)
Coffin, James H. (1806)
Coffin, John H. C. (1815)
Coffin, Selden J. (1838)
Colburn, Warren (1793)
Colburn, Zerah (1804)
Compton, Alfred G. (1835)
Conant, Levi L. (1857)
Cottier, Joseph G. C.
  (1874?)
Courtenay, Edward H. (1803)
Craig, Thomas (1855)
Daboll, Nathan (1750)
Daniels, Archibald L. (1849)
Darby, John (1804)
Davies, Charles (1798)
Davis, Ellery W. (1857)
Davis, John E. (   )
Day, Jeremiah (1773)
Dean, James (1776)
DeForest, Erastus L. (1834)
DeMotte, Harvey C. (1838)
Devarre, Spencer H. (   )
Doolittle, Charles L. (1843)
Downes, John (1799)
Duffield, John T. (1823)
Dunbar, William (1749)
Dutton, W. T. (1852?)
Eddy, Imogen W. (   )
Eimbeck, William (1841)
Ellicott, Andrew (1754)
Ely, Achsah M. (1848)
Engler, Edmund A. (1856)
Esty, William C. (1838)
Farrar, John (1779)
Faunce, Charles M. (1867?)
Fisher, Alexander M. (1794)
Fisher, George E. (1863)
Folger, Walter (1765)
Frary, Hobart D. (1887)
Fry, Joshua (ca.1700)
Gable, George D. (1863)
Garber, Davis (1829?)
Garnett, John (ca.1748)
Gibbs, Josiah W. (1839)
Godfrey, Thomas (1704)
Green, G. W. (1857?)
Green, Gabriel M. (1891)

Shaw, Benjamin S. (1827)
Shrady, George F. (1837)
Sinkler, Wharton (1845)
Smith, Abram A. (1847)
Smith, Andrew H. (1837)
Smith, Eugene A. (1864)
Smith, Job L. (1827)
Smith, Nathan (1762)
Southard, Elmer E. (1876)
Spalding, Lyman (1775)
Starkey, George (1628)
Sternberg, George M. (1838)
Stevens, Edward (ca.1755)
Stewart, David D. (1858)
Stille, Alfred (1813)
Stimson, Lewis A. (1844)
Sutton, Walter S. (1877)
Taylor, Herbert D. (1888?)
Thayer, Charles P. (1843)
Thomson, William H. (1833)
Tiffany, Louis M. (1844)
Trudeau, Edward L. (1848)
Tully, William (1785)
Tyson, James (1841)
Van Rensselaer, Jeremiah
  (1793)
Vaughan, Victor C., Jr.
  (1879)
Wagner, Clinton (1837)
Ward, Samuel B. (1842)
Warren, John (1753)
Warren, Jonathan M. (1811)
Waterhouse, Benjamin (1754)
Weir, James, Jr. (1856)
Wells, William C. (1757)
Wheeler, Claude L. (   )
Wheeler, Ernest S. (1869?)
Whitehead, William R. (1831)
Willard, DeForest (1846)
Williams, Stephen W. (1790)
Williamson, Hugh (1735)
Williston, Samuel W. (1851)
Wilson, Thomas M. (   )
Winthrop, John, Jr. (1605/6)
Withington, Charles F.
  (1852)
Wood, Edward S. (1846)
Wood, Horatio C. (1841)
Woodward, Joseph J. (1833)
Wright, Hamilton K. (1867)
Wyman, Morrill (1812)
Zabriskie, John B. (1805)

METALLURGY
Arents, Albert (1840)
Balbach, Edward, Jr. (1839)
Blake, William P. (1825)
Brinley, Charles A. (1847)

Browne, David H. (1864)
Cheever, Byron W. (1841)
Christy, Samuel B. (1853)
Church, John A. (1843)
Cowles, Eugene H. (1855)
Daniels, Fred H. (1853)
Davenport, Russell W. (1849)
DeChalmot, Guillaume L. J.
  (   )
Douglas, James (1837)
Egleston, Thomas (1832)
Eilers, Frederic A. (1839)
Firmstone, Frank (1846)
Fritz, John (1822)
Gayley, James (1855)
Goetz, George W. (1856)
Havard, Francis T. (1878?)
Hill, Nathaniel P. (1832)
Holley, Alexander L. (1832)
Hunt, Alfred E. (1855)
Johnson, John E., Jr.
  (1860?)
Kirchhoff, Charles W. H.
  (1853)
Lord, Nathaniel W. (1854)
Metcalf, William (1838)
Overman, Frederick (ca.1803)
Pearse, John B. S. (1842)
Peters, Edward D. (1849)
Piggot, Aaron S. (1822)
Poole, Herman (1849)
Raht, August W. (1843)
Raymond, Rossiter W. (1840)
Reese, Jacob (1825)
Ricketts, Pierre D. (1848?)
Spilsbury, Edmund G. (1845)
Stetefeldt, Carl A. (1838)
Tassin, Wirt d. (1868)
Tuttle, David K. (1835)
Wahl, William H. (1848)
Weeks, Joseph D. (1840)
Whitehead, Cabell (1863)

METEOROLOGY; CLIMATOLOGY
Abbe, Cleveland (1838)
Beck, Theodric R. (1791)
Blasius, William (1818)
Blodget, Lorin (1823)
Bowditch, Jonathan I. (1806)
Brocklesby, John (1811)
Butler, Thomas B. (1806)
Chalmers, Lionel (1715)
Coffin, James H. (1806)
Coffin, Selden J. (1838)
Curtis, George E. (1861)
Davis, Walter G. (1851)
Dunbar, William (1749)
Eddy, William A. (1850)

Gibbs, George (1776)
Gilmor, Robert (1774)
Gratacap, Louis P. (1851)
Hall, Frederick (1780)
Hawes, George W. (1848)
Hidden, William E. (1853)
Horton, William (   )
Hubbard, Oliver P. (1809)
Jackson, Charles T. (1805)
Joy, Charles A. (1823)
Kato, Frederick (1866)
Keating, William H. (1799)
Koenig, George A. (1844)
Lewis, Henry C. (1853)
Lord, Nathaniel W. (1854)
Meade, William (   )
Moses, Alfred J. (1859)
Olmsted, Denison, Jr. (1824)
Osborn, Henry S. (1823)
Penfield, Samuel L. (1856)
Robinson, Franklin C. (1852)
Scovell, Josiah T. (1841)
Seybert, Adam (1773)
Seybert, Henry (1801)
Shepard, Charles U. (1804)
Silliman, Benjamin, Sr.
  (1779)
Smith, Edward D. (1778)
Smith, John L. (1818)
Smithson, James L. M. (1765)
Sperry, Francis L. (   )
Taylor, William J. (ca.1833)
Teschemacher, James E.
  (1790)
Trautwine, John C. (1810)
Troost, Gerard (1776)
Vaux, William S. (1811)
Wheatley, Charles M. (1822)
Williams, George H. (1856)
Williams, John F. (1862?)
Williams, Lewis W. (1807?)
Yeates, William S. (1856)

MINING. Also see ENGINEERING
MINING
Buckley, Ernest R. (1872)
Egleston, Thomas (1832)
Emmons, Samuel F. (1841)
Irving, John D. (1874)
Kerr, Mark B. (1860)
Rothwell, Richard P. (1836)
Verrill, Clarence S. (   )
Wheeler, Charles G. (1836)

MORPHOLOGY
Bruce, Adam T. (   )
Graf, Arnold (1870)

MYCOLOGY
Ellis, Job B. (1829)
McIlvaine, Charles (1840)
Morgan, Andrew P. (1836)
Peck, Charles H. (1833)
Schweinitz, Lewis D. v.
  (1780)

**NATURAL HISTORY; NATURE
STUDY**
Abbott, Charles C. (1843)
Adams, Charles B. (1814)
Agassiz, Alexander E. R.
  (1835)
Agassiz, Elizabeth C. C.
  (1822)
Agassiz, Jean Louis R.
  (1807)
Aldrich, Charles (1828)
Apgar, Austin C. (1838)
Atwater, Caleb (1778)
Bachman, John (1790)
Baird, Spencer F. (1823)
Barnard, William S. (1849)
Barton, Benjamin S. (1766)
Bartram, William (1739)
Beal, Foster E. L. (1840)
Beck, Lewis C. (1798)
Bell, John G. (1812)
Bessels, Emil (1847)
Bickmore, Albert S. (1839)
Binney, Amos (1803)
Bland, Thomas (1809)
Boll, Jacob (1828)
Bouve, Thomas T. (1815)
Brace, John P. (1793)
Brooks, William K. (1848)
Brown, Arthur E. (1850)
Buckley, Samuel B. (1809)
Byrd, William (1674)
Catesby, Mark (1683)
Caton, John D. (1812)
Cist, Jacob (1782)
Clinton, DeWitt (1769)
Colman, Benjamin (1673)
Conklin, William A. (1837)
Conrad, Solomon W. (1779)
Cook, Albert J. (1842)
Cooper, William (1797)
Coues, Elliott (1842)
Couper, James H. (1794)
Cuming, Alexander (ca.1690)
Cutting, Hiram A. (1832)
Damon, William E. (1838)

Hays, Isaac (1796)
Heilprin, Angelo (1853)
Hyatt, Alpheus (1838)
James, Uriah P. (1811)
Jefferson, Thomas (1743)
Lacoe, R. D. (   )
Leidy, Joseph (1823)
Loper, Samuel W. (1835)
Lyon, Sidney S. (1807)
Marcou, Jules (1824)
Marsh, Othniel C. (1831)
Meek, Fielding B. (1817)
Newberry, John S. (1822)
Norwood, Joseph G. (1807)
Owen, Richard (1810)
Prout, Hiram A. (1808)
Read, Motte A. (1872)
Redfield, William C. (1789)
Reed, William H. (1848)
Shumard, Benjamin F. (1820)
Simpson, George B. (   )
Troost, Gerard (1776)
Wachsmuth, Charles (1829)
Wagner, William (1796)
Warren, John C. (1778)
White, Charles A. (1826)
Whitfield, Robert P. (1828)
Williams, Henry S. (1847)
Williston, Samuel W. (1851)
Yandell, Lunsford P. (1805)

PATHOLOGY
Blackburn, Isaac W. (1851?)
Brinckerhoff, Walter R.
  (1874)
Carroll, James (1854)
Clark, Admont H. (1888)
Claypole, Edith J. (1870)
Hendrickson, W. F. (1876?)
Herter, Christian A. (1865)
Hiss, Philip H., Jr. (1868)
Hodenpyl, Eugene (1863)
Jackson, John B. S. (1806)
Janeway, Edward G. (1841)
Kinyoun, Joseph J. (1860)
Lazear, Jesse W. (1866)
Meltzer, Samuel J. (1851)
Mooers, Emma W. D. (   )
Ohlmacher, Albert P. (1865)
Ricketts, Howard T. (1871)
Salmon, Daniel E. (1850)
Schoney, Lazarus (1838)
Stewart, Hugh A. (1882)
Van Gieson, Ira T. (1866)
Vaughan, Victor C., Jr.
  (1879)
Weil, Richard (1876)
Wesbrook, Frank F. (1868)

Wiener, Joseph (1828)
Willson, Robert N. (1873)
Wilson, Thomas M. (   )
Woodward, Joseph J. (1833)

PEDIATRICS
Rotch, Thomas M. (1849)

PETROGRAPHY. See PETROLOGY

PETROLOGY; PETROGRAPHY
Iddings, Joseph P. (1857)
Julien, Alexis A. (1840)
Williams, George H. (1856)

PHARMACOLOGY. See PHARMACY

PHARMACY; PHARMACOLOGY
Alpers, William C. (1851)
Avery, Charles E. (1848?)
Battey, Robert (1828)
Bullock, Charles (1826)
Carpenter, George W. (1802)
Carson, Joseph (1808)
Chapman, Nathaniel (1780)
Coxe, John R. (1773)
Durand, Elias (1794)
Fisher, William R. (1808)
Hoffmann, Friedrich (1832)
Jackson, Samuel (1787)
Koch, Waldemar (1875)
Lee, Charles A. (1801)
Maisch, John M. (1831)
Mayer, Ferdinand F. (   )
Meltzer, Samuel J. (1851)
Oldberg, Oscar (1846)
Procter, William (1817)
Remington, Joseph P. (1847)
Rice, Charles (1841)
Rother, Reinhold F. W.
  (1843)
Schlotterbeck, Julius O.
  (1865)
Seabury, George J. (1844)
Squibb, Edward R. (1819)
Stabler, Richard H.
  (ca.1820)
Stewart, David (1813)
Thomas, Robert P. (1821)
Tully, William (1785)
Wilbert, Martin I. (1855)
Wood, Horatio C. (1841)
Zollickoffer, William (1793)

PHILANTHROPY
Everhart, Isaiah F. (1840)
Gibbs, George (1776)
Herrman, Esther (1822?)

About the Compiler

CLARK A. ELLIOTT is Associate Curator of the Harvard University Archives. Previously, Elliott served as an assistant professor of library science at Simmons College in Boston. He is the author of *Biographical Dictionary of American Science: The Seventeenth Through the Nineteenth Centuries* (Greenwood Press, 1979) and is currently coediting a work on *Science at Harvard University: Historical Perspectives.*